T0133076

COMPUTER-AIDED DESIGN,
ENGINEERING, AND MANUFACTURING
SYSTEMS TECHNIQUES AND APPLICATIONS

VOLUME
IV

# OPTIMIZATION
# METHODS FOR
# MANUFACTURING

# COMPUTER-AIDED DESIGN, ENGINEERING, AND MANUFACTURING

## SYSTEMS TECHNIQUES AND APPLICATIONS

### VOLUME IV

# OPTIMIZATION METHODS FOR MANUFACTURING

## EDITOR
## CORNELIUS LEONDES

**CRC Press**

Boca Raton   London   New York   Washington, D.C.

Elsevier Science is acknowledged for permission to reprint the following tables and figures in this book:

**Tables 5.1 through 5.5:** Reprinted from D. Golenko-Ginzburg et al., "Industrial Job-Shop Scheduling with Random Operations and Different Priorities," *International Journal of Production Economics*, 40, 185–195, 1995. **Tables 5.6 through 5.10:** Reprinted from D. Golenko-Ginzburg and A. Gonik, "Using 'Look Ahead' Techniques in Job-Shop Scheduling with Random Operations," *International Journal of Production Economics*, 50:1, 13–22, 1997. **Figures 9.1, 9.4, and 9.18:** Reprinted from S. Ema and E. Marui, "A Fundamental Study on Impact Dampers," *International Journal of Machine Tools and Manufacture*, 34, 3, 1994. **Figures 9.1, 9.2, 9.4 through 9.7, 9.9 through 9.11, and 9.13, and Tables 9.1 through 9.4:** Reprinted from S. Ema and E. Marui, "Damping Characteristics of an Impact Damper and Its Application," *International Journal of Machine Tools and Manufacture*, 36, 3, 1996.

### Library of Congress Cataloging-in-Publication Data

Catalog record is available from the Library of Congress.

# Preface

A strong trend today is toward the fullest feasible integration of all elements of manufacturing, including maintenance, reliability, supportability, the competitive environment, and other areas. This trend toward total integration is called concurrent engineering. Because of the central role information processing technology plays in this, the computer has also been identified and treated as a central and most essential issue. These are the issues which are at the core of the contents of this volume.

This set of volumes consists of seven distinctly titled and well-integrated volumes on the broadly significant subject of computer-aided design, engineering, and manufacturing: systems techniques and applications. It is appropriate to mention that each of the seven volumes can be utilized individually. In any event, the great breadth of the field certainly suggests the requirement for seven distinctly titled and well-integrated volumes for an adequately comprehensive treatment. The seven volume titles are

1. Systems Techniques and Computational Methods
2. Computer-Integrated Manufacturing
3. Operational Methods in Computer-Aided Design
4. Optimization Methods for Manufacturing
5. The Design of Manufacturing Systems
6. Manufacturing Systems Processes
7. Artificial Intelligence and Robotics in Manufacturing

The contributions to this volume clearly reveal the effectiveness and great significance of the techniques available and, with further development, the essential role that they will play in the future. I hope that practitioners, research workers, students, computer scientists, and others on the international scene will find this set of volumes to be a unique and significant reference source for years to come.

**Cornelius T. Leondes**
Editor

# Editor

**Cornelius T. Leondes, B.S., M.S., Ph.D.,** is Emeritus Professor, School of Engineering and Applied Science, University of California, Los Angeles. Dr. Leondes has served as a member or consultant on numerous national technical and scientific advisory boards and as a consultant for a number of Fortune 500 companies and international corporations. He has published more than 200 technical journal articles and edited or co-authored more than 120 books. Dr. Leondes has been a Guggenheim Fellow, Fulbright Research Scholar, and IEEE Fellow. He is a recipient of the IEEE Baker Prize Award and the IEEE Barry Carlton Award.

# Contributors

**Vito Albino**
Università della Basilicata
Potenza, Italy

**Alka Chaudhary**
Meerut College
Meerut, India

**Satoshi Ema**
Gifu University
Gifu-shi, Japan

**Dimitri Golenko-
 Ginzburg**
Ben-Gurion University of the Negev
Beer Sheva, Israel

**Aharon Gonik**
Ben-Gurion University of the Negev
Beer Sheva, Israel

**Rakesh Gupta**
Ch. Charan Singh University
Meerut, India

**Peter Heimann**
Aachen University of Technology
 (RWTH)
Aachen, Germany

**Udo Konradt**
University of Kiel
Kiel, Germany

**Hochang Lee**
Kyung Hee University
Korea

**E. Mauri**
Gifu University
Gifu-shi, Japan

**O. Geoffrey Okogbaa**
University of South Florida
Tampa, Florida

**Hui-Ming Wee**
Chung Yuan Christian University
Chungli, Taiwan

**Bernhard Westfechtel**
Aachen University of Technology
 (RWTH)
Aachen, Germany

**Huanxin Henry Xiong**
Unxipros, Inc.
Eatontown, New Jersey

**MengChu Zhou**
New Jersey Institute of Technology
Newark, New Jersey

# Contents

# 1

# Synchronization of Production Rate with Demand Rate in Manufacturing Systems

Vito Albino
*Università della Basilicata*

O. Geoffrey Okogbaa
*University of South Florida*

## 1.1 Introduction

In the twentieth century, the market competitive pressure has shifted from cost to cost, quality, and responsiveness. The market globalization has forced companies to become more effective and efficient in their operations all over the world (Pfeifer et al. 1994, Porter 1985). Information and communication technology advances are changing the nature of the marketplace (Spar and Bussgang 1996). The virtual market has become a reality. In this context, companies need to synchronize their manufacturing activities with the variable product demand. Several business functions and activities are involved in synchronizing production rate with demand rate.

In this chapter, the problem of synchronization in the manufacturing area is addressed. The process-based organization and the main manufacturing planning and control techniques and systems are considered as the basic support for the matching between production and demand flows.

# 1.2   Markets and Production Systems

The synchronization of production rate with demand rate in manufacturing systems concerns both company strategy and operations. Some preliminary considerations about customer demand and manufacturing, logistics, and distribution systems seem particularly suitable to understand how companies can allocate and integrate resources to ensure an effective synchronization.

## Customer Demand

A customer is an individual or group who buys or uses the products, services, or both produced by a company. It is common to make distinction between external and internal customers. The internal customer is the user of the output of a single process, being the "customer" to the preceding component of the supply chain. In fact, most businesses can get benefits by improving the interface and responsiveness to internal customer needs. However, it is important to recognize that the customer of the overall product supply system is the end consumer. Everyone else in the product supply system is both supplier and customer of a product on its way to the consumer. Therefore, understanding the end consumer's need can provide the key to a new breakthrough in competitive advantage. The customer has always been important for the supplier organization. Today, however, due to an intense domestic and international competition, the need for a different consideration of the customer is critical for survival and success. This consideration must include a clear understanding of the customer's needs and acting as a value-added servicer to meet those needs: in short, establishing a closer interface and actively responding to change. While interim customers are important, the end user is the only true customer and holds the key to competitive advantage (Bockerstette and Shell 1993).

Customers will gravitate to and do business with the manufacturer that provides the highest value. In fact, the price customers will pay for products is directly related to the perceived value of a supplier in relation to its competitors. Value from the customer's perspective comes from an optimum balance of product quality, customer service, cost, and response times.

Among these variables, customers are showing increasing interest for response times. Response times can be basically related to two time variables: the lead time to supply a product to the customer for a specific order and the time to develop a product from concept to final market delivery (Blackburn 1991).

Reducing product-supply lead time is the strategic objective of the time-based manufacturing company. Any time beyond the time necessary to add value to the product represents excess costs and waste to both the supplier and the customer. A customer will pay a premium to the supplier who can supply its product faster and more reliably than the competition. A reduction of the lead time has also a positive impact on product quality, customer service, and cost (Bockerstette and Shell 1993, Stalk and Hout 1990). In consequence of the change from a "seller's market" to a "buyer's market" and of accelerated technological developments, enterprises are forced to reorganize their business processes, to be able to quickly and cost effectively react to fast-changing market demands (Slats et al. 1995).

Every business today competes in two worlds: a physical world of resources that managers can see and touch and a virtual world made of information. The latter has created the world of electronic commerce, a new locus of value creation. Executives must pay attention to how their companies create value in both the physical world and the virtual world. But the processes for creating value are not the same in the two worlds (Rayport and Sviokla 1995). In the physical world, the value chain (Porter 1985) is a model that describes a series of value-adding activities connecting a company's supply side (raw materials, inbound logistics, and production processes) with its demand side (outbound logistics, marketing, and sales). The value chain model treats information as a supporting element of the value-adding process, not as a value source itself.

The value-adding processes that companies must employ to turn raw information into new services and products are unique to the information world. The value-adding steps are virtual in that they are performed through and with information. Creating value in any stage of a virtual chain involves a sequence of five activities: gathering, organizing, selecting, synthesizing, and distributing information

(Rayport and Sviokla 1995). This value-adding process can be the key for the synchronization of production activities with customer demand.

The evolution towards the electronic marketplace is fast as estimated by recent market researches (Forrester Research 1997). The electronic marketplace deals with final consumers and companies. Companies can sell products and buy raw materials, components, and services. Unfortunately, there is an initial bottleneck for the development of the electronic marketplace. Companies wait for customers able to buy in the electronic marketplace and, on the other hand, customers buy in the electronic marketplace if they find sellers.

In fact, some experiences in the fields of telecommunication, computer, and multimedia have shown attractive economic results, in particular when they are based on electronic communities (Armstrong and Hagel 1996, Spar and Bussgang 1996). Also, the implementation of the Trading Process Network by General Electric can be considered an interesting example, whereas the IBM World Avenue online mall has resulted in market failure (Seminerio 1997).

## Manufacturing Systems

A manufacturing system can be defined (Chryssolouris 1992) as a combination of humans, machinery, and equipment that are bound by a common material and information flow. The development of the manufacturing system architecture begins with an assessment of the basic tasks that the factory must perform. A production process consists of a sequence of steps, traditionally assigned to separate departments, that take various inputs (such as parts, components, and subassemblies) and convert them into desired outputs. Each task performed by the various departments in a factory accomplishes one or more of three things: it converts inputs into outputs, it moves material from one location to another, and it transfers information.

Conversion activities add value to the product by transforming materials, components, or subassemblies into a higher level of component or assembly. Material flows refer to the movement of materials from one department (location) to another. Because of the number of conversion steps, the material volumes, and the distances involved in most factories, such movements are both numerous and extremely important. Information transfers serve not only to coordinate conversion steps and material flows, but also provide the feedback necessary to make improvements in the factory's procedures, process technology, and operating characteristics.

A factory has goals with respect to output volume, cost, quality, time, and location of delivery. Many of these goals are broken down into subgoals that define short-term operating targets: product specification, inventory turns, standard costs, times and quantities, and daily shipment requirements. Other goals express longer-term objectives, such as the speed of new product introduction, defect reduction, and the rate of productivity improvement.

The challenge for the manufacturing organization is to integrate and synchronize all these tasks and align them with the company's chosen goals, as well as to do that better than most or all of its competitors (Hayes et al. 1988). Synchronizing departments and integrating operations can require changes in the location and sequencing of equipment, reductions in setup times, faster and more accurate information about production problems, and improvements in process quality. The complexity and uncertainty in manufacturing have led over the years to an array of tools, techniques, programs, and principles for designing different manufacturing architectures, each with its own set of acronyms (Hayes et al. 1988).

In industrial practice, there are five general approaches to structuring manufacturing systems: job shop, project shop, cellular system, flow line, and continuous system. Each of these approaches also refers to different situations of the synchronization of production rate with demand rate (for instance, flow line is suitable for stable demand rate and standard products, and job shop for unstable demand rate and customized products). In the actual manufacturing world, these standard system structures often occur in combinations, or with slight changes. The choice of the structure depends on the design of the parts to be manufactured, the lot sizes of the parts, and the market factors (i.e., the required responsiveness to market changes).

In general, there are four classes of manufacturing attributes (Chryssolouris 1992) that are considered in manufacturing decision-making: cost, time, quality, and flexibility. These attributes are related to the manufacturing system performance and are strongly affected by the approach adopted in the design of the manufacturing system components, in their integration, and in the interaction logic with the environment.

## Logistics and Distribution

Logistic activities within an enterprise can be related to (Slats et al. 1995):

1. flow of goods, including transportation, material handling, and transformation (manufacturing, assembly, packaging, etc.);
2. flow of information, including information exchange regarding orders, deliveries, transportation, etc.; and
3. management and control, including purchasing, marketing, forecasting, inventory management, planning, sales, and after-sales service.

A company's performance can be strongly affected by logistics and distribution processes. The synchronization of production rate with demand rate is strictly related to the effectiveness of logistics and distribution processes. Also, logistics has gained much attention in increasing the efficiency and flexibility of organizations, as logistics costs make up a significant part of total production costs.

Nowadays logistics is seen as a value-adding process that directly supports the primary goal of the enterprise, which is to be competitive in terms of a high level of customer service, competitive price and quality, and flexibility in response to market demands.

To reduce stocks and response times and to increase efficiency, in the large consumer market the logistics function can play the critical role of coordinating the distribution and production planning activities.

The integration of all logistics processes, from the acquisition of raw materials to the distribution of end-customer products, makes up a logistic chain consisting of multiple actors. Developments in telecommunication and information technology (IT) have created many opportunities to increase the integration of these actors, and, as a consequence, to increase the performance of the total logistic chain providing each participant with benefits (from material supplier to end-customer) (Slats et al. 1995). However, when actors are integrated along the supply chain, possible conflicting needs of individuals, functions, and companies can arise in the overall organization, especially when synchronization and cooperation are required between them (Womack and Jones 1994). Finally, the integration stresses the effect of uncertainty that plagues the performance of suppliers, the reliability of manufacturing and transportation processes, and the changing desires of customers (Davis 1993).

## Integrated Manufacturing

The architecture of a manufacturing system (which includes its hardware, its materials and information flows, the rules and procedures used to coordinate them, and the managerial philosophy that underlies them all) largely determines the productivity of the people and assets in the factory, the product quality, and the responsiveness of the organization to customer needs.

The major issue of the architecture design of a manufacturing system is the integration of manufacturing activities. The potential performance impact of integrated manufacturing goes well beyond cost reduction. Integrated manufacturing techniques have been recognized to improve performance in terms of what customers receive and when, i.e., quality and lead time. In addition, by their implementation flexibility can be also achieved (Dean and Snell 1996). Manufacturing activities need to be integrated if performance has to be improved. For instance, short throughput times depend on a coordinated effort of reduction and handling of manufacturing activities variability. Under conditions of reduced inventory or throughput time, interdependence among resources increases. To reduce processing times, the actual activities performed by various functions must be coordinated and integrated (Duimering et al. 1993).

Integrated manufacturing is driven by the widespread adoption of information technology (Computer-Integrated Manufacturing) and management techniques and approaches (Just-in-time and Total Quality Management) which can be referred to the lean production (Womack et al. 1990).

The concept of world-class manufacturing (Gunn 1987, Schonberger 1986) has been proposed to point out how performance can be enhanced by adopting integrated manufacturing. Among the key attributes of world-class manufacturers, being able to quickly and decisively respond to changing market conditions is particularly important (Hayes et al. 1988). Manufacturing can achieve this attribute by integrating product and process design, production planning and control, manufacturing, logistics, and distribution as well as suppliers and customers. The integration approach can be based on information technology and organization. Computer-integrated manufacturing and lean production may be considered the most known approaches emphasizing information technology and organization, respectively (Mariotti 1994).

## Computer-Integrated Manufacturing

A computer-integrated manufacturing (CIM) system can be defined as a computer-based integration of all components involved in the production of an item. It starts with the initial stages of planning and design and encompasses the final stages of manufacturing, packaging, and shipping, combining all existing technologies to manage and control the entire business (Sule 1994). A CIM system includes components such as computer-aided design, computer-aided process planning, database technology and management, expert systems, information flow, computer-aided production control, automated inspection methods, process and adaptive control, and robots. All of these elements work together using a common database. Data are entered from the operations of the entire plant: from manufacturing, receiving and shipping, marketing, engineering, and designated individuals within other departments who are related to the operations of the plant. The information is continuously processed, updated, and reported to people on an as-needed basis in an appropriate format to assist in making decisions.

The level of automation applied within a manufacturing enterprise has a dramatic effect on the production-planning and -control function, and then on the synchronization of production rate with demand rate. The higher the degree of automation, the greater the potential of simplification of the production-planning function can be. However, as the degree of automation grows, the production-control function becomes the key to the success of the manufacturing facility. Thus, the increasing automation of the manufacturing process for productivity improvement needs a very sophisticated computerized and integrated production-planning and -control system, able to react to problems in very short time intervals (Bedworth and Bailey 1987). The synchronization of production rate with demand rate has to be assured also by an effective order-handling system. In fact, the order-handling system constitutes the link between the production and sale areas. Such a system does not always belong to production planning and control (Scheer 1994). In an inventory-based serial manufacturing system, for example, the problem of customer order handling can be separated from production order planning and control. However, the more customer-specific demands affect the manufacturing process, the closer the link between these two systems must be.

In the case of customer-demand-oriented production (individual orders, variant production) cost information must be established in the course of order acceptance. This preliminary pricing necessitates access to primary production data (bills of materials, work schedules, equipment data) and then order handling cannot be separated from production planning.

The use of computers in manufacturing is effective only after the creative assimilation of their potential for converting to new organizational structures. The computer-based integration needs the redesign of the organization before implementing flexible technology (Duimering et al. 1993). In fact, flexible technology alone does not address the causes of manufacturing problems. A correct approach should begin by examining and eventually redesigning the interdependencies that exist among functional units within the overall manufacturing organization. Then, investments in flexible technologies should be considered as the last resort.

## Lean Production

The concept of lean production has been introduced by Womack et al. (1990) to point out the difference in the car industry between the western mass production and the Japanese production techniques. "Lean production is lean because it uses less of everything compared with mass production, half the human effort in the factory, half the manufacturing space, half the investment in tools, half the engineering hours to develop a new product in half the time. Also, it requires keeping far less than half the needed inventory on site, results in many fewer defects, and produces a greater and ever growing variety of products" (Womack et al. 1990).

From the point of view of synchronization between production and demand rates, lean production systems are customer driven, not driven by the needs of manufacturing. All activities are team-based, coordinated, and evaluated by the flow of work through the team or the plant, rather than by each department meeting its plan targets in isolation. The whole system involves fewer actors (including suppliers), all integrated with each other (Lamming 1993). The system is based on stable production volumes, but with a great deal of flexibility (Jones 1992). Based on several experiences, Womack and Jones (1994) observe that applying lean techniques to discrete activities can improve performance, but if individual breakthroughs can be linked up and down the value chain to form a continuous value stream that creates, sells, and services a family of products, the performance of the whole can be raised to a dramatically higher level. A new organizational model is then proposed to join value-creating activities, the lean enterprise. This is a group of individuals, functions, and legally separate but operationally synchronized companies. However, the benefits of lean production have shown some limits (Albino and Garavelli 1995, Cusumano 1994). Due to global competition, faster product development, and increasingly flexible manufacturing systems, a large variety of products are competing in markets. Despite the benefits to consumers, this phenomenon is making it more difficult for manufacturers and retailers to predict which of their goods will sell and to synchronize production rate with demand rate.

As a result, inaccurate forecasts are increasing, and along with them the costs of those errors. Manufacturers and retailers alike are ending up with more unwanted goods that should be marked down even as they lose potential sales because other articles are no longer in stock. In industries with highly volatile demand, like fashion apparel, the costs of such stock-outs and markdowns can actually exceed the total cost of manufacturing.

To reduce the impact of the inaccuracy of forecasts, many managers have implemented popular production-scheduling systems. But quick-response programs (Blackburn 1991), just-in-time (JIT) inventory systems, manufacturing resource planning, and the like are simply not up to the task. With a tool like manufacturing resource planning, for example, a manufacturer can rapidly change the production schedule stored in its computer when its original forecast and plan prove to be incorrect. A new scheduling doesn't help, though, if the supply chain has already been filled based on the old one.

Similarly, quick response and JIT address only a part of the overall picture. A manufacturer might hope to be fast enough to produce in direct response to demand, virtually eliminating the need for a forecast. But in many industries, sales of volatile products tend to occur in a concentrated season, which means that a manufacturer would need an unjustifiably large capacity to be able to make goods in response to actual demand. Using quick response or JIT also may not be feasible if a company is dependent on an unresponsive supplier for key components (Fisher et al. 1994).

From the strategic point of view, Hayes and Pisano (1994) observe that simply improving manufacturing by, for example, adopting JIT or TQM is not a strategy for using manufacturing to achieve competitive advantage. Neither is aspiring to lean manufacturing, continuous improvement, or world-class status. In today's turbulent competitive environment, a company more than ever needs a strategy that specifies the kind of competitive advantage that it is seeking in its marketplace and articulates how that advantage has to be achieved. Then, companies need strategies for building critical capabilities to achieve competitive advantage.

## Competitiveness and Time-Based Capabilities

Starting from the 1970s, quality has become an important prerequisite for the competition. Determining what the customer needs and wants, which attributes add value to the product/service, and creating operations capabilities to satisfy those requirements are some characteristics of world-class companies. Customer-driven manufacturing has been taken one step further by several excellent manufacturing organizations. Defining both internal and external customers, everyone within the organization has a customer. Everyone's goal is to provide its own customers with quality products and services. Then, there is a necessary progression from quality to delivery, flexibility, and/or cost.

While timing and staging of priorities may vary from company to company, a firm should achieve multiple capabilities. For instance, as more and more manufacturers achieve quality and delivery goals, the importance of both cost and flexibility capabilities will increase dramatically. Being good at only one element will probably not lead to success. Today, however, manufacturers understand that the customer cares little about the length of the manufacturing process and much about how long it will take to receive his product (Bockerstette and Shell 1993). They also understand that continually shrinking product life cycles can no longer support the length of the design process typical of most engineering functions today. Further, they realize that many of the problems experienced in manufacturing on the shop floor are heavily influenced by the work of process and product designers.

Manufacturers have discovered the mistake of trying to achieve efficiency and quality by separating equipment types into different areas of a facility, and they have since then undertaken a massive desegregation effort to recast production around products rather than around functional departments. Then, the process-based approach can be pursued to support the reorganization of manufacturing activities (Roth et al. 1992).

Time may sometimes be the most important performance. The ways leading companies manage time—in production, in sales and distribution, in new product development and introduction—can be a powerful source of competitive advantage (Blackburn 1991, Stalk and Hout 1990). The improvement of time-based performance is essential in the competitive race to respond faster to customers while reducing costs and improving quality. It is a race in which the only winning strategy is to do it right the first time.

To compete successfully, the manufacturer must know exactly what is happening in the plant and must be able to manage it. The closer to real time this information, the better the  manufacturer's ability to compete and respond (Hakanson 1994). In the time-based competition, the initial focus is on a first-level time-based capability, i.e., dependable on-time delivery. After dependable delivery, delivery speed or reduced cycle time until time-to-market comes. A list of time-based capabilities is shown in Figure 1.1. Then, the capability to produce a high-quality product, on time and faster than the competition, can represent a formidable competitive advantage.

| Time-based capabilities |
| :---: |
| Reliability of delivery time |
| Prompt handling of customer complaints |
| Delivery speed/reduced lead times |
| Rapidly confirm order delivery |
| Speed of new product introductions |
| Rapidly handle custom orders |

**FIGURE 1.1**  Time-based capabilities.

# 1.3   Synchronizing by Process Organization

Forced by market and technology evolution, companies are focusing on both technological and organizational aspects to improve performance and then competitiveness. In fact, as several experiences have dramatically shown, a change only in either technological aspects (Evans et al. 1995, Hammer and Champy 1993) or organizational aspects is unable to provide significant and robust improvements in performance. For instance, the simple use of information technology is not able to solve synchronization problems between production and demand rates in manufacturing systems. Performance improvements can be obtained if the organization is modified and adapted to new information technologies. Similarly, the simple implementation of a production planning and control technique cannot improve significantly manufacturing system performance. Focusing primarily on the organization aspects seems particularly suitable to increase the effectiveness of the technological and organizational changes (Duimering et al. 1993).

## Process Organization

All manufacturing organizations, aiming at synchronizing manufacturing rate with demand rate, must cope with a certain amount of variability, which is caused by market, suppliers, competition, government regulations, and so forth. The choices an organization makes to handle variability within these external domains have implications on how the organization functions internally. The same logic applies internally. At every organizational level, each unit's activities may generate undesirable variability for other units. Organizations must make sure that the amount of variability generated by one unit matches the capability of other units to handle that variability. An effective management of variability requires a strong degree of communication and coordination among interdependent units. Such coordination is quite difficult to achieve when interdependent units are separated in the organization (Duimering et al. 1993). Galbraith (1973) suggests that when there is a high degree of uncertainty among interrelated tasks, organizations should try to reduce uncertainty as well as associated information processing and coordination costs by grouping these tasks together. The principle of grouping highly interdependent activities applies at various organizational levels (from manufacturing processes to group units and departments).

Reducing the impact of environmental variability, and reducing internal processing variability allow the reduction of production throughput times. Then, organizations must create an integrated manufacturing system, which is essentially an issue of organizational rather than of technological system design. From this point of view, an integrated manufacturing system would be one in which interrelated organizational tasks and activities are effectively coordinated with each other resulting in shorted throughput times.

Fast and unpredictable changes in market and technology can often force companies to fast organization redesign, which is usually focused on higher integration and coordination among activities and tasks. In this context, business process reengineering (Hammer 1990), core process redesign (Kaplan and Murdock 1991), and business process redesign (Davenport and Short 1990) are some of the approaches proposed to gain competitiveness by undertaking a change of organization's internal processes and controls from a traditional vertical, functional hierarchy to a horizontal, cross-functional, team-based, flat structure that focuses on customers delighting. The evolution would consist in moving to teams of people who bring their specialisms and abilities to specific processes, change projects or technical support projects characterized by highly interdependent activities. A process is defined as a set of structured activities designed to produce an output for a customer or a market. Process design is focused on how the organization works rather on what the organization produces. Then a process is a specific and logic sequence of activities having well-defined inputs and outputs (Davenport 1993).

A high reliance is placed on an open, informal culture with excellent communications within the process team backed by a degree of information and communication technology. Teams are networked using information technology and the organization links strategy, technology, and people into a model that ensures flexibility and responsiveness to change.

The reduction of time to market, cycle time, and costs as well as the improvement of quality can be obtained by the reengineering of a company's activities. The major benefits can be grouped into four categories:

1. cost (reduction in work effort required to perform a process, elimination of unnecessary tasks, more effective ways of performing tasks);
2. quality (reduced error rate, improved performance matching client needs and expectations, improved and innovative products and services);
3. time (faster access to information and decision-making, elimination of waiting time between steps); and
4. working environment (improved morale, teamwork and commitment, working conditions).

## Process Identification and Integration

Processes are customer-oriented and must include only activities that add value and satisfy customer needs. Customers can be both internal and external to the organization and the responsibility of each process has to be clearly defined. Different organization functions can be involved in a process. Then a strong coordination among organization functions has to be embedded in process operations (Omachonu and Ross 1994). Teams are important to perform processes; in fact, coordination and integration among human resources is the major issue in process design. Information technology applications should be oriented to integrate and coordinate activities within a process and between processes and also to avoid automation islands. For each process design, inputs, tools and techniques, and outputs should be identified. For each input and output, provenience and destination should be determined.

Processes frequently cut "horizontally" across the "vertical" functional boundaries of the marketing, production, accounting, etc. departments of the "traditional" organizations. Even if the organization is not process-oriented, processes exist and probably have not been documented, and no one is responsible for them. They need to be identified. Also, investments in training and information technologies, that are usually departmental, need to be applied to the processes that actually deliver the outputs sought. The process performances are

1. effectiveness, i.e., the extent to which the outputs of the process meet the needs and expectations of customers (timeliness, reliability, accuracy, and usability);
2. efficiency, i.e., the extent to which resources are minimized and waste is eliminated in the pursuit of effectiveness (processing time, waiting time, and resource cost per unit); and
3. adaptability, i.e., the flexibility of the process to handle future, changing customer expectations and today's individual, special customer requests (average time for special customer requests, percent of special requests turned down).

In a process-oriented organization, both process improvement and innovation should be pursued. For instance, a company analyzing its process of order management can eliminate redundant or no value-added operations to obtaining process improvement. Another company can decide to be connected with customers by a computer network to assure a computerized order management. This process innovation can allow a faster response to customer orders and a sale personnel reduction.

Process innovation is based on the integration of information technology applications and changes of organization and information flow. However, organization changes can involve difficult steps during their implementation.

Process organization and information technology have become inextricably linked. The creative use of IT (e.g., Electronic Data Interchange links) could be a key element of a process reengineering solution, while IT tools (e.g., process modeling) could be used to support the actual work of reengineering. The IT main objectives are

1. to increase the speed of the work flow;
2. to improve the reliability of the transferred data;

3. to decentralize; and
4. to support and improve the communication with customers and suppliers.

Then, IT tools can be effective if they can support "network" relationships with customers, suppliers and business partners, enable new patterns of work, and facilitate the synchronization of the production flow with the demand. In fact, as mentioned earlier, the electronic marketplace in which companies conduct a major portion of business electronically is rapidly becoming a reality. This new way of doing business is possible because of an array of technologies (including such computer/communications technologies such as electronic data interchange, electronic mail, fax messaging, and electronic funds transfer) that support the inter-enterprise movement of information. These technologies are able to eliminate paper from business-to-business transactions. However, the most important issue is that they enable companies to reengineer their fundamental processes.

Among the different IT tools available, a short description of the most used and promising is hereafter provided.

People in different places, but locally grouped, can be linked on a Local Area Network (LAN) to share files and exchange documents. LANs can be also linked together via Wide Area Networks (WANs). Using LANs and WANs, information and communication technologies can be linked to help organizations radically change the way they work. The linkages can be achieved in different ways: phone lines, mobile phones, satellites, hosts, and Integrated Service Digital Networks (ISDNs). There is, for instance, an increasing use of ISDN for video conference.

Electronic Data Interchange (EDI) is a tremendous cost and time-saving system. Since the transfer of information from computer to computer is automatic, there is no need to recode information. For instance, purchase order costs can be strongly reduced. EDI can also support improvements in customer service. The quick transfer of business documents and the marked decrease in errors during transmission allow to fill orders faster. Someone can send stocks as the EDI system reports, and automatically bill the client. It cuts days, even weeks, from the order fulfillment cycle and can ensure the synchronization of the production rate with the demand rate.

The workflow software allows a previously manual process, with defined steps of information retrieval, manipulation and onward distribution, to be automated. The software manages the flow of work at each point of the cycle. Such software is usually adapted using flexible software that allows a high degree of adaptability to current/new processes. The groupware software differs from the workflow software in that it is not tied to a specific process, but allows different people working in different locations to communicate and work together.

Smart/intelligent agents are software packages that can look at a variety of databases that are continually updated, and then alert the user on any changes that need particular attention, or filter out irrelevant data (to the user's specifications). They can take new data and automatically update the user's own database as well as those of others he or she needs to interact with. They certainly help to overcome the information overload that can be faced by people using IT systems linked to different databases and other users.

Some other tools are available. For instance, the Internet allows effective searches for goods and services.

## Synchronizing by Processes

The major processes in a company (Davenport 1993) can be classified as:

1. operation processes, such as product development, customer acquisition, production, logistics, order management, and after-sale customer support; and
2. management processes, such as performance monitoring, information management, assets management, human resource management, and resource planning.

Some key processes can be particularly effective to achieve the synchronization of production rate with demand rate. In fact, order management, production, logistics, and resource planning are considered processes whose design/redesign can provide significant improvements in synchronization performance.

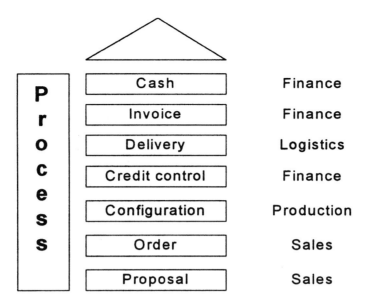

**FIGURE 1.2**   Order management process and its interactions with business functions.

## Order Management Process

This process includes all the activities between the order acceptance and the customer payment. The major goals are to reduce the process lead time and cost, and to increase the customer satisfaction.

Several business functions are involved in performing activities related to the order management process as shown in Figure 1.2. Information technology can enhance the activity and resource coordination and facilitate communications. Among the different applications that can support the process, multimedia, electronic marketplace, and market analysis and forecasting at a microlevel are the most promising.

## Production Process

This process is related to all the activities that transform raw materials in finished products. It has strong interdependencies with engineering, logistics, and order process. Company performances in terms of cost, time, and quality are affected by the coordination level in all manufacturing activities. The major issue is how to create a flow of products without defects and synchronized with the customer request. Manufacturing system layout (for instance, cellular manufacturing), process technology, planning and control techniques, and computer-aided manufacturing applications can enhance performance. However, in several cases (Schonberger 1982 and Womack et al. 1990) the most important benefits have been achieved by modifying work organization through simplification and waste elimination.

Also, teams have shown to be highly effective to increase worker commitment and reactive response to problems (Cutcher-Gershenfeld et al. 1994). These characteristics are essential to improve flexibility and productivity.

## Logistics Process

This process is related to the materials movements inside the shop and between it and customers and suppliers. This process also concerns the supply chain management and very strongly contributes to the effectiveness of the synchronization of production activities with customer demand. A full integration of the supply chain involves the extension of the internal management philosophy (not strict control) to suppliers and customers—to create strategic partnerships which often mean a supplier number reduction (Evans et al. 1995). A deep understanding of customer requirements will change the way the business operates, i.e., from pushing products to selling desired goods.

The strategies within the supply chain must have the common goal of improving the performance of the chain as viewed from the final customer's perspective. An overall strategic intent is therefore necessary with mutually defined goals and rules of operations within the supply chain. This is believed to be most effectively achieved with an integrated logistics information system providing a rapid information channel. This should be engineered in parallel with a lean materials channel to provide a rapid and transparent supply chain (Evans et al. 1995).

Several applications support the logistic process, ranging from material/resource position identification to warehouse automation and advanced resource planning. Information and communication technology provides new tools able to increase the process productivity and effectiveness such as satellite-based traffic control, barcode diagram, and cellular communication.

Barcode and scanning techniques have made it easier to track the physical progress of goods while the exchange of information through electronic data interchange workflow and document management have streamlined administration.

Large companies, whose operations have been centralized, are increasingly looking at standardization and globalization of their software solutions. These enable companies to install software in each national center for integrated accounting, manufacturing, and distribution packages. Good communications are vital at every stage and they have to extend beyond the organization to suppliers. For instance, this has become vital to push the principles of (and responsibilities for) just-in-time manufacturing techniques further down the line.

### Resource Planning Processes

This management process can contribute significantly to improve time and cost performance of manufacturing systems and of the whole company. In particular, by strengthening interactions among all the participants to planning decisions, personnel involvement and learning process are more effective. In fact, group decision making and problem solving increase the response timeliness and effectiveness creating opportunities for innovation in procedures. Simulation tools and scenario analysis are other traditional tools that can support resource planning. Artificial intelligence techniques promise to provide significant improvements in this process, especially for complex businesses. Some of the most important manufacturing planning and control techniques and systems are reported and analyzed in the next section.

## 1.4 Synchronizing by Manufacturing Planning and Control (MPC) Techniques and Systems

In the manufacturing environment, inventory, scheduling, capacity, and resource management requires planning and control. As manufacturing facilities grow, planning and control become extremely complex, and efficiency and effectiveness of manufacturing must be assured by planning and control techniques and systems. The synchronization of the production rate with the demand rate is one of the most important problems that manufacturing companies try to solve by means of planning and control activities. However, the needs of different firms call for different activities and each company adopts different techniques and systems with diverse degrees of detail for planning and control.

In any firm, manufacturing planning and control (MPC) encompass three different aspects or phases. In the first phase, the overall manufacturing plan for the company is created. It must be stated in production terms such as end items or product options. In the second phase, the detailed planning of material and capacity needs to support the overall plans is defined. In the third and final phase, these plans are executed at the shop floor and purchasing level.

### A General Framework

A general framework for manufacturing planning and control systems is shown in Figure 1.3. This framework can be useful to evaluate and compare alternative systems (Vollmann et al. 1992). The scheme represents a set of functions that must be performed, coordinated, and managed. It does not indicate how

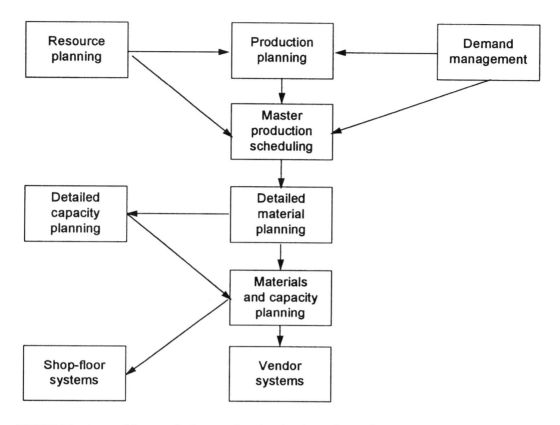

**FIGURE 1.3**   A general framework of a manufacturing planning and control system.

the functions are to be performed or which technique has to be adopted. However, each of these functions has to be accomplished in any firm. In this scheme no reference is made to the emphasis that each module will have in a specific company.

Different firms can also use different systems or techniques. Moreover, all systems can be improved. Once a fairly complete integrated system is in place, it is easier to evaluate and implement new procedures. The specific procedures must be evaluated on an ongoing basis, considering both possible alternatives and what is currently considered as important. In some instances, activities can be combined, performed less frequently, or reduced in importance.

For many of the newer approaches to manufacturing planning and control, an upgrade of the existing techniques and systems is not enough. It is also necessary to consider the broad management and sociotechnical environments in which manufacturing planning and control systems operate. Sometimes it is necessary to make major changes in managerial practices and in the evaluation systems for manufacturing performance to realize improvement (Vollmann et al. 1992). This is particularly true for the new approaches based on the intensive use of information technologies.

Choosing a system and making it work requires a careful evaluation of human resource capabilities (Aggarwal 1985). For instance, even if executives can ignore the complex design, huge input requirements, and large investments each of these systems needs, most of them cannot overlook the real constraints in terms of the working habits of employees who must operate them. In fact, some executives have tried to relay to their employees:

1. the perpetual need for updating the system in accordance with actual happenings;
2. the importance of precise record keeping for receipts, issues, returns, rejects, and so forth; and

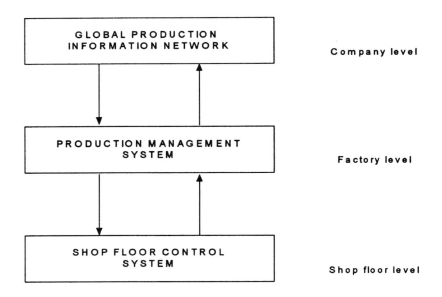

**FIGURE 1.4**  Production information systems.

3. the strict discipline required for the movement of materials, physical inventory and storage control, and placement and disbursement of parts, components, and subassemblies.

Successful companies have developed in-house competence in working with these systems. Initially, they have used experts to train employees, after which each company has struggled on its own to solve the problems that surface constantly.

## Production Information Systems

Facing drastic environmental changes in production activities, such as the change from small variety and large volume to large variety and small volume production, the increasing quantity of information for production management requires information systems in order to make better and quick decisions.

Production information systems represent the backbone for manufacturing planning and control activities. Planning and control techniques are based on specific information requirement and processing and often the advance in such techniques is related to information technology improvements. The production information system can be divided (Usuba et al. 1992) into three levels hierarchically as shown in Figure 1.4. The first and top level is the global production information network for all company activities. The second level is the production management system for activities at the factory. The third and lowest level is the shop floor control system for information flow and activities on the shop floor.

The global production information network support functions are essential for optimizing global business logistics so that the allocation of work and the exchange of information among business operations located in different areas can be handled effectively (Usuba et al. 1992). The main functions are related to supply and demand ordering, distribution, procurement, and technical information management.

The production management system exists primarily to plan and control the efforts of employees and the flow of materials within a factory. Materials in manufacturing are the parts (raw and in-process), the assemblies, and the finished products. As materials move from one operation to the next, their status changes and their progress needs to be monitored. Management must ensure that everything runs smoothly and efficiently. If parts are unavailable or if certain work centers become bottlenecks, the associated loss in productivity must be considered a production control problem.

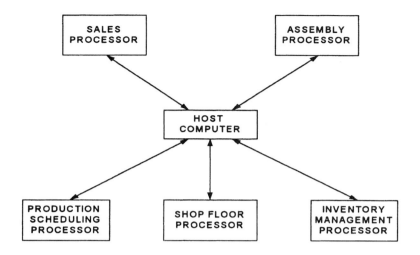

**FIGURE 1.5**    Information flows and processing for a real-time computerized production control.

The management control is accomplished through the information system that monitors the production activity, assisting in identifying problems as they occur and communicating the corrective decisions to those who will implement them. At the shop floor level, production control requires that information about production operations be made available to decision makers in a timely fashion. This means that data have to be gathered from all segments of the production operation: receiving, shipping, warehousing, manufacturing, maintenance, and so on.

Two important aspects (Bedworth and Bailey 1987) of the production control, rooted in the production information system, are

1. the shop floor information input; and
2. production control information outputs.

The large volume of these information flows usually requires the computerization of the production control. The value of fast and accurate information flow suggests that the cost of extensive computerization is often justifiable as it provides evident benefit to the synchronization of production rate with demand rate (Bedworth and Bailey 1987). A simplified scheme of information flows and processing for a real-time computerized production control is reported in Figure 1.5.

## Techniques

In this section the most important techniques adopted to support manufacturing planning and control are described. These techniques can effectively contribute to ensure the synchronization between the production and demand flows.

### Demand Management

A highly integrative activity that significantly contributes to the synchronization of production rate with demand rate is demand management. Through demand management all potential demands on manufacturing capacity are collected and coordinated. This activity manages day-to-day interactions between customers and the company (Vollmann et al. 1992). Starting from an accurate planning of all externally and internally generated demands, capacity can be better planned and controlled, timely and honest customer order promises are possible, and physical distribution activities can be significantly improved. Demand management concerns forecasting, order entry, order-delivery-date promising, customer order service, physical distribution, and other customer-contact-related activities. Demand management also takes in consideration other sources of demand for manufacturing capacity, including service-part demands, intra-company requirements, and pipeline inventory stocking.

For many firms, planning and control of demand quantities and timings are a day-to-day interactive dialogue with customers. For other firms, particularly in the process industries, the critical coordination is in scheduling large inter- and intra-company requirements. For still others, physical distribution is critical, since the factory must support a warehouse replenishment program that can differ significantly from the pattern of final customer demand.

Demand management is a gateway module in manufacturing planning and control providing the link to the marketplace. Activities performed here provide coordination between manufacturing and the marketplace, sister plants, and warehouses. In addition to the synchronization and communication between market and manufacturing, a key activity in the demand management module is to assure the completeness of demand information. All sources of demand for manufacturing resources must be identified and incorporated in the production and resource-planning processes, including, for instance, spare part demand, inter-company transfers, and pipeline buildups.

The interactions between demand management and master production scheduling (MPS) are frequent and detailed. Details vary significantly between make-to-stock (MTS), assemble-to-order (ATO), and make-to-order (MTO) environments (Groenevelt 1993).

In make-to-stock environments there are very few actual customer orders since demand is generally satisfied from inventory. Thus, the master production scheduling provides inventory to meet forecasted customer orders.

In the assemble-to-order environment, the most important task of MPS is to provide viable customer promise dates. Usually, there are customer orders already booked for several periods into the future. The master production scheduler uses the available-to-promise concept for each module or customer option to manage the conversion from forecasts to booked orders.

Still different demand management/MPS interactions characterize the firm in a make-to-order environment even though there is a relatively larger backlog of customer orders. Some orders can be in progress even though they are not completely specified and engineered. In this case the master production scheduler is concerned with controlling such customer orders as they progress through all steps in the process. This also needs engineering activities. Those activities impact on resource requirements as well as on manufacturing. All this has to be coordinated with customers as the orders become completely specified. This is also referred to as Flexible Manufacturing.

In each company environment, the objective of the demand management module is always to bridge the firm and the customer. However, companies differ for the types of uncertainty that affect them. Master production scheduling and demand management can facilitate the buffering against uncertainty. In the MTS case, uncertainty is largely related to the demand variations around the forecast at each of the inventory locations. In this case, levels of safety stock (and/or safety lead time) must be set in order to provide the required service levels. In the ATO case, the uncertainty affects not only the quantity and timing of customer orders but product mix as well. Safety stocks can be used. In MTO environments, the uncertainty is often related to how much of the company's resources will be required as the engineering is finally completed and exact requirements are determined.

Distribution activities are planned using information developed in the demand management function. Distribution requirements planning (DRP) provides the basis for tying the physical distribution system to the manufacturing planning and control system. Customer delivery promise dates, inventory resupply shipments, interplant shipments, and so on are used to develop short-term transportation schedules. Information used for master production schedules can be integrated with distribution planning and used to plan and control warehouse resupply. Then, a DRP system supports management anticipate future requirements in the field, closely match materials supply to demand, effectively deploy inventory to meet customer service requirements, and rapidly adjust to the variability of the marketplace.

DRP is a link between the marketplace, demand management, and master production scheduling. The link is effected through time-phased information on inventories and through materials and shipping plans that coordinate activities in these modules. Finished-good inventories are often positioned in a complicated physical system, consisting of field warehouses, intermediate distribution centers, and a central supply. The most important data for DRP are the planned timings and quantities for replenishing inventories throughout the physical distribution system. These data take into account currently available field inventories and forecasts.

DRP uses detailed records for individual products at locations as close to the final customer as possible. The record for a single stock-keeping unit (SKU) at a field warehouse is essential. In fact, products need to be identified also in terms of their locations. Quantities and times are insufficient data to effectively synchronize the production rate with the demand rate.

Plans derived from the DRP information and from the resultant shipping requirements are the basis for managing the logistics system (Jarvis and Ratliff 1992). By planning future replenishment needs, DRP establishes the basis for more effective transportation dispatching decisions. These decisions are continually adjusted to reflect current conditions. Long-term plans help to determine the necessary transportation capacity. As actual field demands vary around the forecasts, DRP continually makes adjustments, sending products from the central warehouse to those distribution centers where they are most needed. If insufficient total inventory exists, DRP provides the basis for deciding on allocations. Some policies to manage such a situation are providing stock sufficient to last the same amount of time at each location, or favoring the "best" customers, or even accurately saying when delivery can be expected.

Safety stocks are also adopted in DRP. In distribution, not only are planners concerned about how much uncertainty there is but also where it is. Less is known about where to put safety stocks (Vollmann et al. 1992). One principle is to carry safety stocks where there is uncertainty, i.e., close to the customer. If the uncertainty from several field locations could be aggregated, it would require less safety stock than having stock at each field location. In this case, the concept of a "national level" safety stock (Brown 1977) has been proposed. Some central stock that can be sent to field locations as conditions warrant, or to permit transshipments between field warehouses is the basic concept. This added flexibility would provide higher service levels than maintaining multiple field safety stocks.

It is in demand management that service levels and resultant safety stocks can be defined. The requisite degree of flexibility for responding to mix or product design changes is set here as well. This is performed through the determination of buffer stocks and timings. The master scheduler is responsible for maintaining the required level of buffer stocks and timings. The company's service levels are realized through conversion of day-to-day customer orders into product shipments. Careful management of actual demands can provide the stability needed for efficient and effective production. Booking actual orders also serves to monitor activity against forecasts. As changes occur in the marketplace, demand management can and should routinely pick them up indicating when managerial attention is required.

Monitoring activities of demand management fall into two broad categories concerning the overall market and the detailed product mix. The most appropriate activity for production planning is related to overall market trends and patterns. The second activity concerns managing the product mix for master production scheduling and customer order promising. A day-to-day conversion of specific demands to manufacturing planning and control actions requires managing the mix of individual products.

Actual customer demands must be converted into production actions regardless of whether the firm manufactures MTS, MTO, or ATO products. Details may vary depending on the nature of the company's manufacturing/marketing conditions. In MTO environments, the primary activity is controlling customer orders to meet customer delivery dates. While firms often perform this function the same way for ATO products, communication with the final assembly schedule may also be needed to set promise dates. In both these environments, the communications from the customer (a request) and to the customer (a delivery date) can be named order entry, order booking, or customer order service.

In a MTS environment, since materials are usually in stock, the customer is more often served from inventory. If inventory is insufficient for a specific request, the customer must be told when materials will be available or, if there is allocation, told what portion of the request can be satisfied. Conversion of customer orders to manufacturing planning and control actions in the MTS environment triggers resupply of the inventory from which sales are made. This conversion largely recurs through forecasting, since the resupply decision anticipates customer orders. In this case, demand forecasting is essential for demand management.

Forecasts of demand are an important input to MPC systems. Short-term forecasting techniques are usually the most useful to day-to-day decision-making in manufacturing planning and control. However, other decision problems both in manufacturing and other functional areas of the firm require different approaches to forecasting. Among these, there are long-run decisions involving, for instance, investments for new plants or resources. In these cases, forecasts of aggregate levels of demand are required.

Therefore, a careful analysis about the choice of the most suitable forecasting technique is necessary. Accuracy, simplicity of computation, and flexibility to adjust the rate of response (Bedworth and Bailey 1987) can be considered the main criteria to select the technique. Accuracy in forecasting is a measure of how well a forecasted condition matches the actual condition when it occurs. In general, high accuracy requirements do not allow simplicity of computation. Therefore, a trade-off between the two criteria usually exists. If time is important, the computation has to be simple. The flexibility in adjusting the rate of response is a measure of the ability of a forecast in adapting to changing conditions. For instance, the forecast technique has to be able to adjust model parameters any time a change in the forecasting environment occurs.

A further general consideration in forecasting is the lead time, defined here as the elapsed time between making the forecast and implementing its result. One problem with a long lead time is that forecast errors grow as the lead times increase.

Another concept, related to lead time, is forecasting horizon, i.e., the elapsed time from when the forecast is acted on to when the effects of those actions take place. If the forecasting horizon is wide, large forecast errors are likely to occur. When the cost of forecast error is high, the demand for accuracy is great. Errors due to long lead times or forecast horizons must be reduced. In many MPC problems, forecasts for many thousand of items on a weekly or monthly basis have to be made: then a simple, effective, low-cost procedure is needed. Since the resultant decisions are made frequently, any error in a forecast can be compensated for in the decision next time.

How often to gather data is another relevant issue. In sales forecasting, data-gathering frequency is usually dictated by the forecast requirements. For instance, data cannot be brought in at a slower rate than the maximum allowable lead time.

Forecasting techniques can be classified into two types: quantitative and qualitative. Quantitative techniques are applicable when some quantifiable data are available about the past performance and it is assumed that the same pattern will continue in the foreseeable future. Time-series analysis and projection, and causal and econometric models are the most known quantitative techniques. Qualitative forecasting techniques are adopted when data are not available and forecasting has to be based on the best estimates of acknowledged experts in that specific activity or product. Subjective estimates and Delphi techniques are well-known approaches for qualitative forecasting.

While each technique has strengths and weaknesses, every forecasting situation is limited by constraints like time, funds, competencies, or data. Balancing advantages and disadvantages of techniques with regard to a situation's limitations and requirements is an important management task.

A very effective and practical guide that can help managers sort out their priorities when choosing a technique and enable them to combine methods to achieve the best possible results is provided by Georgoff and Murdick (1986). The authors have compiled a chart that profiles the 20 most common forecasting techniques and rates their attributes against 16 important evaluative dimensions.

This chart is still a powerful management tool because no dramatic breakthroughs in technique development have occurred during the last years. Efforts to improve forecasts are now focused on searching for better approaches to technique selection and combination. In MPC problems, quantitative forecasting techniques and formalized procedures are usually adopted assuming that past conditions that produced the historical data will not change. However, managers know very well that they cannot rely exclusively on past information to estimate the future activity. In some cases, data are not available. This occurs, for example, when a new product is introduced, a future sales promotion is planned, a new competitor appears, or a new legislation affects a company's business. In these cases new knowledge needs to be included in the forecasting system.

For aggregate forecasts, causal relationships and the statistical tools of regression and correlation can be used. In fact, causal models are those that relate the firm's business to indicators that are more easily forecast or are available as general information. Both quantitative and qualitative forecasting methods can also be utilized for medium-run decisions, such as the annual budgeting process.

In manufacturing planning and control, a linkage between the forecasting system and the order entry system is essential. Sometimes, if demand information is not available, sales data or shipment data can be used even if there is a difference between sales, shipments, and demand.

Sales forecasts usually have many different decision-making purposes in a company. There is often the need of a significant coordination because independent forecasts are rarely brought together within the organization to ensure that they are consistent.

Some methods, such as specialized bills of material (super bill) and pyramid forecasting, are available to integrate and force consistency between forecasts prepared in different parts of the organization (Vollmann et al. 1992).

It is well known that long-term or product-line forecasts are more accurate than detailed forecasts. There are several reasons for aggregating product items in both time and level of detail for forecasting purposes, although this has to be done with caution. For example, aggregating individual products into families, geographical areas, or product types facilitate the forecasting task but must be done in ways that are compatible with the planning systems. However, forecast errors are unavoidable. Many efforts to reduce them are wasted, since they will not go away. It is better to improve systems for responding to forecast errors.

For MTS demand management, providing adequate inventory to meet customers' needs throughout the distribution system and to maintain desired customer service levels requires detailed forecasts. In fact, in MTS situations the need is for item-level forecasts by location and time period.

For many firms, product-mix ratios remain fairly constant over time. In such situations, managerial attention can directly focus on forecasting overall sales. Percentage forecasting can be used to solve the individual item forecasting problem.

In the ATO environment, accurate promise dates to customers are the major concern. This requires MPS stability and predictability. The master scheduler can help to achieve this by using the concept of time fences (Vollmann et al. 1992). Setting the time fence and managing the hedge units must take into account both economic trade-off and current conditions. If the time fence is set too early, inventory will be carried at higher levels in the product structure, which often decreases alternative uses of basic materials and so reducing flexibility. If the time fence is set too late, time to evaluate whether the mix ratio is changing might not be enough.

In the MTO environment, the basic activity concerns the control of customer orders after they are entered into the system. Even though the number and timing of customer orders for some time into the future can be known, there can still be much uncertainty concerning these orders in terms of engineering and/or process technology details.

## Inventory Planning and Control

Basically, inventory allows to decouple successive operations or anticipate changes in demand. In this paragraph, only the inventory of final products is considered. This is influenced by external factors and product demand is usually subject to random variations. Several incentives and reasons motivate the carrying of final product inventory (Masri 1992).

In terms of customer service, maintaining final product inventories ensures the availability of a product at the time the customer requests it and minimizes the customer's waiting time or the product's delivery time. To provide effective customer service through product availability, carrying inventories at different post-manufacturing stages of the supply chain is required. In terms of demand-supply decoupling, inventories of final products serve as buffers to protect against variations on both the demand and the supply of the last stage of the supply chain. Inventory on the demand side protects

| Order frequency | Order quantity | |
| --- | --- | --- |
| | Fixed (Q) | Variable (S) |
| Variable (R) | Q,R | S,R |
| Fixed (T) | Q,T | S,T |

Q: order a fixed quantity (Q); S: order up to a fixed expected opening inventory quantity (S); R: place an order when the inventory balance drops to (R); T: place an order every (T) periods.

FIGURE 1.6    Inventory decision rules.

against fluctuations such as a sudden surge in demand by ensuring product availability. Similarly, inventory on the supply side acts as a buffer against such uncertainties as raw material shortage or manufacturing labor force strikes.

Incentives to pursue economies of scale in manufacturing, transportation, and ordering often result in carrying and storing inventories in amounts that are in excess of the quantities immediately needed. The inventory planning function is primarily concerned with the determination of appropriate levels of inventory. These are determined by minimizing the costs associated with inventories. These costs are often conflicting. Then, optimal inventory levels correspond to trade-off that is acceptable by management. Inventory-related costs are numerous and need to be accurately defined to enable managers to correctly carry out the inventory planning and control function.

Investment in inventory is not the only cost associated with managing inventories, even though it may be the most evident. Three other cost elements have to be considered: cost of preparing an order for more inventory, cost of keeping that inventory on hand until a customer requests it, and cost involved with a shortage of inventory. When demand exceeds the available inventory for an item, the incurring cost is more difficult to measure than the order preparation or inventory carrying costs. In some cases, shortage costs may equal the product's contribution margin when the customer can purchase the item from competing firms. In other cases, it may only involve the paperwork required to keep track of a backorder until a product becomes available. However, this cost may be relevant in cases where significant customer goodwill is lost. Basically, only two decisions need to be made in managing inventory: how much to order (size) and when to order (timing). These two decisions can be made routinely using any one of the four inventory control decision rules shown in Figure 1.6.

A common measure of inventory performance, inventory turnover, relates inventory levels to the product's sales volume. Inventory turnover is computed as annual sales volume divided by average inventory investment. High inventory turnover suggests a high rate of return on inventory investment. Nevertheless, though it does relate inventory level to sales activity, it doesn't reflect benefits of having the inventory. To consider a major benefit of inventory, firms can use customer service to evaluate their inventory system performance. One common measure of customer service is the fill rate, wihich is the percentage of units immediately available when requested by customers. The level of customer service can be also measured by the average length of time required to satisfy backorders, or percentage of replenishment order cycles in which one or more units are back ordered.

A number of order quantities should be evaluated to determine the best trade-off between ordering and inventory carrying costs. The well-known model to determine the economic order quantity provides the lowest-cost order quantity directly (Masri 1992). Common assumptions for the determination of order quantities are fixed demand rate and constant replenishment lead time. These assumptions are rarely justified in actual operations. Random fluctuations in demand for individual products occur because of variations in the timing of consumers' purchases of the product. Likewise, the replenishment lead time often varies because of machine breakdowns, employee absenteeism, material shortages, or transportation delays in the factory and distribution operations. To cope with these uncertainty sources, safety stocks are created. Two criteria are often used to determine safety stocks: the probability of stocking out in any given replenishment order cycle, or the desired level of customer service in satisfying product demand immediately out of inventory (the fill rate) (Masri 1992).

## Production Planning

The process of production planning provides a plan for how much production will occur in the next time periods, during an interval of time called "planning horizon" that usually ranges between six months to two years. Production planning also determines expected inventory levels, as well as the work force and other resources necessary to implement the production plans. Production planning is based on an aggregate view of the production facility, the demand for products, and even of time (using monthly time periods, for example).

The selection of the aggregation level is important and needs to be defined. Basically, the aggregation level of the products (one aggregate product, a few product categories, or full detail) and the aggregation level of the production resources (one plant or multiple resources) are the most important decisions to be made (Thomas and McClain 1993). A forecast of demand needs to be developed by product categories as an input to the production plan. The quality of the forecast varies widely in practice. Still some forecast, with an estimate of its accuracy, is always needed.

The production plan provides a direct and consistent dialogue between manufacturing and top management, as well as between manufacturing and the other functions. Then, the plan must be necessarily in terms that are meaningful to the firm's no-manufacturing executives. The production plan states the mission manufacturing must accomplish if the firm's overall objectives are to be met. Manufacturing management takes care of how to accomplish the production plan in terms of detailed manufacturing and procurement decisions. Conceptually, production planning should precede and direct MPC decision-making providing the basis for making the more detailed set of MPC decisions. The production plan has to be stated in dollars or aggregate units of monthly or quarterly output. Some firms refer to the output of individual factories or major product lines. Still other firms state the production plan in terms of total units for each product line. Measures that relate to capacity (such as direct labor-hours and tons of product) are also used by some firms.

An important linkage exists between production planning and demand management. This module quantifies every source of demand against manufacturing capacity, such as interplant transfers, international requirements, and service parts. The match between actual and forecast demand is monitored in the demand management module. As actual demand conditions do not correspond to forecast, the necessity for revising the production plan increases.

As shown in Figure 1.3, production planning is also linked with resource planning. This activity encompasses long-range planning of facilities. Extended production plans are translated into capacity requirements, usually on a gross or aggregate basis. In some firms, the unit of measure might be dollar output rates; in others, it might be labor-hours, machine-hours, key-facility-hours, tons of output, or some other output measure. The need is to plan capacity, at least in aggregate terms, for a horizon at least as long as it takes to make major changes.

Resource planning is directly related to production planning, since, in the short term, the available resources represent a set of constraints to production planning. In the longer run, to the extent that production plans call for more resources than available, financial considerations are needed. If the planning of plant and equipment, sometimes referred to as facility planning, takes a very long time period, these facility resources are assumed fixed in the production plan while manpower and inventory resources are assumed adjustable (Blocher and Chand 1992).

Much of the very near term production plan is constrained by available material supplies. Current levels of raw material, parts, and subassemblies limit what can be produced in the short run, even if other resources are available. This is often hard to assess unless information links from the detailed material planning and shop status data base are effective. The linkage of production planning with the master production scheduling will be analyzed in the next paragraph. The production-planning process requires the monitoring of performance. Deviations in output as well as changes in the master production scheduling usually call for adjustments in the production plan. Then, a frequent evaluation of production situation and comparison to plan are recommended.

Production planning is really important to firms facing time-varying demand commonly caused by seasonal fluctuations. Firms facing fairly constant demand patterns might prefer to go directly to the next planning level (master production scheduling) as little can be achieved at the production planning level.

Firms having time-varying demand might choose to reduce its effect through the use of marketing efforts. This option could be viewed as demand smoothing. Instead of demand smoothing (or occasionally in addition to demand smoothing) the firm faced with time-varying demand will either adjust production rates to respond to the changes in demand or keep constant the production rate and build inventories during low-demand periods to be used in the high-demand periods.

If the production rate is maintained constant and large inventories are used to absorb the demand fluctuations (a "production smoothing" strategy), inventory holding costs and costs due to possible customer dissatisfaction become high. On the contrary, if the production rate is continually adjusted to be synchronized with demand rate fluctuations (a "chase" strategy), overtime and work-force adjustment costs, and poor facility utilization arise. There are many ways to change the production rate, including subcontracting, hiring, and overtime, and the specific approaches to be used depend on the available alternatives. Then, in the case of time-varying demand, firms usually look for this trade-off (Blocher and Chand 1992). Using combinations of the chase and smoothing strategies, managers face a very large number of production plans to meet demand (Thomas and McClain 1993).

The production plan must be disaggregated into specific products and detailed production actions. Moreover, if the production plan has been based on a single facility, the problem of determining which facility will produce which products, in which quantities, is an important prerequisite for the planning at each facility. In fact, managers must make day-to-day decisions on a product and unit basis rather than on the overall output level. Some mathematical approaches have been proposed to solve these disaggregation problems (Bedworth and Bailey 1987, Shapiro 1993, Vollmann et al. 1992). Even though there has been some progress in both theory and practice, to date the number of applications is limited. An interesting review and discussion of these approaches are reported in McKay et al. (1995) and in Thomas and McClain (1993).

An approach to disaggregation that closely parallels the managerial organization for making these decisions is the hierarchical production planning (HPP). Actually, HPP is an approach to aggregate capacity analysis that is based upon disaggregation concepts and can accommodate multiple facilities (Bitran and Tirupati 1993). The approach incorporates a philosophy of matching product aggregations to decision-making levels in the organization. Thus, the approach is not a single mathematical model but utilizes a series of models, each model at the organization level where it can be formulated. Since the disaggregation follows the organization structure, managerial input is possible at each stage. In Figure 1.7 a scheme of the approach is reported.

Another principle in HPP is that it is only necessary to provide information at the aggregation level appropriate to the decision. Thus, for instance, it is not necessary to use detailed part information for the plant assignment decisions. Finally, it is necessary to schedule only for the lead time required to change decisions. That means detailed plans can be made for periods as short as the manufacturing lead times. The mentioned references can provide useful insights and further details about HPP.

## Master Production Scheduling

The master production schedule (MPS) is an anticipated build schedule for manufacturing end products (or product options). A master production schedule provides the basis for making customer delivery promises, utilizing plant capacity effectively, attaining the firm's strategic objectives as reflected in the production plan, and resolving trade-off between manufacturing and marketing.

The MPS takes into account capacity limitations, as well as desires to fully utilize capacity. This means some items may be built before they are needed for sale, and other items may not be built even though the marketplace demand them. These are stated in product specifications, and in part numbers for which bills of material exist in the MPS. Since it is a build schedule, it must be stated in terms used to determine component-part needs and other requirements. Specific products in the MPS can be end items or groups of items such as models. In this case, the exact product mix will be determined by a final assembly schedule (FAS), which is not ascertained until the latest possible moment. If the MPS is stated in terms of product groups, special bills of material (planning bills) for these groups are used. The production plan constrains the MPS since the sum of the detailed MPS quantities must always equal the whole

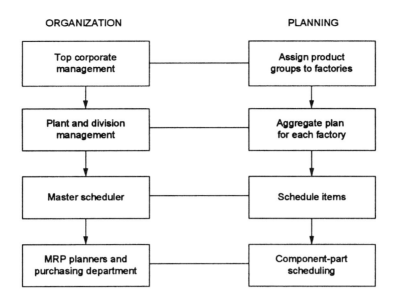

**FIGURE 1.7** A scheme for hierarchical planning.

quantity established by the production plan. The MPS drives the rough-cut capacity planning which involves an analysis of the MPS to determine the existence of manufacturing facilities that represent potential bottlenecks in the production flow.

At an operational level, the most important decisions concern how to construct and update the MPS. This involves processing MPS transactions, maintaining MPS records and reports, having a periodic review and update cycle (rolling through time), processing and responding to exception conditions, and measuring MPS effectiveness on a routine basis. On a day-to-day basis, marketing and production are coordinated through the MPS in terms of order promising. This is the activity by which customer order requests receive shipment dates. When customer orders create a backlog and require promise dates that are unacceptable from a marketing viewpoint, trade-off conditions are established for making changes.

Three classic types of MPS approaches have been identified depending on the environment: make-to-stock, make-to-order, and assemble-to-order. Each approach depends on the unit used for the MPS: end items, specific customer orders, or some group of end items and product options, respectively.

The MTS company produces in batches, carrying finished good inventories for most, if not all, of its end items. The MPS determines how much of and when each end item is to be produced. Firms that make to stock usually produce consumer products as opposed to industrial goods, but many industrial goods, such as supply items, are also made to stock. As MPS unit, all use end items, but many tend to group these end items into model groupings until the latest possible time in the final assembly schedule.

The MTO company, in general, does not carry end-item inventory and builds each end item as needed. This form of production is often used when there is a very large number of possible product configurations, and, thus a small probability of anticipating a customer's exact needs. In the MTO company, the MPS unit is typically defined as the particular end item or set of items comprising a customer order. In the ATO firm a large number of possible end-item configurations, all made from combinations of basic components and subassemblies, are possible. Order lead times are often shorter than total manufacturing lead times so production must be started in anticipation of customer orders. Due to the large number of end-item types, forecasting end-item demands are extremely difficult, and stocking end items is very risky. As a result, the ATO firm starts to produce basic components and subassemblies, but, in general, does not start final assembly until a customer order is received.

| | Week number | | | | | | | | | | | |
|---|---|---|---|---|---|---|---|---|---|---|---|---|
| | 1 | 2 | 3 | 4 | 5 | 6 | 7 | 8 | 9 | 10 | 11 | 12 |
| Forecast | 50 | 50 | 50 | 50 | 50 | 50 | 50 | 50 | 50 | 50 | 50 | 50 |
| MPS | 50 | 50 | 50 | 50 | 50 | 50 | 50 | 50 | 50 | 50 | 50 | 50 |
| Available | 10 | 10 | 10 | 10 | 10 | 10 | 10 | 10 | 10 | 10 | 10 | 10 |
| On hand | 10 | | | | | | | | | | | |

FIGURE 1.8   Simple scheme of MPS.

| | Week number | | | | | | | | | | | |
|---|---|---|---|---|---|---|---|---|---|---|---|---|
| | 1 | 2 | 3 | 4 | 5 | 6 | 7 | 8 | 9 | 10 | 11 | 12 |
| Forecast | 25 | 25 | 25 | 25 | 25 | 25 | 75 | 75 | 75 | 75 | 75 | 75 |
| MPS | 50 | 50 | 50 | 50 | 50 | 50 | 50 | 50 | 50 | 50 | 50 | 50 |
| Available | 35 | 60 | 85 | 110 | 135 | 160 | 135 | 110 | 85 | 60 | 35 | 10 |
| On hand | 10 | | | | | | | | | | | |

FIGURE 1.9   Production smoothing strategy in MPS.

| | Week number | | | | | | | | | | | |
|---|---|---|---|---|---|---|---|---|---|---|---|---|
| | 1 | 2 | 3 | 4 | 5 | 6 | 7 | 8 | 9 | 10 | 11 | 12 |
| Forecast | 25 | 25 | 25 | 25 | 25 | 25 | 75 | 75 | 75 | 75 | 75 | 75 |
| MPS | 25 | 25 | 25 | 25 | 25 | 25 | 75 | 75 | 75 | 75 | 75 | 75 |
| Available | 10 | 10 | 10 | 10 | 10 | 10 | 10 | 10 | 10 | 10 | 10 | 10 |
| On hand | 10 | | | | | | | | | | | |

FIGURE 1.10   Chase strategy in MPS.

Then, the ATO firm usually does not use end items as MPS unit. This is stated in planning bills of material. The MPS unit (planning bill) has as its components a set of common parts and options. The option usages are based on percentage estimates, and their planning in the MPS incorporates buffering or hedging techniques to maximize response flexibility to actual customer orders.

Basic techniques for MPS are based on time-phased records. These are used to show the relationships between production output, sales forecast, and expected inventory balance. Using time-phased records as a basis for MPS preparation and maintenance means that they can be easily computerized, and they are consistent with material requirements planning techniques.

In Figure 1.8 a highly simplified scheme of a MPS involving an item with an initial inventory of 10 units, sales forecast of 50 units per week, and MPS of 50 units per week as well. The MPS row states the timing for completion of units available to meet demand. In the example, the record covers a 12-week period (planning horizon) with total sales forecast equal to 600 units. The total MPS is also 600 units.

The planned available inventory balance (available) is shown in the third row of the record in Figure 1.8. The "Available" row represents the expected inventory position at the end of each week in the 12-week schedule. It results from adding to the initial inventory of 10 units the MPS of 50 units per week, and subtracting the sales forecast of 50 units per week. Any negative values in the "Available" row represent expected backorders. The main reason to hold a positive planned inventory balance is the variability in demand and production.

In Figures 1.9 and 1.10 a different sales forecast from that of Figure 1.8 is presented. The overall result is the same: total sales of 600 units as well as total production of 600 units during the 12-week period. The two MPS perform two extreme strategies: production smoothing in Figure 1.9 and chase in Figure 1.10.

Obviously, there are many possible MPS plans between these two extremes. The goal is to find that plan that represents a suitable trade-off between costs and benefits. For instance, batch manufacturing can be adopted. In this case, a trade-off between carrying inventory and incurring possible back orders has to be found. Some adjustment to the MPS can be required to reflect actual conditions. However, high costs are typically associated with making production changes. The MPS should be buffered from overreaction with changes made only when essential. A stable MPS permits stable component schedules with improved performance in plant operations. Too many changes in the MPS are costly in terms of reduced productivity. However, too few changes can lead to poor customer service levels and increased inventory. In some cases, the delivery date of customer orders can be negotiated. If the company has a backlog of orders for future shipments, the order promising task is to determine the promise date.

Some approaches are available for order promising (Vollmann et al. 1992) that allow companies to operate with reduced inventory levels. In fact, rather than carry safety stocks to buffer uncertainties, companies can "manage" the promise dates. In the ATO environment, safety stocks may be used for the options to absorb variations in the mix. When no safety stock is used for the common parts, buffer is provided for product mix variances but not for variances in the overall MPS quantity. In this case, the order promising needs a check for each option included in the order to accept a customer order.

The final assembly schedule is the schedule used to plan and control final assembly and test operations. It determines the types of end products to be built over some time period. The MPS states an anticipated build schedule. The FAS is the actual build schedule. In fact, MPS generally uses forecasts or estimates of actual customer orders; the FAS represents the as late as possible adjustment that can be made to the MPS. Even in the MTS environment, to increase flexibility the MPS can be stated in options or groups of items. In this way, the final commitment to end items can be made as late as possible.

The time horizon for the FAS is only as long as dictated by the final assembly lead time. A critical activity that parallels developing material plans is developing capacity plans. The MPS is a primary data source for capacity planning. Insufficient capacity quickly leads to problems in delivery performance and work-in-process inventories. On the other hand, excess capacity might be a needless expense. While the resource planning is directly linked to the production planning module, the MPS is the primary information source for rough-cut capacity planning. A particular MPS's rough-cut capacity requirements can be estimated by several techniques: capacity planning using overall planning factors (CPOF), capacity bills, or resource profiles (Vollmann et al. 1992). These techniques provide information for modifying the resource levels or material plan to ensure the MPS execution.

## Material Requirements Planning (MRP)

The output of the MPS process is a master schedule for final assembly/production. The master schedules for all components that feed into the end item need to be determined. These master schedules are usually defined in terms of when the components need to be released to manufacturing or a vendor to satisfy the end-item master schedule. One procedure for determining these release times (and the quantities to be released) is material requirements planning (Bedworth and Bailey 1987). Narrowly defined, a material requirements planning system consists of a set of logically related procedures, decision rules, and records (alternatively, records may be viewed as inputs to the system) designed to translate a master production schedule into time-phased net requirements and the planned coverage of such requirements for each component inventory item needed to implement this schedule (Orlicky 1975). Then, MRP is concerned with both production scheduling and inventory control. It provides a precise scheduling (priorities) system, an efficient and effective materials control system, and a rescheduling mechanism for revising plans as changes occur. It keeps inventory levels at a minimum while assuring that required materials are available when needed. The major objectives of an MRP system are simultaneously (Tersine 1985) to:

1. ensure the availability of materials, components, and products for planned production and for customer delivery;
2. maintain the lowest possible level of inventory; and
3. plan manufacturing activities, delivery schedules, and purchasing activities.

In the MRP logic, the concept of dependent demand is fundamental (Vollmann et al. 1992). The demand (gross requirements) for subassemblies or components depends on the net requirements

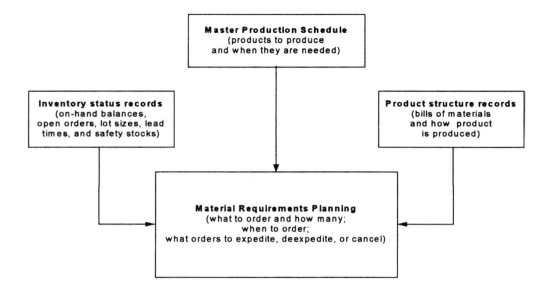

**FIGURE 1.11**   MRP inputs.

for their use in the end item. To do correct calculations, bill of material, inventory, and scheduled receipt data are all necessary. With these data, the dependent demand can be exactly calculated. It does not need to be forecasted. On the other hand, independent demand items are subject to demand from outside the firm and their needs have to be forecasted.

The three major inputs of an MRP system are the master production schedule, the inventory status records, and the product structure records. The master production schedule outlines the production plan for all end items. The product structure records contain information on all materials, components, or subassemblies required for each end item. The inventory status records contain the on-hand and on-order status of inventory items. A flow diagram of MRP inputs is represented in Figure 1.11.

MRP takes the master production schedule for end items and determines the gross quantities of components required from the product structure records. Gross requirements are obtained by "exploding" the end-item product structure record into its lower level requirements. The exploding process is simply a multiplication of the number of end items by the quantity of each component required to produce a single end item. The explosion determines what, as well as how many, components are required to produce a given quantity of end items. By referring to the inventory status records, the gross quantities will be netted by subtracting the available inventory items. The actual order quantity for an item may be adjusted to a suitable lot size, or it may simply be the net requirement. Order releases are scheduled for the purchase or the manufacture of component parts to assure their availability in a synchronized way with the total manufacturing process.

Considering when components are scheduled to be purchased or produced and the lead times for their supply, MRP time-phases orders by lead time offsetting or setbacks. For purchased components, lead time is the time interval between placement of the purchase order and its availability in inventory. For manufactured items, it is the time interval between the release of the work order and its completion. To make sure that the component is available when needed, the planned order is offset (moved back) so that it is placed at the beginning of the lead time. Normally, all components of an assembly are planned to be available before the start date and therefore are set back at least to the beginning of their lead time period. Thus, the material requirements for each component are phased over time in a pattern determined by lead times and parent requirements. MRP plans orders (planned order releases) for purchasing and shop scheduling for the quantity of items that must be available in each time period to produce the end items. The planned order release provides the quantity and time period in which work orders are to be released

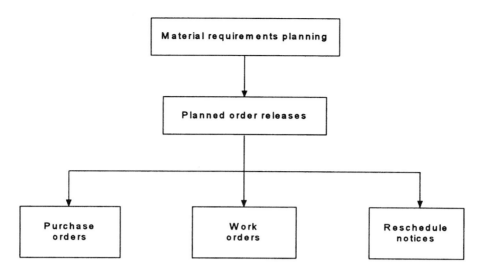

**FIGURE 1.12**   MRP outputs.

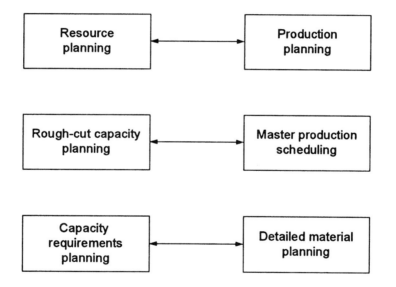

**FIGURE 1.13**   Long and medium range capacity planning.

to the shop or purchase orders placed to suppliers. A schematic of planned order releases as MRP outputs is contained in Figure 1.12.

## Capacity Requirements Planning

A critical activity that parallels developing material plans is developing capacity plans. Capacity planning decisions (Vollmann et al. 1992) are usually adopted starting from an overall plan of resources, proceeding to a rough-cut evaluation of a particular MPS's capacity needs, and moving to a detailed evaluation of capacity requirements based on detailed material plans. As shown in Figure 1.13, resource planning and rough-cut capacity are linked with production plan and MPS, respectively. Resource planning and rough-cut capacity planning have been briefly described previously. There has to be a correspondence between the capacity required to execute a given material plan and the capacity made available to execute the plan. Without this correspondence, the plan either will be impossible to execute or inefficiently executed.

Capacity requirements planning (CRP) is a commonly used technique adopted by companies that prepare detailed material plans using time-phased material requirements planning records (Vollmann et al. 1992). Each capacity planning technique's goal is to evaluate capacity required by either the production plan, the MPS, or the detailed material plan, so timely actions can be taken to balance the capacity needs with the available capacity. Once a feasible capacity plan has been developed, it needs to be monitored to see whether the actions were correct and sufficient. Capacity requirements planning is only applicable in companies using time-phased MRP records for detailed material planning and shop-order-based shop scheduling systems. CRP is unnecessary under JIT operations since minimal work-in-process levels mean that there is no need to estimate the impact on capacity requirements of partially processed work.

## Production Activity Control

Production activity control (PAC) concerns the execution of detailed material plans. It deals with the planning and release of individual orders to both factory and suppliers. PAC also takes care, when necessary, of detailed scheduling and control of jobs at the work centers on the shop floor (shop-floor monitoring and control systems, scheduling, and dispatching), and it takes care of purchasing and vendor scheduling (vendor scheduling and follow-up). A PAC system can reduce work-in-process inventories and lead times as well as improve vendor performance (Vollmann et al. 1992). Then, an effective PAC system can ensure the synchronization of production rate with demand rate. The shop-floor monitoring and control systems (SFMCS) involves the planning, release, and tracking of open orders that are on the shop-floor plus the coordination of several production and material-handling operations (Mabert 1992).

The American Production and Inventory Control Society Dictionary defines shop-floor control as "a system for utilizing data from the shop floor as well as data processing files to maintain and communicate status information on shop orders and work centers." Different variables can affect the characteristics of a SFMCS. For example, the type of product manufactured influences the shop layout, material flow, and manufacturing process. The objectives of top management control and influence the type of reporting system. Available capacity of manpower and/or equipment controls type and amount of information collected on the shop activities. In many manufacturing organizations SFMCS is computerized.

The shop floor monitoring and control system communicates priorities between the firm's planning system and the shop floor, evaluates job progress on the shop floor, and gathers shop performance information for the management control. SFMCS functions can be grouped into three areas: shop planning, shop monitoring, and shop controlling. Shop planning deals with the preparation of shop packet information, and the determination of a job's due date/release to the floor. Shop monitoring involves the collection of data on the status of a job, as it moves on the shop floor, and on the work center performance. Shop controlling determines the priorities of open-order jobs and communicates these priorities to the shop floor in order to coordinate production activities. Scheduling can be defined as the allocation of resources over time to perform a certain collection of tasks (Okogbaa and Mazen 1997).

Detailed scheduling of the various elements of a production system is crucial to the efficiency and control of operations. Orders have to be released and translated into one or more jobs with associated due dates. The jobs often are processed through the machines in a work center in a given sequence. Queueing may occur when jobs have to wait for processing on machines that are busy; preemption may occur when high-priority jobs arrive at busy machines and have to proceed at once (Pinedo 1992). Then, the scheduling process has to interface with the shop floor control. Events that happen on the shop floor have to be taken into account as they may have a considerable impact on the schedules.

Given a collection of jobs to be processed in the shop floor, scheduling means to sequence the jobs, subject to given constraints, in such a way that one or more performance criteria are optimized. Three primary objectives apply to scheduling problems: due dates, i.e., to avoid late job completion; flow times, i.e., to minimize the time a job spends in the system from when a shop order is open until it is closed; and work center utilization, i.e., to fully utilize the capacity of expensive resources. These three objectives often conflict and trade-offs have to be determined.

Another issue related to scheduling is the shop structure. Repetitive and cellular manufacturing systems are characterized by a flow shop structure; namely, all jobs tend to go through a fixed sequence of the same routing steps. Other structures are more of a job shop nature involving custom-made products. Each particular job has its specific routing, and the processing time required at each work center can be highly variable. Scheduling complexity and constraints in a flow shop can be significantly different from those in a job shop. Also, various forms of uncertainties, such as random job processing times, machine breakdowns, rush orders, and so forth may have to be dealt with.

Classical scheduling research has addressed two fundamental types of scheduling problems: static scheduling and dynamic scheduling. The static scheduling problem consists of a fixed set of jobs to be run. Typical assumptions are that the entire set of jobs arrive simultaneously and all work centers are available at that time. In dynamic scheduling problems new jobs are continually added over time. The research on the scheduling problem has produced extensive literature and even a general overview of the developed algorithms and techniques is not possible. Some references (Baker 1974, French 1982, Lawler et al. 1993, Okogbaa and Mazen 1997, Rachamadugu and Stecke 1994, Shapiro 1993) can provide a useful and complete guide for each manufacturing situation.

Real-life scheduling problems can be very different from the mathematical models studied by researchers in academia and industrial research centers (Pinedo 1992). Some of the differences between the real problems and the theoretical models regard the low level of knowledge about what will happen in the near future, the resequencing problems, the preferences in resource utilization, or the variable weight of a scheduling objective. Dispatching is a procedure that selects a job for processing on a machine that has just become available. The selection is usually based on logical decision rules, called dispatching rules, that determine the priority of each job. The priority of a job is calculated as a function of such parameters as the job's processing time and due date, the length of the queue in which the job is waiting, and the length of the queue at the next work center along the job's route. Once the priorities for all candidate jobs are determined, the jobs are sorted and the job with the highest priority is chosen. Dispatching rules are usually designed so that the priorities are easy to be calculated using up-to-date information pertaining to the jobs and/or work centers. Computational ease and real-time decision making are attractive features of dispatching (Bhaskaran and Pinedo 1992).

A noticeable shortcoming is the inherent myopic nature of dispatching. This priority-based scheduling approach suffers from the same disadvantages as many greedy optimization schemes; it can result in suboptimal schedules over a long time horizon. Consequently, dispatching is largely used in short-term scheduling where disturbances such as machine failures and rush jobs can have a significant impact on the schedule. In a longer time horizon, where one may assume a more stable system, detailed scheduling approaches based on advanced optimization techniques are preferable. As for scheduling problems, it is not possible in this chapter to overview the literature on dispatching problems. Some helpful references (Bhaskaran and Pinedo 1992, Blackstone et al. 1982, Lawler et al. 1993) are then provided.

## Vendor Scheduling and Follow-Up

The vendor scheduling and follow-up aspects of PAC are similar to the shop-floor scheduling and control systems. In fact, from the vendor's perspective, each customer is usually one of the demand sources, from the customer's standpoint, the objectives of vendor scheduling are the same as those of internal work center scheduling: keep the orders lined up with the correct due dates from the material plan. This means that a continually updated set of relative priority data has to be sent to the vendor. A typical approach to providing this information is a weekly updated report reflecting the current set of circumstances in the customer's plant and, sometimes, the final customer requirements that dictate them. Increasingly, computerized communication is used to transmit this information (Vollmann et al. 1992). Since the vendor follow-up system is often concerned with changes to the customer schedule, there must be limits to the amount of change the vendor will be asked to accommodate. Contractual agreements with the vendor typically define types and degree of changes that can be accepted, time frames for making changes, and so on.

# Systems

The implementation of the MPC techniques, either singly or in concert with others, requires a clear assess of the company organization and environment. Any MPC implementation usually replaces informal planning and control systems with formal systems. In this section, the most known systems used for planning and control manufacturing activities are presented and compared referring to their basic logic and philosophy. In actual cases, improvements in manufacturing performance can also be obtained by combining the use of different systems.

## Manufacturing Resource Planning

In a broader perspective, overall production planning activities are the result of the interaction of many different functions, such as general management, marketing, finance, manufacturing, and engineering. Proper tools and ability to use them are required to correctly perform and integrate all these activities to achieve the established goals.

For manufacturing managers, obtaining the right materials in the right quantity and at the right time is crucial. Managing the materials flow in a manufacturing company requires the balance of three objectives: maximizing customer service, minimizing inventory investment, and optimizing facility utilization. The manufacturing planning process starts with production planning. The MPS determines the top-level shop priorities therefore providing the direct input to MRP. The subsequent output is the input to CRP and PAC. The control phase requires follow-up actions for both materials and capacity. Unfortunately, customer demand is irregular, vendors do not always deliver on time, machines break down, people miss work. Disruptions in a manufacturing environment require a continuous control based on a fast updating about shop-floor situation, inventory status, and orders portfolio.

When a closed-loop system is introduced based on a real-time control tying MPS, MRP, CRP, and PAC, a closed-loop MRP system is obtained (Vollmann et al. 1992). This can be considered as a company-wide system able to support planning and control of company operations. For this system, Wight (1979) coined the term MRP II where MRP means "manufacturing resource planning." MRP II is a business planning and control system that makes a connection between aggregate-level planning and detailed plans. In MRP II both aggregate and detailed production planning are included and the planning approach considers capacity at several levels of aggregation (Thomas and McClain 1993). Figure 1.13 shows three levels of capacity planning: resource planning, rough-cut capacity planning, and capacity requirements planning. In MRP II these are used in three integrated levels of production planning: production planning, master production scheduling, and material requirements planning. Different patterns exist in different implementations of MRP II. The pattern shown in Figure 1.14 is illustrative.

The actual software for the implementation of MRP II systems usually has three major characteristics: it is designed in a modular way, integrates production and inventory plans into the financial planning and reporting system, and has a simulation capability to answer "what-if" questions. Some improvements in the business planning and control systems can be obtained by extending their influence to all enterprise (Enterprise Resource Planning) or supply chain (Supply Chain Synchronization) resources (Gumaer 1996, Martin 1995).

## Just-In-Time (JIT) Systems

The just-in-time (JIT) production system was first developed and promoted by Toyota Motor Company as "Toyota Production System" (Shingo 1989) and has been adopted by many companies around the world. The JIT production system is a reasonable way of making products because it completely eliminates unnecessary production elements to reduce costs. The basic and simple idea in such a production system is to produce the kind of units needed, at the time needed, in the quantities needed (Monden 1992). Even though the system's most important goal is cost reduction, three sub-goals must be first achieved. They are

1. quantity control, which enables the system to adjust to daily and monthly demand fluctuations in quantity and variety;

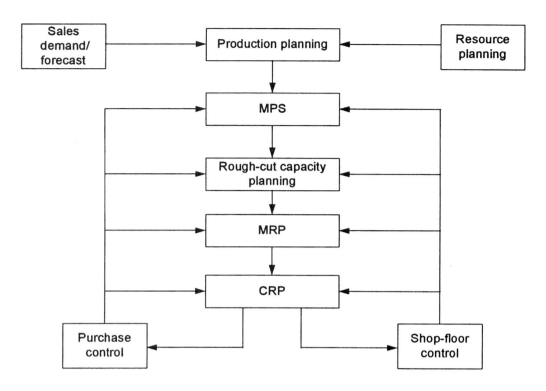

**FIGURE 1.14**  A simple scheme of a manufacturing resource planning system.

  2. quality assurance, which ensures that each process will supply only defect-free units to subsequent processes; and
  3. respect of the worker, which is the base to increase human resources productivity necessary for attaining the system's cost objectives.

A special feature of the JIT production system is that the primary goal cannot be achieved without realizing these subgoals, and vice versa. A continuous flow of production adapting to demand changes in quantity and variety is obtained by producing the needed units in the needed quantities at the needed time, by never allowing defective units from a preceding process to flow into and disrupt a subsequent process, by varying the number of workers based on demand changes, and by capitalizing on worker suggestions. To implement these basic concepts, the JIT system includes the following subsystems and methods (Monden 1992):

  1. the kanban system to maintain JIT production;
  2. a production-smoothing approach to allow adaptation to demand changes;
  3. shortened setup times that reduce production lead times;
  4. standardizing operations to attain line balancing;
  5. machine layouts that promote multiple-skilled workers and the flexible work-force concept;
  6. small-group improvement activities and a worker suggestion system to reduce the work force and increase the morale, respectively;
  7. a visual control system to achieve autonomous defect control; and
  8. a "functional management" system that promotes company-wide quality control, and so forth.

The kanban system is an information system that controls the production quantities in and between work centers. The kanban is usually a card. The two kinds used are the withdrawal and the production-ordering kanban. A withdrawal kanban tells the downstream work center to withdraw a certain quantity, while a production-ordering kanban tells the upstream work center what quantity to produce.

Production smoothing is the most important condition of kanban-based production for minimizing slacks in terms of work force, equipment, and work-in-process in each upstream work center including suppliers, and also for realizing JIT delivery of each product under a variety of products.

To prevent production rate fluctuations caused by the variability of the demanded quantity, an effort must be made to minimize production fluctuation in the final assembly line. The most difficult point in implementing smoothed production or reducing lead time is setup. Setup times must be shortened in order to increase process flexibility and reduce inventory investments.

The JIT system covers more than MPC activities and does impact all areas of the MPC framework (Vollmann et al. 1992) shown in Figure 1.3. JIT systems permit greatly streamlined execution on the shop floor and in purchasing, eliminating large portions of standard shop-floor control systems. A significant reduction of work-in-process, lead times, and costs of detailed shop scheduling is achieved.

Production plans and master production schedules will be required to provide smoothed shop operations. In many cases, this also requires a rate-based MPS. This drive toward more stable, level, daily-mix schedules is possible if many of the required JIT activities such as setup time reduction are implemented. To the extent that lead times are sufficiently reduced, many firms that had to provide inventories in anticipation of customer orders (MTS firms) now find themselves more like MTO or ATO companies able to respond to customer orders.

A general overview of the JIT system literature is reported in Groenevelt (1993) while the contribution of Schonberger (1982, 1986) can be considered the most useful for actual JIT implementation. In the JIT concept, the initial factor that activates the system is the customer demand. This initial motive will, in turn, function throughout the plant as a "pull system." Each subsequent process pulls the necessary units from its preceding process just in time and the ultimate subsequent process is a customer in the marketplace. However, the JIT system will not use only pull systems; it needs to use the "push system" at the same time. The push system refers to a central planning system in which the central production planning office instructs all processes including suppliers before the production starts. In other words, central plans will be pushed to each process simultaneously whereas kanban system (pull system) operates as a fine-tuning production control system. An interesting description of the pull system in the most known JIT environment, the Toyota MPC system, is reported in Monden (1992).

## Integrating MRP and JIT

The basic difference between pull and push systems is that a pull system initiates production as a reaction to current demand, while a push system initiates production in anticipation of future demand (Karmarkar 1989). These methods are not mutually exclusive, and each has its benefits and costs. The best solution is often a hybrid that uses the strengths of both approaches.

Pull methods tend to be cheaper because they do not require computerization (hardware and software) and leave control and responsibility at a local level. In this way significant incentives for lead-time management are possible. MRP systems are effective at materials planning and coordination and support inter-functional communication and data management.

The most suitable method for production control depends on the nature of the production process. In Figure 1.15 various manufacturing control approaches are proposed for different processes. In particular, continuous flow, repetitive batch, dynamic batch, and custom engineering are the process types considered.

In the continuous-flow process, the production process is dedicated to one or a few similar products. Production is continuous and level so that the lead time for production is uniform and predictable. For a continuous-flow process, ongoing materials planning is not essential and JIT supply techniques are appropriate. Orders are released without changes each week, so a rate-based approach can be adopted. At the shop-floor level, JIT materials-flow controlled by a pull release (kanban, for example) is effective.

| | | Materials Planning | Order Release | Shop Floor |
|---|---|---|---|---|
| Low | Pull: Continuous Flow | JIT | Rate Based | JIT-Pull |
| Lead-Time | Hybrid Push-Pull: Batch, Repetitive | JIT-MRP | Pull or MRP | Pull |
| Variability | Hybrid Push-Pull: Batch, Dynamic | MRP | MRP | Order Scheduling |
| High | Push: Custom Engineering | MRP | Order Scheduling | Operation Scheduling |

**FIGURE 1.15**  Types of production control.

In the batch and repetitive process, some part of the production system may resemble a continuous-flow process while other parts involve multiple products produced in batches. Lead times are fairly constant and predictable. The product mix is relatively constant. In a repetitive manufacturing environment with fairly stable but varying schedules, materials planning can be a hybrid of MRP and JIT methods. Order release may require MRP calculations if changes are frequent or if it is necessary to coordinate with long lead times or complex materials supply. Pull methods are suitable on the shop-floor.

In the batch and dynamic process, production is in batches and the output mix and volume can vary; customer orders arrive on a weekly and monthly basis. The load on the facility changes; bottlenecks can shift and lead times become variable. For more dynamic and variable contexts, such as job-shop manufacturing, MRP becomes crucial for planning and release. Pull techniques cannot cope with increasing demand and lead-time variability. The shop-floor control requires higher levels of tracking and order-scheduling complexity.

In the custom-engineering process, there is no regularity in production patterns because of low-volume, complex-engineered products, or with custom manufacturing. The load on the facility can vary widely and lead-time management requires a high level of analysis and detail. In such complex environments, sophisticated push methods are required. Where these are too expensive, poor time performance, large inventories, and plenty of tracking and expediting are the only alternatives.

As stressed before, there are possible modifications of existing MRP II systems, which add pull elements and remove some of the problems related to the system's lack of responsiveness. Some such modifications are "synchro-MRP," "rate-based MRP II," and "JIT-MRP" (Baudin 1990, Karmarkar 1989). These systems are appropriate for continuous-flow or repetitive processes, where production has a level rate and lead times are constant. In such situations, the order release and inventory management functions are not essential. The facility can be designed to operate in a JIT manner so that all materials that enter the facility flow along predictable paths and leave at predictable time intervals. Work is released by a pull mechanism, so there is no WIP buildup on the floor. In short, MRP is used mainly for materials coordination, materials planning, and purchasing and not for releasing orders. The shop floor is operated as a JIT flow system.

## Optimized Production Technology

Synchronized manufacturing, as achieved by the optimized production technology (OPT) system for production management, was first introduced in the United States at the end of the 1970s (Goldratt 1980). OPT encompasses many functions in the MPC framework (Vollmann et al. 1992) but combines them in a way that allows us to plan both materials and capacity at the same time. A basic principle of OPT is that only the bottleneck operations (resources) are of critical scheduling concern because production output is limited by them. Increased throughput can only be obtained by higher capacity utilization of the bottleneck facilities, and increased batch sizes are one way to increase utilization.

OPT calculates different batch sizes throughout the plant depending upon whether a resource is a bottleneck or not. This has several implications on the manufacturing planning and control. In fact, a batch size for one operation on a part could be different for other operations on the same part. OPT is designed to do order splitting.

The key to lot sizing in OPT is distinguishing between a transfer batch (that quantity that moves from operation to operation) and a process batch (the total lot size released to the shop). Any differences are held as work-in-process inventories in the shop.

The lot-sizing issue is closely related to a scheduling approach in OPT called drum-buffer-rope. The name comes from the bottleneck defining the schedule (the drum), the "pull" scheduling in nonbottleneck operations (the rope), and the buffers at both the bottleneck and finished goods (but not at nonbottlenecks). The basic concept is to move material as quickly as possible through non-bottleneck resources until it reaches the bottleneck. There, work is scheduled for maximum efficiency using large batches. Thereafter, work again moves at the maximum speed towards finished-goods stocks. What this means for lot sizing is very small transfer batches to and from the bottleneck, with a large process batch at the bottleneck. Then, JIT operating conditions are used everywhere except at the bottleneck. In fact, OPT states that utilization of the bottleneck is critical and reduced utilization of nonbottleneck resources costs nothing. In the traditional cost accounting view, people should always be working. If these people are at nonbottleneck resources, the net result of their work will increase work-in-process and create problems for scheduling at other resources. Under OPT, as with JIT, it is quite all right not to work if there is no work to do. In fact, working in this situation will cause problems.

A difference between OPT and JIT can be stressed. JIT techniques and philosophy are more appropriate to assembly and repetitive industries. OPT is best suited to a job shop environment where manufacturing situations keep shifting (Ronen and Starr 1990). The essence of the drum-buffer-rope philosophy is then to provide synchronization in manufacturing activities by (Umble and Srikanth 1990) developing the MPS so that it is consistent with the constraints of the system (drum), protecting the throughput of the system from the unavoidable reductions through the use of buffers at a relatively few critical points in the system (buffer), and tying the production at each resource to the drum beat (rope).

For a firm with an operating MPC system, OPT can be considered as an enhancement. OPT is an example of separating the vital few from the trivial many, and thereafter providing a mechanism to exploit this knowledge for better manufacturing planning and control. In Figure 1.16 some principles of OPT are summarized.

---

1. **Balance flow, not capacity**
2. **The level of utilization of a non-bottleneck is connected not to its own potential but to some other constraint in the system**
3. **Utilization and activation of a resource are not synonymous**
4. **An hour lost at a bottleneck is an hour lost forever**
5. **Bottlenecks determine both throughput and inventory in the system**
6. **The transfer batch may not be equal to the process batch**
7. **The process batch should be variable, not fixed**
8. **Schedules should be established by considering all constraints simultaneously. Lead times are the result of a schedule**

---

**FIGURE 1.16**   Main OPT principles.

While the details of OPT remain proprietary, it is known (Morton and Pentico 1993) to involve four basic stages:

1. determine the bottleneck resource in the shop;
2. schedule to use the bottleneck resource most effectively;
3. schedule the remainder of the shop up to the bottleneck; and
4. schedule the remainder of the shop after the bottleneck.

Although OPT was an extremely successful and widely used system, it suffered a number of legal problems. However, a number of newer systems incorporate and expand on many of OPT principles. A distinction between the OPT strategic managerial principles (BIG OPT) and the OPT scheduling mechanism (SMALL OPT) has to be made.

The BIG OPT approach may be applied to almost all types of processes without using any software. It is most suitable to use the SMALL OPT scheduler for the job shop environment or very complex assembly lines. Management by constraints (or theory of constraints) is the broad way to define BIG OPT concepts, and it is definitely an advanced step beyond what is normally considered OPT (Ronen and Starr 1990).

The concept of a bottleneck has been generalized into "constraint," which includes marketplace constraints. In fact, one goal is to have the shop output constrained by the marketplace, not by internal constraints over which the firm has more control. Constraints are explicitly identified, and following the drum-buffer-rope concepts, they are buffered with inventory and are operated continuously. Moreover, jobs are closely evaluated to find any that can be alternatively routed, even if the solution is more expensive for the work so routed. The goal is always to break a constraint or bottleneck condition, and thereafter identify the next constraint. Continuous improvement is an integral part of the theory of constraints philosophy. Moreover, the path for the improvement is directed by the theory and always follows the constraints.

OPT presents several difficulties in implementation. In general, it is not easy to use OPT. Companies need to understand basic finite scheduling concepts. They also need education, top-management support, and a willingness to unlearn some ingrained habits. Finally, many OPT results are quite counterintuitive (Vollmann et al. 1992).

# 1.5   More Advanced Approaches for Synchronization

The increasing complexity of the production environment is basically caused by the high innovation rate in manufacturing processes and by the market globalization. As the complexity increases, the manufacturing management requires more advanced approaches. In this section artificial intelligence and multimedia tools able to support manufacturing planning and control are presented as a way to face complexity by creating intelligent and flexible manufacturing systems. Also, the approach of global manufacturing is introduced to stress how manufacturing can be worldwide integrated to successfully compete in the global market.

## Artificial Intelligence and Multimedia

In manufacturing systems, artificial intelligence and multimedia tools are applied to different issues such as manufacturing planning and control. These tools can aspire to support several kinds of problem-solving and decision-making tasks (Dattero et al. 1989) providing manufacturing systems with intelligence (Dagli 1994).

Intelligent manufacturing systems make use of different technologies. In the following section, only expert systems, neural networks, and multimedia are shortly described in terms of their potential contribution to a more effective synchronization between production and demand flows.

### Expert Systems

Expert systems are suitable tools for manufacturing planning systems because these systems can support decision making to solve problems that are dynamic and have multiple criteria (Badiru 1992). These types of problems occur in quality, material shortages, vendor deliveries, forecast errors, and in the timing for new product roll-outs. Expert systems represent an attractive and reasonable set of tools able to provide alternatives for intelligent managerial choice. Expert systems are usually adopted in interactive decision support systems (DSS). For example, when a shortage of some item occurs, the system needs to find out which customer orders are impacted. If the shortage condition cannot be resolved, the expert systems have to be employed to analyze the consequences. The first step is to determine what factory orders are also tied to that order. Next, it is necessary to examine how the shortfall is to be resolved and what rescheduling actions are possible for the other resource areas. It is also necessary to find out which customer orders can be satisfied and, perhaps, to reorder the priorities for the resulting set of components so that revenue goals can be met. At the same time, it is important to meet as many shipping deadlines as possible and to use the resources of the factories as wisely as possible.

In planning, scheduling, and dispatch/control function, several algorithms and rules, often based on heuristics, have been developed and implemented by expert systems, ideally giving complete computer control. However, it has been recognized that humans and computers have different strengths and weaknesses and therefore should act as a team (Badiru 1992, Morton and Pentico 1993). Computerized scheduling strengths include a very fast computation, the capability to consider many factors, and a very reliable response. Computerized scheduling weaknesses include difficulty in learning from experience and no broad common sense. Human strengths include a very broad base of common sense, intuitive but poorly understood thinking abilities, and an ability to strategically summarize knowledge. All these abilities often work to allow experts to react quickly and accurately to glitches and problems. Human weaknesses include variability in response and difficulty in transferring their expertise to others. Human-machine interactive systems should be able to have both sets of strengths and fewer of the weaknesses.

Expert systems in scheduling fit the DSS interactive scheduling models. They generally emphasize heavily the automatic scheduling mode. There is typically a forward or backward chaining "inference engine," which implements rules (modularized heuristics) depending on the current conditions in the shop. Also, there is typically a knowledge base (sophisticated database). An expert system for scheduling can be considered a complex version of an interactive system for scheduling. These complexities result in the sophistication of the data management, manipulation of the decision rules by the inference engine, and sometimes in complex operations research heuristics used to form the rules. However, while prototypes of a number of scheduling systems have been built, tested, and demonstrated, very few have been actually implemented with success (Morton and Pentico 1993).

## Artificial Neural Networks

Recently, artificial neural networks (ANNs) have been receiving increasing attention as tools for business applications. In the business administration, ANNs can be used for solving a very broad range of problems. The function-optimizing approach is used for solving the traveling salesman problem or scheduling problems, the function-building approaches are used for credit assessment and classification of companies, and for predicting stock market evolution and exchange rates. Based on the results shown in the literature, it seems a likely supposition that function-building approaches are fundamentally better suited to supporting MPC than function-optimizing approaches (Corsten and May 1996, Zhang and Huang 1995).

In the MPC framework, particular decision situations arise, which are characterized by the fact that there is missing or uncertain information on cause-effect correlation or that there is no efficient solution algorithm. ANNs can provide the solution for the traveling salesman problem, a classic example of a problem without an efficient solution algorithm, and they can learn unknown cause-effect correlation if there is a sufficient amount of available data. This prerequisite is especially fulfilled in the shop floor because of the common use of data collection systems (Cheung 1994).

ANNs seem not suitable for administrative functions such as customer order processing or inventory control (Corsten and May 1996). In Figure 1.17 some areas of possible ANN applications in the MPC framework are shown. The master production schedule is based on customer orders or demand forecasts, depending on the production environment. ANNs can support master production scheduling in the case of no complex demand forecasts. In contrast, rough-cut capacity planning, which determines the capacity needs, can be adequately done by conventional methods, so that the use of ANNs seems inappropriate.

| Planning | master production scheduling | forecast |
|---|---|---|
| | | rough-cut capacity planning |
| | quantity planning | material requirement planning (demand driven/consumption driven) |
| | | order planning |
| | capacity planning | calculation of the capacity utilization |
| | | capacity adjustment |
| Control | order releasing | availability check |
| | | scheduling |
| | steering and control of the processes | capacity and order tracking |
| | | short-term interventions into running processes |

**FIGURE 1.17** Areas of possible applications of ANNs in a MPC system.

At the level of quantity planning, for demand-driven material requirements planning (i.e., the bill of material is exploded to give the gross requirements of the parts), conventional procedures are sufficient. In the case of consumption-driven material requirements planning, ANNs can help in forecasting the material requirements. The task of order planning is to determine the lot sizes for production and procurement. In both cases ANNs can be useful.

The initial task of capacity planning is to calculate the capacity utilization, using the data from time planning. Following that, a comparison of required and available capacity takes place. If bottlenecks occur, appropriate measures have to be taken to adjust capacity demand and supply. In general this task can be supported by ANNs. Order release transfers orders from the planning to the realization phase. The check of resource availability does not represent an area of potential application. Differently, ANNs seem to be appropriate for scheduling, which is usually a problem without an efficient solution algorithm.

About the steering and control of production processes, even though the problems differ as far as content is concerned, regarding the structural similarity it seems quite possible that ANNs have potential in supporting such processes.

## Multimedia

Multimedia tools can support several production management applications, e.g., in customer service operations, sales efforts, product documentation, after sales service and trouble shooting. In fact, the complexity of production management itself together with the complexity of its environment has increased significantly. The products, markets, and ideas have become more complex. Every interaction involves a large amount of people, which explodes the quantity of links for an efficient cooperation. Production processes are technically more and more demanding, too. There is a need for concurrent engineering to speed-up the process of designing and introducing new products into the market. In this context, the need to enhance information exchange in quantity and quality is evident.

Multimedia can offer a significant contribution in this direction. The categories of the use of multimedia-based solutions can be classified (Eloranta et al. 1995) into five groups:

1. tools to support and simplify the production planning mechanisms via multimedia;
2. tools for better customer service;
3. tools for better management;
4. tools for better and faster learning processes; and
5. tools to rationalize and simplify existing practices.

The future systems in production management will have database and multimedia-oriented features in order to perform different production management tasks. These can be classified (Eloranta et al. 1995), for instance, as:

1. transaction processing;
2. operational planning;
3. analysis and learning;
4. documentation; and
5. training.

Transaction processing means entering basic data, updating simple records, and collecting data on actual activities. Entering a customer order, receiving materials to stock, or collecting data on hours worked are examples of simple transactions. Operational planning, analysis, and learning are complex tasks involving analysis, risk-taking, decision making, and alternative solutions evaluation. Graphics and windows help planners to make production plans interactively and take into account the factory floor situations and flexibility. The whole production process can be simulated and animated to evaluate how the plan works in practice. Voice mail is easier and faster to use than written mail. Messages and questions concerning the practical execution of the plan can be sent, for example, to foremen. Answers will be sent back with the same media.

A new application domain for multimedia capabilities can be found in documentation and training. For instance, the whole after-sales business or the technical documentation management are affected by the development of multimedia. Also, job training can turn from reading and listening into virtual hands-on experimentation, as multimedia capabilities are permitting.

## Global Manufacturing Planning and Control

The last two decades have been characterized by the growing globalization of economic and social scenarios. In particular, multinational corporations (MNCs), due to their global market, have been experimenting a tight competition taking the advantage of their contemporary presence in different countries (Porter 1990). Companies could better achieve the exploitation of these advantages by the coordination of a network of geographically dispersed subsidiaries. The coordination may concern some as well as all the value chain activities, such as research and development, product and process design, engineering, marketing, supply, manufacturing, distribution, and so on.

Global manufacturing planning and control refers to the problem of coordinating a MNC's manufacturing activities. Since such manufacturing activities involve the entire network made up of the MNC's subsidiaries, coordination should extend to the global supply network and distribution channels and it must also deal with the attendant logistics issues. Specific problems and techniques related to the synchronization of production rate with demand rate in global manufacturing are reviewed by Pontrandolfo and Okogbaa (1997).

# 1.6   Some Examples

The general overview and description of the main processes, systems, and techniques that companies use to synchronize demand rate with production rate has been provided in the previous paragraphs. However, as several times it has been underlined, each company has developed its own processes, systems, and techniques. This usually results in hybrid approaches developed through an evolutionary process driven by goals and urgent problems. However, theoretical frameworks can be useful in providing some guidelines to managers involved in manufacturing decisions. It seems interesting to describe two case examples based on actual manufacturing situations.

## Magneti Marelli's MPC System

Magneti Marelli is an automotive market supplier. This company supplies engine control systems to car manufacturers in Europe and in America (USA and Brazil). The plant located in Bari (Italy) manufactures fuel injection systems. This plant is organized by operation units, each of them composed of several

| | | | Week # | 23 | | | | | Week # | 24 | | |
|---|---|---|---|---|---|---|---|---|---|---|---|---|
| | Mo | Tu | We | Th | Fr | Sa | Mo | Tu | We | Th | Fr | Sa |
| Module #1 | CCC 2500 | CCC 2500 | CEE 2100 | EFF 2100 | FFQ 2100 | GGG 2500 | CCC 2500 | CCC 2500 | HEE 2100 | MMF 2100 | FFQ 2100 | PGG 2100 |
| Module #2 | AAA 3000 | ADD 3000 | DDD 3000 | CCC 3000 | CCC 3000 | CCC 3000 | AAA 3000 | ADD 3000 | DDD 3000 | III 3000 | CCC 3000 | CCC 3000 |
| Module #3 | BBB 3000 | BBB 3000 | BBB 3000 | BBB 3000 | BBB 3000 | AAA 3000 | BBB 3000 | BBB 3000 | BBB 3000 | BBB 3000 | BBB 3000 | AAA 3000 |

| | | | Week # | 25 | | | | | Week # | 26 | | |
|---|---|---|---|---|---|---|---|---|---|---|---|---|
| | Mo | Tu | We | Th | Fr | Sa | Mo | Tu | We | Th | Fr | Sa |
| Module #1 | CCC 2500 | CCC 2500 | HEE 2100 | ENF 2100 | FFQ 2100 | PGG 2100 | CCC 2500 | CCC 2500 | HEE 2100 | EFF 2100 | FFQ 2100 | PGG 2100 |
| Module #2 | AAA 3000 | ADD 3000 | DDD 3000 | CCC 3000 | CCC 3000 | CCC 3000 | AAA 3000 | ADD 2500 | DDD 3000 | CCC 3000 | CCC 3000 | CCC 3000 |
| Module #3 | BBB 3000 | BBB 3000 | BBB 3000 | BBB 3000 | BBB 3000 | AAA 3000 | BBB 3000 | BBB 3000 | BBB 3000 | BBB 3000 | BBB 3000 | AAA 3000 |

Letters (A, B, and so on) represent different types of end items (fuel injectors).

**FIGURE 1.18** Smoothed MPS for the PICO operation unit with quantities and types of end item detailed for each shift (three shifts per day).

elementary technology units (UTE). Each operation unit is involved in the fabrication and assembly of end items. A particular operation unit, named PICO, consists of UTE (modules) for assembling and UTE (composed of some cells) for component fabrication. Each module can assemble all types of end items. Cells can supply components to all modules.

The management approach is based on lean production principles. Some JIT manufacturing techniques have been applied to improve productivity and customer service and integrated with existing MPC methods.

Orders are received by customers based on their MRP systems. Different shipment dates may be required by a customer even for the same type of end items and a MPS with a planning horizon of three months is determined each month. No mix specifications are included in the MPS being end-item types similar for manufacturing processes and for components. At the level of each operation unit, smoothed production is obtained and a more detailed MPS with a planning horizon of four weeks (including one frozen week) is provided each four weeks.

Mix specifications are determined by each operation unit's planners at the beginning of each week for the next four weeks based on the most updated available information. Only quantities and types of end item for the next week are considered frozen. MPS for a single operation unit is shown in Figure 1.18.

MRP is adopted to determine quantities and timing for suppliers. At the beginning of each week, based on the consumptions estimated by the MRP for that week, raw materials and components are requested to suppliers in order to replenish inventory. Production control is assured by a kanban system. Modules produce following the MPS of the operation unit. Cell manufacturing is driven by the consumption of parts in the downstream stock.

In Figure 1.19, material and information flows through the PICO operation unit are depicted.

## Industrie Natuzzi's MPC System

Industrie Natuzzi manufactures upholstered furniture. It is the world leader of the leather-cover sofa market. This company sells products all over the world. In 1996 more than 90% of products was sold abroad. Given the competitive pressure, a large assortment of sofa types is available and also new products are continually created. All production activities are located within a limited geographical area in the south of Italy and include the manufacturing of some components and the assembling of end items. The main factory consists of manufacturing shops and assembling areas, all connected by an automated material-handling system.

Final customer orders are collected by dealers located in different countries. Products are containerized and transported either directly to dealers in the case of a large order from a single dealer, or to warehouses

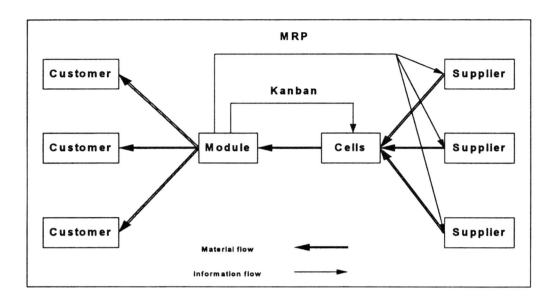

**FIGURE 1.19**   Material and information flows through the PICO operation unit.

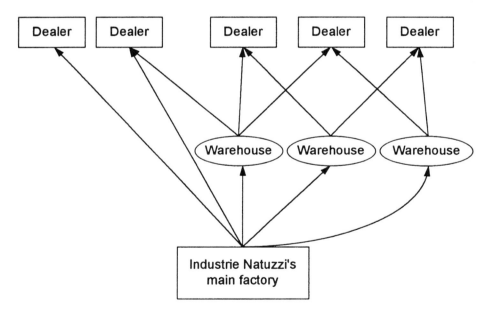

**FIGURE 1.20**   A simplified scheme of the distribution system.

located in some geographical areas in the case of small orders from different dealers (Figure 1.20). Shipment dates are determined by assuming that an order lead time of no more than six weeks is usually accepted. Oversea transportation times can range between two to four weeks.

A hybrid between an ATO and a MTO environment is adopted. Smoothed production is pursued and a MPS with a planning horizon of two weeks is provided each week. Basically, two supply policies are adopted for the main components. Polyurethane foams and wood frames are purchased weekly following the planned orders for the next week.

The suppliers' manufacturing process for the leather requires a long lead time. Moreover, because a large investment is required, inventory is held based on the forecasted production for the next two

| Activity | Lead time (weeks) |
|---|---|
| Polyurethane foam order | 1 |
| Wood frame order | 1 |
| Leather-supplier manufacturing process | 4 |
| Leather-supply order | 2 |
| Upholster order | 4-6 (depending on destination) |
| Upholster manufacturing process | 2 |

**FIGURE 1.21** Lead times.

weeks; this purchasing policy results in about 5% of stock-out. In Figure 1.21 the main lead times are reported.

The transportation by containers requires that all end items scheduled for the same container should be available at the same time at the end-item warehouse of the main factory. Manufacturing disruptions, order changes, and the leather-inventory policy often cause the delay of some end items and, then, the delay of the related container. This situation calls for an effective shop-floor monitoring and control system based on a real-time information system able to provide support for process rescheduling and for changes in container filling.

## Acknowledgments

We are grateful to the Magneti Marelli S.p.A. and to the Industrie Natuzzi S.p.A. for the kind permission to present their manufacturing planning and control systems. We are also grateful to A. Claudio Garavelli (Università della Basilicata), Pierpaolo Pontrandolfo (Politecnico di Bari), and Luciano Attolico (Magneti Marelli S.p.A.) for their suggestions and support. However, we remain responsible for any errors.

## References

Aggarwal S. C., 1985, MRP, JIT, OPT, FMS? Making Sense of Production Operations Systems, *Harvard Business Review*, Vol. 63, No. 5, pp. 8–16.

Albino V. and Garavelli A. C., 1995, A methodology for the vulnerability analysis of just-in-time production systems, *International Journal of Production Economics*, Vol. 41, pp. 71–80.

Armstrong A. and Hagel III J., 1996, The Real Value of ON-LINE Communities, *Harvard Business Review*, Vol. 74, No. 3, pp. 134–141.

Badiru A. B., 1992, *Expert Systems Applications in Engineering and Manufacturing*, Prentice-Hall, Englewood Cliffs, New Jersey.

Baker K. R., 1974, *Introduction to Sequencing and Scheduling*, J. Wiley & Sons, New York.

Baudin M., 1990, *Manufacturing System Analysis: With Application to Production Scheduling*, Prentice-Hall, Englewood Cliffs, New Jersey.

Bedworth D. D. and Bailey J. E., 1987, *Integrated Production Control Systems. Management, Analysis, Design,* 2nd ed., J. Wiley & Sons, New York.

Bhaskaran K. and Pinedo M., 1992, Dispatching, in Salvendy G. (ed.), *Handbook of Industrial Engineering*, 2nd ed., J. Wiley & Sons, New York, pp. 2182–2198.

Bitran G. R. and Tirupati D., 1993, Hierarchical production planning, in Graves S. C. et al. (eds.), *Handbooks in Operations Research and Management Science*, Vol. 4, Elsevier, Amsterdam, pp. 523–568.

Blackburn J. D., 1991, *Time-Based Competition. The Next Battle Ground in American Manufacturing*, Irwin, Homewood, Illinois.

Blackstone J. H., Phillips D. T., and Hogg G. L., 1982, A State-of-the-Art Survey of Dispatching Rules for Manufacturing Job Shop Operations, *International Journal of Production Research*, Vol. 20, No. 1, pp. 27–45.

Blocher J. D. and Chand S., 1992, Resource planning for aggregate production, in Salvendy G. (ed.), *Handbook of Industrial Engineering*, 2nd ed., J. Wiley & Sons, New York, pp. 2098–2115.

Bockerstette J. A. and Shell R. L., 1993, *Time Based Manufacturing*, McGraw-Hill, New York.

Brown R. G., 1977, *Materials Management Systems,* J. Wiley & Sons, New York.

Cheung J. Y., 1994, Scheduling, in Dagli C. H. (ed.), *Artificial Neural Networks for Intelligent Manufacturing*, Chapman & Hall, London, pp. 159–193.

Chryssolouris G., 1992, *Manufacturing Systems. Theory and Practice*, Springer-Verlag, New York.

Corsten H. and May C., 1996, Artificial Neural Networks for Supporting Production Planning and Control, *Technovation*, Vol. 16, No. 2, pp. 67–76.

Cusumano M. A., 1994, The Limits of "Lean," *Sloan Management Review*, Vol. 35, No. 4, pp. 27–32.

Cutcher-Gershenfeld J., Nitta M., Barrett B., Belhedi N., Bullard J., Coutchie C., Inaba T., Ishino I., Lee S., Lin W.-J., Mothersell W., Rabine S., Ramanand S., Strolle M., and Wheaton A., 1994, Japanese team-based work systems in North America: explaining the diversity, *California Management Review*, Vol. 37, No. 1, pp. 42–64.

Dagli C. H., 1994, Intelligent manufacturing systems, in Dagli C. H. (ed.), *Artificial Neural Networks for Intelligent Manufacturing*, Chapman & Hall, London.

Dattero R., Kanet J. J., and White E. M., 1989, Enhancing manufacturing planning and control systems with artificial intelligence techniques, in Kusiak A. (ed.), *Knowledge-Based Systems in Manufacturing*, Taylor & Francis, London, pp. 137–150.

Davenport T. H., 1993, *Process Innovation. Reengineering Work through Information Technology*, Harvard Business School Press, Boston, Massachusetts.

Davenport T. H. and Short J. E., 1990, The new industrial engineering: information technology and business process redesign, *Sloan Management Review*, Vol. 31, No. 4, pp. 11–27.

Davis T., 1993, Effective supply chain management, *Sloan Management Review*, Vol. 34, No. 4, pp. 35–46.

Dean J. W. Jr and Snell S. A., 1996, The strategic use of integrated manufacturing: an empirical examination, *Strategic Management Journal*, Vol. 17, No. 6, pp. 459–480.

Duimering P. R., Safayeni F., and Purdy L., 1993, Integrated manufacturing: redesign the organization before implementing flexible technology, *Sloan Management Review*, Vol. 34, No. 4, pp. 47–56.

Eloranta E., Mankki J., and Kasvi J. J., 1995, Multimedia and production management systems, *Production Planning and Control*, Vol. 6, No. 1, pp. 2–12.

Evans G. N., Towill D. R., and Naim M. M., 1995, Business process re-engineering the supply chain, *Production Planning and Control*, Vol. 6, No. 3, pp. 227–237.

Fisher M. L., Hammond J. H., Obermeyer W. R., and Raman A., 1994, Making supply meet demand in an uncertain world, *Harvard Business Review*, Vol. 72, No. 3, pp. 83–93.

Forrester Research, 1997, *The Forrester Report,* http://www.forrester.com.

French S., 1982, *Sequencing and Scheduling: An Introduction to the Mathematics of the Job-Shop*, Horwood, Chichester.

Galbraith J., 1973, *Designing Complex Organizations*, Addison-Wesley, Reading, Massachusetts.

Georgoff D. M. and Murdick R. G., 1986, Manager's guide to forecasting, *Harvard Business Review*, Vol. 64, No. 1, pp. 110–120.

Goldratt E., 1980, *OPT-Optimized Production Timetable*, Creative Output, Milford, New York.

Groenevelt H., 1993, The just-in-time system, in Graves S. C. et al. (eds.), *Handbooks in Operations Research and Management Science*, Vol. 4, Elsevier, Amsterdam, pp. 629–670.

Gumaer B., 1996, Beyond ERP and MRP II. Optimized planning and synchronized manufacturing, *IIE Solutions*, Vol. 28, No. 9, pp. 32–35.

Gunn T. G., 1987, *Manufacturing for Competitive Advantage*. Becoming a World Class Manufacturer, Ballinger Pub., Cambridge, Massachusetts.

Hakanson W. P., 1994, Managing manufacturing operations in the 1990's, *Industrial Engineering*, Vol. 26, No. 7, pp. 31–34.

Hammer M., 1990, Reengineering work: don't automate! obliterate, *Harvard Business Review*, Vol. 68, No. 4, pp. 104–112.

Hammer M. and Champy J., 1993, *Re-engineering the Corporation: A Manifesto for Business Revolution*, Brealey Pub., London.

Hayes R. H. and Pisano G. P., 1994, Beyond world-class: the new manufacturing strategy, *Harvard Business Review*, Vol. 72, No. 1, pp. 77–86.

Hayes R. H., Wheelwright S. C., and Clark K. B., 1988, *Dynamic Manufacturing*, The Free Press, New York.

Jarvis J. J. and Ratliff H. D., 1992, Distribution and logistics, in Salvendy G. (ed.), *Handbook of Industrial Engineering*, 2nd ed., J. Wiley & Sons, New York, pp. 2199–2217.

Jones D. T., 1992, Beyond the Toyota production system: the era of lean production, in Voss C.A. (ed.), *Manufacturing Strategy. Process and Content*, Chapman & Hall, London.

Kaplan R. S. and Murdock L., 1991, Core process redesign, *The McKinsey Quarterly*, 2, pp. 27–43.

Karmarkar U., 1989, Getting control of just-in-time, *Harvard Business Review*, Vol. 67, No. 5, pp. 122–131.

Lamming R., 1993, *Beyond Partnership: Strategies for Innovation and Lean Supply*, Prentice-Hall, Englewood Cliffs, New Jersey.

Lawler E. L., Lenstra J. K., Rinnooy Kan A. H. G., and Shmoys D. B., 1993, Sequencing and scheduling: algorithms and complexity, in Graves S. C. et al. (eds.), *Handbooks in Operations Research and Management Science*, Vol. 4, Elsevier, Amsterdam, pp. 445–522.

Mabert V. A., 1992, Shop floor monitoring and control systems, in Salvendy G. (ed.), *Handbook of Industrial Engineering*, 2nd ed., J. Wiley & Sons, New York, pp. 2170–2181.

Mariotti S., 1994, Alla ricerca di un'identità per il "post-fordismo," *Economia e politica industriale*, n. 84, pp. 89–111.

Martin R., 1995, *Turbocharging MRP II Systems with Enterprise Synchronization, IIE Solutions*, Vol. 27, No. 11, pp. 32–34.

Masri S. M., 1992, Finished goods inventory planning in Salvendy G. (ed.), *Handbook of Industrial Engineering*, 2nd ed., J. Wiley & Sons, New York, pp. 2064–2086.

McKay K. N., Safayeni F. R., and Buzacott J. A., 1995, A review of hierarchical production planning and its applicability for modern manufacturing, *Production Planning and Control*, Vol. 6, No. 5, pp. 384–394.

Monden Y., 1992, Just-in-time production system, in Salvendy G. (ed.), *Handbook of Industrial Engineering*, 2nd ed., J. Wiley & Sons, New York, pp. 2116–2130.

Morton T. E. and Pentico D. W., 1993, *Heuristics Scheduling Systems with Applications to Production Systems and Project Management*, J. Wiley & Sons, New York.

Okogbaa O. G. and Mazen a.-N., 1997, *Production Scheduling: An Epistemological Approach*, Working paper.

Omachonu V. K. and Ross J. E., 1994, *Principles of Total Quality*, St. Lucie Press, Delray Beach, Florida.

Orlicky J., 1975, *Material Requirements Planning*, McGraw-Hill, New York.

Pfeifer T., Eversheim W., König W., and Weck M. (eds.), 1994, *Manufacturing Excellence: The Competitive Edge*, Chapman & Hall, London.

Pinedo M., 1992, Scheduling, in Salvendy G. (ed.), *Handbook of Industrial Engineering*, 2nd ed., J. Wiley & Sons, New York.

Pontrandolfo P. and Okogbaa O. G., 1997, *Global Manufacturing: A Review and a Framework for the Planning in a Global Corporation*, Working paper.

Porter M. E., 1985, *Competitive Advantage*, The Free Press, New York.

Porter M. E., 1990, *The Competitive Advantage of Nations*, The Free Press, New York.

Rachamadugu R. and Stecke K. E., 1994, Classification and review of FMS scheduling procedures, *Production Planning and Control*, Vol. 5, No. 1, pp. 2–20.

Rayport J. F. and Sviokla J. J., 1995, Exploiting the virtual value chain, *Harvard Business Review*, Vol. 73, No. 6, pp. 75–85.

Ronen B. and Starr M. K., 1990, Synchronized manufacturing as in OPT: from practice to theory, *Computers & Industrial Engineering,* Vol. 18, No. 4, pp. 585–600.

Roth A. V., Giffi C. A., and Seal G. M., 1992, Operating strategies for the 1990s: elements comprising world-class manufacturing, in Voss C.A. (ed.), *Manufacturing Strategy. Process and Content,* Chapman & Hall, London, pp. 133–165.

Scheer A.-W., 1994, CIM. *Towards the Factory of the Future,* Springer-Verlag, Berlin.

Schonberger R. J., 1982, *Japanese Manufacturing Techniques. Nine Hidden Lessons in Simplicity,* The Free Press, New York.

Schonberger R. J., 1986, *World Class Manufacturing. The Lessons of Simplicity Applied,* The Free Press, New York.

Seminerio M., 1997, *IBM Shuts the Doors on World Avenue Online Mall,* http://www3.zdnet.com/zdnn/content/zdnn/0610/zdnn0005.html, June 10th.

Shapiro J. F., 1993, Mathematical programming models and methods for production planning and scheduling, in Graves S. C. et al. (eds.), *Handbooks in Operations Research and Management Science,* Vol. 4, Elsevier, Amsterdam, pp. 371–443.

Shingo S., 1989, *A Study of Toyota Production System from an Industrial Engineering Viewpoint,* Productivity Press, Cambridge, Massachusetts.

Slats P. A., Bhola B., Evers J. J. M., and Dijkhuizen G., 1995, Logistic chain modelling, *European Journal of Operational Research,* Vol. 87, No. 1, pp. 1–20.

Spar D. and Bussgang J. J., 1996, Ruling the net, *Harvard Business Review,* Vol. 74, No. 3, pp. 125–133.

Stalk G. Jr. and Hout T. M., 1990, *Competing Against Time. How Time-based Competition is Reshaping Global Markets,* The Free Press, London.

Sule D. R., 1994, *Manufacturing Facilities: Location, Planning, and Design,* 2nd ed., PWS-Kent Publishing Co., Boston, Massachusetts.

Tersine R. J., 1985, *Production/Operations Management: Concepts, Structure, and Analysis,* 2nd ed., North-Holland, New York.

Thomas L. J. and McClain J. O., 1993, An overview of production planning, in Graves S. C. et al. (eds.), *Handbooks in Operations Research and Management Science,* Vol. 4, Elsevier, Amsterdam, pp. 333–370.

Umble M. M. and Srikanth M. L., 1990, *Synchronous Manufacturing: Principles for World Class Excellence,* South-Western Pub., Cincinnati, Ohio.

Usuba K., Seki T., and Kawai A., 1992, Production information systems, in Salvendy G. (ed.), *Handbook of Industrial Engineering,* 2nd ed., J. Wiley & Sons, New York, pp. 2025–2063.

Vollmann T. E., Berry W. L., and Whybark D. C., 1992, *Manufacturing Planning and Control Systems,* 3rd ed., Irwin, Homewood, Illinois.

Wight O. W., 1979, MRP II—manufacturing resource planning, *Modern Materials Handling,* September, pp. 78–94.

Womack J. P. and Jones D. T., 1994, From Lean Production to the Lean Enterprise, *Harvard Business Review,* Vol. 72, No. 2, pp. 93–103.

Womack J. P., Jones D. T., and Roos D., 1990, *The Machine That Changed the World,* Rawson, New York.

Zhang H.-C. and Huang S. H., 1995, Applications of neural networks in manufacturing: a state-of-the-art survey, *International Journal of Production Research,* Vol. 33, No. 3, pp. 705–728.

# 2

# Integrated Product and Process Management for Engineering Design in Manufacturing Systems[1]

Peter Heimann
*Aachen University of Technology
(RWTH)*

Bernhard Westfechtel
*Aachen University of Technology
(RWTH)*

Worldwide competition has increased the pressure to optimize business processes in manufacturing enterprises. To produce innovative designs of high-quality products rapidly at low costs, methods such as concurrent and simultaneous engineering have been developed recently. Moreover, there is pressing need for sophisticated software tools.

We describe a system for *integrated product and process management* which has been designed in the face of these challenges. The system was developed in the SUKITS project, an interdisciplinary project involving both computer scientists and mechanical engineers. It supports management of complex, dynamic development processes; provides managers with detailed and accurate views; offers them sophisticated tools for planning, analyzing, and controlling development processes; maintains workspaces for executing development tasks (including document and version management, release control, and tool activation); and supplements formal cooperation with annotations for spontaneous, ad hoc communication, both synchronously (conferences) and asynchronously (messages).

---

[1]This work was partially supported by the German Research Council (DFG).

## 2.1  Introduction

Worldwide competition has increased the pressure to optimize business processes in manufacturing enterprises. Optimization is performed in three dimensions: costs, time-to-market, and quality. In particular, this challenges the capabilities of *engineering design*, which is concerned with the design of the products as well as the processes of manufacturing. To produce innovative designs of high-quality products rapidly at low costs, methods such as concurrent and simultaneous engineering have been developed recently. Moreover, there is pressing need for sophisticated software tools.

Tools can provide suppport at different levels. First, there are tools for carrying out individual design activities such as creating a drawing, generating a numerical control (NC) program, or running a simulation. Second, these activities have to be coordinated, i.e., dependencies between activities have to be managed.

*Management tools* are concerned with both products and processes of engineering design. The products of engineering design activities are stored in documents such as CAD designs, manufacturing plans, NC programs, or simulation traces. The engineering design process consists of the steps which are carried out to produce these products. In the sequel, we will use the terms "product" and "process" without qualification, referring to engineering design rather than manufacturing.

*Integration* is a key requisite to the success of management tools for engineering design. We distinguish between

- integrated product management, which takes care of the dependencies between engineering design documents such that consistency between different representations can be maintained;
- integrated process management, which coordinates the execution of activities such that they are orchestrated in an optimal way; and
- integrated product and process management, which is responsible for supplying activities with data, but also takes into account that the process structure may depend on the product structure.

We do not provide a general overview on product and process management for engineering design. Instead, we focus on a specific system [28, 30], describe the ways how management is supported, and motivate the underlying design decisions. The system has been developed in a project called SUKITS [9], which is introduced briefly in Section 2.2. Subsequently, the conceptual model for integrated product and process management is described (Section 2.3). User interface and realization of the SUKITS system are the topics of Section 2.4. Section 2.5 reports on applications and experiences. Section 6 briefly summarizes our contributions.

## 2.2  The SUKITS Project

The SUKITS project (*Software and Communication Structures in Technical Systems*, [9]), which was funded by the German Research Council (DFG), was carried out at the Technical University of Aachen (RWTH) from 1991–1997. SUKITS is a joint effort of computer scientists and mechanical engineers to improve tool support for engineering design management. We describe the application domain considered and the approach to providing tool support.

### Application Domain

SUKITS focuses on *engineering design* for mechanical engineering. Engineering design covers all activities that are concerned with designing products and their manufacturing processes; alternatively, we will also use the term *development*. Later phases of the product life cycle such as production planning, production, and disposal go beyond the scope of our work.

More specifically, we investigated the following scenarios:

- *Development of mental parts.* This includes both single parts and assembly parts. In the case of single parts, different versions of CAD designs are prepared (ranging from draft initial designs to detailed final designs), a manufacturing plan is created, NC programs for manufacturing steps defined in the manufacturing plan are generated, simulations are performed, etc. In the case of assembly parts, the design is decomposed into multiple documents (design hierarchy), a part list is constructed, and an assembly plan describes how to assemble the single parts.

- *Development of plastic parts.* Here, we investigated a specific kind of production, namely *injection moulding* (melted plastic material is injected into a mold from which it is removed after having cooled down). Here, the development process consists of steps such as creating a list of requirements (concerning both geometrical data and properties of the plastic material), material selection, design of the plastic part, process and mechanical simulations, and finally the design of the mold used for production.

These scenarios have complementary properties:

- When a metal part is developed, it is often not known a priori which activities have to be carried out and which documents have to be produced. This does not only apply to assembly parts where the part hierarchy is usually determined during the course of the development process. Furthermore, even in the case of single parts the manufacturing steps may be known only when the manufacturing plan is created, and only then may we know which NC programs are required. Finally, it may turn out that new tools have to be constructed, the position of the workpiece has to be described by additional drawings, etc.

- On the other hand, both the documents to be created and the activities to be carried out are frequently known beforehand when a plastic part is constructed. Injection molding is usually concerned with single parts, e.g., a plastic cup; sometimes, the plastic part is composed of a small number of individual parts which are known a priori. However, due to complex interactions—e.g., between the selected material and the moulding process—the activities may have to be iterated due to feedback, making the development process more complicated and less predictable than it may appear at first glance.

## Integration Approach

Tool support for product and process management is achieved by providing an integrating infrastructure which is called *management system*. Figure 2.1 shows the coarse architecture of this system:

- The *communication system* provides uniform services—e.g., database access and file transfer—for communication in a heterogeneous network [7, 14].

- The *management database* contains data about products, processes, and resources (humans and tools).

- *Wrappers* are used to integrate tools with the management system. A wrapper is responsible for supplying some tool with product data (documents) and for activating it with appropriate parameter settings (e.g., tool options).

- Tools are activated via a uniform *front end*. The front end supports process management by displaying agendas (lists of tasks assigned to engineers) and by establishing a detailed work context for each task (including input and output documents as well as tools to operate on these documents).

- The *management environment* consists of tools which support project managers. In particular, a project manager may plan development tasks and supervise their execution. Furthermore, a project manager may define the project team and assign roles to its members. Finally, the management environment also includes product management tools for version and configuration management.

- Finally, the *parameterization environment* is used to adapt the management system to a specific application domain. To make the management system usable in a certain domain (which may also lie outside mechanical engineering), domain-specific knowledge and domain-specific components have to be added to the domain-independent core of the management system. Domain-specific knowledge is expressed by a schema (which defines, e.g., types of documents and dependencies). Domain-specific components are the wrappers mentioned above.

Note that tools are integrated *a posteriori* with the management system. This means that existing tools are reused. This is extremely important because the alternative—development of new tools from scratch, also called *a priori integration*—is not economically feasible. On the other hand, integration of tools

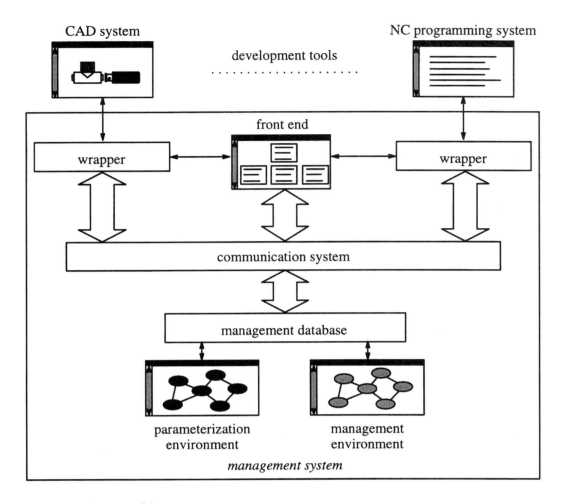

**FIGURE 2.1**    Structure of the management system.

which were developed by different vendors, run under different operating systems, and rely on different data management systems is inherently difficult to achieve.

In contrast, we decided to develop the components of the management system anew. Overall, this results in a *hybrid integration approach*: a priori integration within the management system, and a posteriori integration of existing tools with the management system. When we started the SUKITS project in 1991, our analysis of existing management tools or environments revealed severe deficiencies which could not be overcome by a posteriori integration. For example, classical project management tools are stand-alone tools which do not offer any interface for integration with other tools or environments. Furthermore, they do not handle the dynamics inherent in development processes. For these reasons, we have developed new, tightly integrated tools for management which remove these deficiencies.

## 2.3   Management Model

The management system is based on a formal model which is called *management model*. The model is composed of the following parts:

- The *product management model* is concerned with the products of the development process and serves as the basis for version and configuration management of interdependent documents.

- The *process management model* describes development tasks and their relationships and provides the foundation for planning, analysis, and execution of tasks.
- Finally, the *resource management model*, which will only be discussed briefly, covers both human and tools required for executing development tasks.

All of these models are *generic*, i.e., they are reusable in different application domains. The sections following describe parameterization of generic models in order to obtain *specific models* for a certain application domain. The formal specification of the management model is addressed briefly at the end.

## Product Management Model

Product management has been carried out in different engineering disciplines, including mechanical, electrical, and software engineering. Accordingly, research has been conducted under different umbrellas, e.g., engineering data management [18], product data management [25], CAD frameworks [12], and software configuration management [2]. As noticed in [5], the problems to be solved are largely domain-independent. For example, this becomes evident when comparing work in the CAD domain [15] to research in software engineering [2].

The product management model developed in SUKITS [28–30] integrates version control, configuration control, and consistency control for heterogeneous engineering design documents in a uniform conceptual framework:

- *Management of heterogeneous documents*. Documents such as designs, manufacturing plans, or NC programs are managed which are created by heterogeneous development tools.
- *Representation of dependencies*. Rather than managing collections of unrelated documents, product management represents their mutual dependency relationships.
- *Management of configurations*. Documents and their dependencies are aggregated into configurations which may be nested.
- *Version management*. Product management keeps track of multiple versions into which a document evolves during its lifetime. Configurations are versioned as well.
- *Consistency control*. The representation of dependencies lays the foundations for consistency control between interdependent documents. Note that versioning is taken into account, i.e., dependencies relate document versions and therefore define which versions must be kept consistent with each other.

In the following, we discuss each of these topics in turn.

### Documents and Configurations

Product management deals with *documents*, which are artifacts created and used during the development process (designs, manufacturing plans, NC programs, part lists, simulation results, etc.). A document is a logical unit of reasonable size typically manipulated by a single engineer. According to the constraints of a posteriori integration, we do not make any assumptions regarding the internal structure of documents. This approach is called *coarse-grained* because a document is considered an atomic unit. Physically, a document is typically represented as a file or a complex object in an engineering database.

What is actually modeled as an atomic unit is application-dependent. For example, we may either model each NC program for manufacturing a single part as one document, or we may consider the whole set of NC programs as a single document. In the latter case, a document corresponds to a set of files. Which alternative is selected, depends on whether or not we need to refer to individual NC programs at the management level (e.g., with respect to task assignments).

Documents are related by manifold *dependencies*. Such dependencies may either connect components belonging to the same working area (e.g., dependencies between designs of components of an assembly part), or they may cross working area boundaries (e.g., dependencies between designs and manufacturing plans). Product management records these dependencies so that consistency between interdependent documents can be controlled.

Related documents (e.g., all documents describing a single part) are aggregated into *configuration*. In addition to components, configurations contain dependencies as well. Documents (*atomic objects* at the coarse-grained level) and configurations (*complex objects*) are treated uniformly. Vertical relationships between configurations and their components are called *composition relationships*. In particular, a component of some configuration may be a subconfiguration (nested configurations). Therefore, dependencies between subconfigurations can be modeled in the same way as dependencies between documents.

### Versions

During its evolution history, documents evolve into multiple *versions*. Since documents and configurations are handled uniformly, configurations are versioned too. Versions may be regarded as snapshots recorded at appropriate points in time. The reasons for managing multiple versions of an object (instead of just its current state) are manifold: reuse of versions, maintenance of old versions having already been delivered to customers, storage of back-up versions, or support of change management (e.g., through a diff analysis figuring out what has changed compared to an old version).

A version *v* represents a state of an evolving object *o*. The latter is called *versioned object* (or simply object). A versioned object serves as a container of a set of related versions and provides operations for version retrieval and creation of new versions. We distinguish between *document objects* and *configuration objects*, whose versions are denoted as *document versions* and *configuration versions*, respectively.

According to the distinction between objects and their versions, the product management model distinguishes between an *object plane* and a *version plane* (Figure 2.2). The version plane refines the object

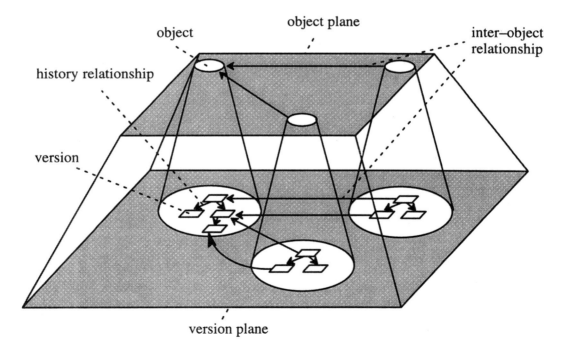

**FIGURE 2.2**   Object and version plane.

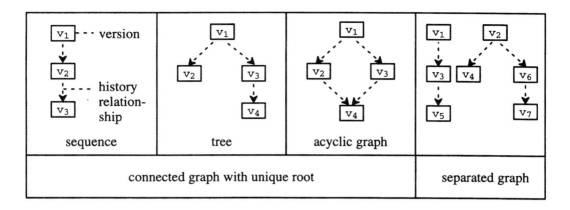

**FIGURE 2.3**    Structures of version graphs.

plane: each object is refined into its versions, and each relationship between two objects is refined into relationships between corresponding versions. Furthermore, history relationships between versions of one object represent its evolution. While the object plane provides an overview by abstracting version-independent structural information, the version plane accurately represents actual versions of documents and configurations as well as their mutual relationships.

Objects, versions, and their interrelationships are formally represented by *graphs*. In order to structure the information contained in such graphs, the product management model distinguishes between different kinds of interrelated subgraphs, namely, version graphs, configuration version graphs, and configuration object graphs. Each subgraph provides a view on the underlying database representing the evolution history of an object, the structure of a configuration version, and the structure of a configuration object, respectively.

A *version graph* (Figure 2.3) consists of versions which are connected by *history relationships*. A history relationship from $v_1$ to $v_2$ indicates that $v_2$ was derived from $v_1$ (usually by modifying a copy of $v_1$). In simple cases, versions are arranged in a sequence reflecting the order in which they were created. Concurrent development of multiple versions causes branches in the evolution history (version tree). For example, $v_2$ may have been created to fix an error in $v_1$, while work is already in progress to produce an enhanced successor $v_3$. Merging of changes performed on different branches results in directed acyclic graphs (dags), where a version may have multiple predecessors. Finally, a version graph may even be separated if multiple branches have been developed in parallel from the very beginning. The product management model allows for all structures shown in Figure 2.3; it merely excludes cycles in history relationships. Variants (alternative versions) may either be represented by branches or by separated subgraphs.

A *configuration version graph* represents a snapshot of a set of interdependent components. Thus, it consists of component versions and their dependencies. The lower part of Figure 2.4 shows three versions of a configuration Shaft which comprises all documents describing a shaft. Versions are denoted by the name of the versioned object, a dot, and a version number. The initial configuration version contains versions of a design, a manufacturing plan, and two NC programs (e.g., for turning the left and the right half of the shaft, respectively). The manufacturing plan depends on the design, and the NC programs depend on both the design and the manufacturing plan. In the second configuration version, the geometry of the shaft was modified, and accordingly all dependent documents were modified, as well. In the last version, it was decided to produce the shaft from a specifically designed raw part in order to minimize consumption of material (previously, a piece of some standard metal bar was used). To this end, a raw part design was added to the configuration, and the dependencies were extended accordingly.

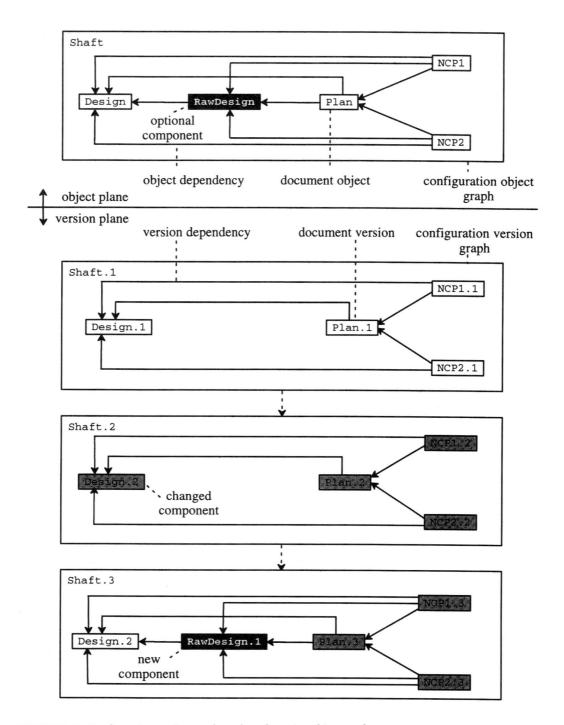

**FIGURE 2.4**    Configuration version graphs and configuration object graphs.

A *configuration object graph* represents version-independent structural information in the object plane. The following constraint must hold between a configuration object graph and the corresponding configuration version graphs in the version plane: For each version component (dependency) contained in a configuration version graph, a corresponding object component (dependency) must exist in the configuration object graph. More precisely, each configuration version graph must be mapped by a graph monomorphism into the corresponding configuration object graph. This constraint guarantees that the version plane actually refines the object plane. Furthermore, an injective mapping excludes that multiple versions of some object are contained in one configuration (version consistency). The upper part of Figure 2.4 shows a configuration object graph which satisfies these constraints.

Configuration object graphs support management of configuration families. Although the members of configuration families should have much in common, it is too restrictive to enforce identical structures. Rather, structural variations have to be permitted (in the example given above, the raw part design is optional). Thus, a *structural variant* is a certain subgraph of the configuration object graph. Such a variant can be specified with the help of variant attributes. To each component and dependency, a variant description is attached which refers to these attributes and controls its visibility. Then, a variant is specified by assigning values to the variant attributes and evaluating the variant descriptions. After having fixed the configuration structure, versions of components still have to be selected in order to obtain a completely specified configuration version.

## Consistency Control

Product management assists in maintaining *consistency* between interdependent documents. Here, configuration versions play a crucial role. Instead of managing versions of documents individually and independently, configuration versions define sets of component versions among which consistency has to maintained. By recording configuration versions, consistent configurations may easily be reconstructed later on, e.g., when some error needs to be fixed or a requirements change has to be accommodated. Guessing consistent combinations of component versions a posteriori is inherently difficult and error-prone.

More specifically, the product management model allows to differentiate between the following aspects of consistency:

- *Internal consistency* refers to the local consistency of a version. Internal consistency does not take relationships to other versions into account. For example, a version of an NC program is denoted as consistent if it consists of legal statements of the NC programming language.

- *External consistency* refers to the consistency of a dependent version with respect to a certain master. For example, a version of an NC program is consistent with a version of a CAD design if it conforms to the geometrical data specified in the design.

- Finally, *component consistency* refers to the consistency of a version component with respect to a certain configuration version. Component consistency implies that the corresponding version is internally consistent, and that it is externally consistent with all master components.

Note that component consistency is a context-dependent notion. Thus, some version $v$ may be consistent with respect to a configuration version $c_1$ and inconsistent with respect to another configuration version $c_2$.

The product management model provides a framework for consistency control between interdependent documents. However, we have to keep in mind that product management is performed at a *coarse-grained level* and consistency checks need to be carried out at the fine-grained level, as well. To this end, appropriate tools are required which have to be interfaced with the product management system.

## Heterogeneity

To support *a posteriori integration*, the product management database is divided into a structural part, which is composed of the various kinds of graphs explained above, and a contents part, which stores the contents of document versions (Figure 2.5). The structural part contains references to the contents of document versions. Typically, the contents part is stored in the file system under control of the management system. To operate on a document version, the corresponding file (or set of files) is checked out into a workspace. In case of write access, the modified version is checked back into the contents part of the product management database when the modifications are finished.

Note that this model of a posteriori integration works fine for tools offering an interface for importing and exporting data. Export and import may include conversions between different data formats.

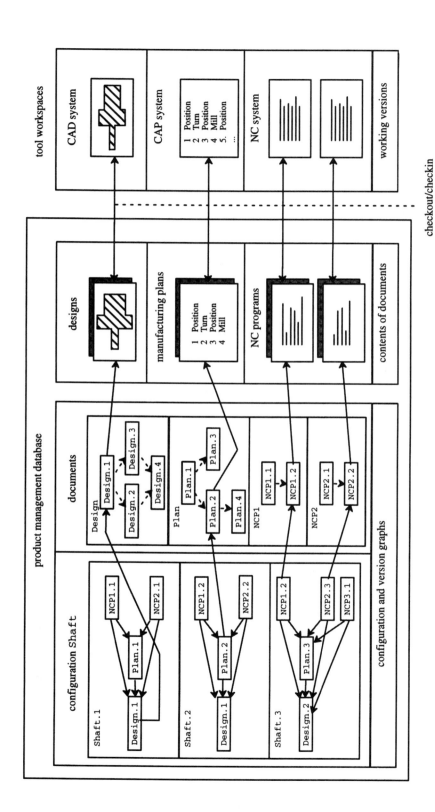

**FIGURE 2.5**    A posteriori integration.

However, closed tools which do not offer any data exchange cannot be integrated in this way. Then, it is merely possible to maintain references to data which are completely controlled by the tool.

## Product Management Operations

So far, we have mainly described the static structure of the product management database. Below, we briefly discuss operations provided for product management. These operations modify the contents of version graphs, configuration version graphs and configuration object graphs:

- *Version graphs.* A version may be created in two ways. Either, it may be created from scratch as a new root version, or it may be derived from a predecessor version. In the latter, predecessor and successor are connected by a successor relationship, and the contents of the successor is copied from the predecessor. In addition, merging creates a new version from multiple predecessors. This can be done either manually or with the help of some merge tool. If a version is deleted, all adjacent successor relationships are deleted as well.
- *Configuration object graphs.* Components and dependencies may be inserted and deleted. Deletion is only allowed if there is no corresponding component/dependency in any configuration version graph.
- *Configuration version graphs.* Similar operations are offered on configuration version graphs. Components and dependencies can only be inserted if there are corresponding elements of the configuration object graph.

Operations on the product management database have to respect a lot of *constraints*. These can be expressed partially by *state transition diagrams*. The diagrams of Figure 2.6 define the states of document and configuration versions, respectively. For each operation, it is defined in which state it may execute and which state will follow the execution.

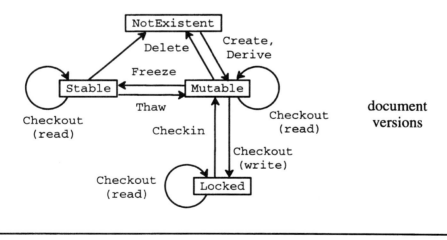

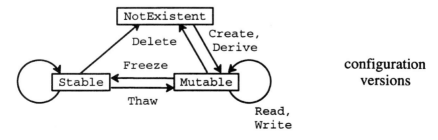

**FIGURE 2.6** State transition diagrams for document and configuration versions.

In order to provide for uniform versioning of documents and configurations, the state transition diagrams have been unified as far as possible. In particular, both diagrams share the states Stable and Mutable. Changes can only be performed in state Mutable.

However, there are still some differences between the state transition diagrams: Since document versions are manipulated outside the product management database, Checkout and Checkin are needed for data exchange. Note that a document version can only be checked out once for writing. On the other hand, configuration versions are manipulated in place, i.e., within the product management database. Therefore, Checkout and Checkin are not provided at the configuration level.

## Process Management Model

Like product management, process management has been dealt with in different disciplines, e.g., software process management [4, 10] or workflow management [11]. More recently, the parallels between these fields have been recognized, and various attempts have been undertaken to foster cooperation [17, 24].

The process management model [28, 30] formalizes planning, analysis, control, and execution of development tasks. It is characterized by the following features:

- *Product-centered approach*. A configuration of interdependent components is interpreted as a net of tasks connected by data flows (product-centered approach). Each task is responsible for developing a version of some document which is consistent with the respective versions of its master documents.

- *Interleaving of planning and execution*. Since the task net depends on the product structure, which is not known beforehand, planning cannot be performed once and for all when development starts. Rather, planning and execution are interleaved seamlessly.

- *Simultaneous engineering*. Concurrent and simultaneous engineering aim at shortening development cycles by turning away from phase-oriented development [1, 21]. In particular, we focus on simultaneous engineering: Tasks connected by producer-consumer relationships (data flows) can be executed in parallel; execution is controlled through a sophisticated release strategy.

### Overview

In order to clarify the goals and the scope of our process management model, we introduce a small taxonomy which is used to classify our approach. Subsequently, process management is discussed more specifically, focusing on the integration between product and process management and the dynamics of development processes.

Process management is concerned with *development tasks* which may be classified as follows:

- *Technical tasks* involve the creation of documents such as CAD designs, manufacturing plans, or NC programs. They are carried out by engineers who are skilled in working areas such as product design, process design (of the manufacturing process), or production planning.

- *Management tasks* include planning, analysis, control, and monitoring of the overall development process. These activities require managerial skills which are quite different from those needed for technical tasks.

In spite of these disparate characteristics, the process management model does not assume that technical and management tasks are performed by different employees. As to be explained, an employee may play multiple roles and may even act as manager and engineer simultaneously.

Tasks can be managed at different levels of granularity:

- At the *coarse-grained level*, the development process is decomposed into tasks such as "create design," "create manufacturing plan," "perform simulation," etc. Each of these tasks involves creation or modification of a small number of documents. They can be aggregated into complex tasks such as "develop single part," but they are not refined any further at the coarse-grained level.

- At the *fine-grained level*, decomposition proceeds down to small process steps which describe in detail how a coarse-grained task is carried out. For example, such process steps may be described in some design method. Carrying this to the extreme, refinement is performed down to the level of single commands offered by development tools.

With respect to the number of persons involved, we distinguish between support for the *personal process* and for *cooperation* between multiple persons, respectively. Furthermore, cooperation can be classified as follows:

- *Formal cooperation* is regulated by rules which are used to coordinate activities. These rules determine the inputs and outputs of development tasks, the order in which tasks are carried out, flow of documents, etc. Formal cooperation is planned and controlled by management.
- *Informal cooperation* takes place spontaneously, without being controlled by management. For example, engineers working on interrelated tasks may communicate to clarify design decisions, discuss alternatives, reveal potential trouble spots, etc.

Although there is no sharp borderline between formal and informal cooperation, the distinction is useful to clarify the focus of a certain approach to process management. For example, workflow management systems primarily support formal cooperation, while groupware tools focus on informal cooperation.

Finally, let us consider integration between product and process management:

- In the case of *product-centered process management*, process management is performed on top of product management by enriching the product structure with process data. The development process is modeled as a partially ordered set of operations on objects, where each operation corresponds to a task. This approach seems natural and is simple to understand.
- In the case of *separate process management*, products and processes are represented by separate, but interrelated structures. Separation allows to select among alternative processes to develop a product, i.e., the development process is not implicitly determined by the product. On the other hand, separate process management is more complex because product and process structures have to be combined and kept consistent with each other.

Within the spectrum outlined above, we may now put our process management model in its place:

- The model covers both technical and management tasks.
- Process support focuses at the coarse-grained level.
- Its main emphasis lies on formal cooperation. However, informal cooperation is covered as well, and it is also integrated with formal cooperation.
- Because of its conceptual simplicity, we follow a product-centered approach to process management.

## Product-Centered Approach

In our product-centered approach, a configuration version is enriched with process data, resulting in a *task net*. For each version component, there is a corresponding *task* which has to produce this component. Dependencies between version components correspond to horizontal task relationships which represent both *data* and *control flow*. Thus, a task has *inputs* defined by its incoming data flows, and it produces a single *output*, namely the corresponding version component. If a flow relationship starts at $t_1$ and ends at $t_2$, $t_1$ and $t_2$ are denoted as *predecessor* and *successor task*, respectively. Furthermore, tasks can also be connected by (vertical) *composition relationships*: version components referring to configuration versions result in hierarchies of task nets. The leaves of the composition hierarchy are called *atomic tasks*; otherwise, a task is *complex*. Finally, source and target of a composition relationship are called *supertask* and *subtask*, respectively.

Note that flow relationships and dependencies have opposite directions. In configuration versions, we have drawn dependencies from dependent to master components, respectively. On the other hand, it is

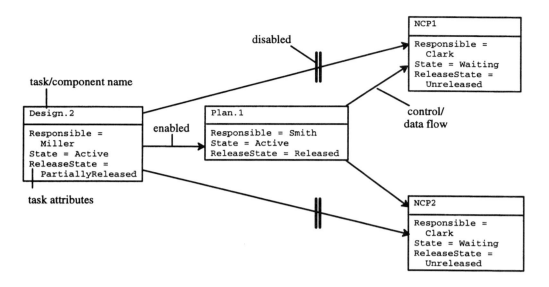

**FIGURE 2.7**    Example of a task net.

more natural to choose the opposite direction in task nets (this is done in the figures given below). Then, the flow relationships indicate the order in which tasks may be executed (control flow) and the direction in which data are transmitted from producer to consumer tasks (data flow). Therefore, we may view a task net as a combination of a *PERT chart* and a *data flow diagram*.

An example of a task net is given in Figure 2.7. The task net corresponds to a configuration version which contains one design, one manufacturing plan, and two NC programs. Each task is named by the corresponding version component, i.e., we refrain from introducing explicit task names such as CreateDesign or CreatePlan. Tasks are decorated with attributes, e.g., for representing task states, release states of outputs, and employees responsible for task execution. The design task and the planning task have already produced outputs (bound version components), while both NC programming tasks are still waiting for their execution (unbound version components).

## Dynamics of Development Processes

It is widely recognized that development processes are highly *dynamic*. Due to their creative nature, it is rarely possible to plan a development process completely in advance. Rather, many changes have to be taken into account during its execution. Figure. 2.8 gives examples of influencing factors by which these changes may be caused. Some of them have been studied thoroughly in the SUKITS project, namely product evolution, simultaneous engineering, and feedback (see right-hand side of the figure). These are discussed below.

*Product evolution* results in incremental extensions and structural changes of task nets. Let us consider again the task net of Figure 2.7. When starting development of a single part, it may only be known that one design and one manufacturing plan have to be created. Then, the initial task net only contains these tasks, which are known a priori. Which NC programs must be written, is determined only when the manufacturing plan has been worked out. Figure 2.7 shows a snapshot of the task net after the corresponding tasks have been inserted. Later on, the manufacturing plan may still be changed, resulting in a different set of NC programs and corresponding modifications to the task net. Another example was given in Figure 2.4, where a raw design was added to minimize consumption of raw material in the manufacturing process.

The "classical," conservative rule for defining the execution order states that a task can be started only after all of its predecessors have been finished. However, enforcing this rule significantly impedes parallelism. In order to shorten development time, the conservative rule needs to be relaxed so that tasks may be executed concurrently even if they are connected by flow relationships. *Simultaneous engineering* [1] denotes development methods to achieve this.

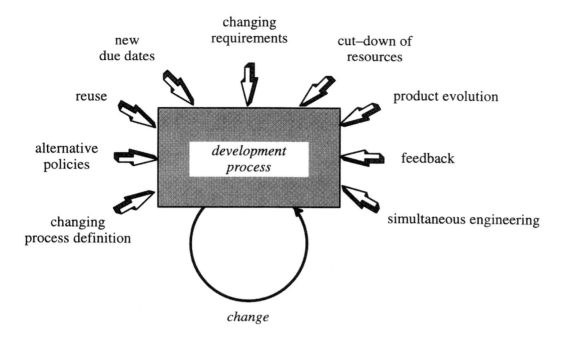

**FIGURE 2.8** Dynamics of development processes.

Simultaneous engineering is supported by *prereleases* of intermediate results. Since a task may release preliminary versions of its output, successors may start execution when their predecessors are still active. Prereleases serve two purposes. First, the overall development process may be finished earlier because of overlapping task execution. In particular, work located on the critical path may be accelerated. Second, wrong decisions can be detected much earlier by taking feedback from later phases in the lifecycle into account (e.g., design for manufacturing).

Versions may be prereleased *selectively* to successor tasks. An example is given in Figure 2.7, where the design task has the release state PartiallyReleased. Attributes attached to flow relationships control which successors may access a selectively released version. In Figure 2.7, double vertical bars indicate *disabled* flows along which outputs may not be propagated. NC programming tasks may not yet access the design because the geometry still has to be elaborated in detail. On the other hand, the flow to the planning task is *enabled* because the manufacturing plan can already be divided into manufacturing steps based on a draft design.

Due to simultaneous engineering, the *workspace* of a task is highly dynamic. The workspace consists of input documents, the output document, and potentially further auxiliary documents which are only visible locally. All components of the workspace are subject to version control. Multiple versions of inputs may be consumed sequentially via incoming data flows. Similarly, multiple versions of the output document may be produced one after the other. Finally, versions of auxiliary documents may be maintained as well.

The workspace of a sample task is illustrated in Figure 2.9. For inputs, the *current version* denotes the version which is currently being used by the responsible engineer. A *new version* may have been released already by the predecessor task (see input Design). When it is consumed, the current version is saved as an *old version*, and the new version replaces the current one. For outputs, we distinguish between a *working version*, which is only locally visible, and a *released version*, which is available to successor tasks. When the working version is released, it is frozen and replaces the previously released version. Any subsequent update to the working version triggers creation of a successor version. Finally, a current and an old version are maintained for auxiliary documents.

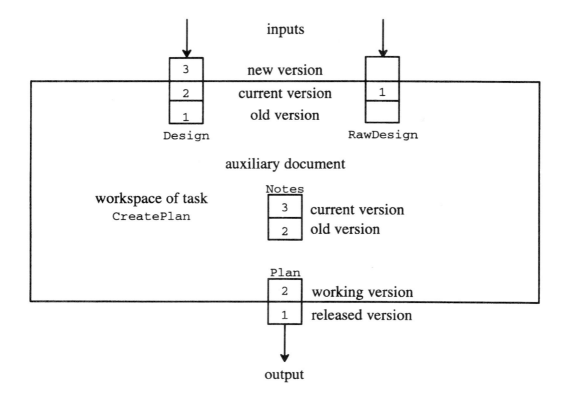

**FIGURE 2.9**   Workspace of a task.

Imports of inputs and exports of outputs are controlled explicitly by the responsible engineer who invokes operations for *consuming* inputs and *producing* outputs. Since only stable versions may be released, the import part of a workspace remains stable as long as no Consume operation is executed. In this way, the engineer may control the time of upgrading to a new input version (and may still access the previously current version as an old version, e.g., to perform a diff operation in order to analyze the differences). Similarly, changes to outputs are visible only when the working version is released. Thus, versions are used to achieve *loose coupling* between activities of multiple engineers.

Start and termination of tasks are controlled by activation and termination conditions, respectively. The *activation condition* requires that all essential inputs are available, i.e., they are (pre-)released by predecessor tasks. "Essential" weakens the activation condition by enforcing availability of only those inputs that are needed to start working. On the other hand, the *termination condition* is much more restrictive and conservative. All predecessor tasks must have terminated and released their output to the successor task; furthermore, all inputs must be up to date, i.e., the final version of each input document must have been consumed. These restrictions prevent premature termination, which could require reactivation of a task when its inputs are changed later on. Note that the conservative termination condition does not slow down the work; rather, it alerts the engineer of possibly changing inputs and mereley delays a state transition.

However, reactivation of terminated tasks cannot be excluded completely because of *feedback* occurring in the development process. Feedback is raised by a task $t_1$ which detects some problem regarding its inputs. This problem is caused by a task $t_2$ which precedes $t_1$ in the task net. In general, $t_2$ may be a transitive predecessor of $t_1$. If $t_2$ has already terminated, it has to be reactivated.

The impact of feedback may be hard to predict, and it may vary considerably from case to case. $t_2$ may raise cascading feedback to a predecessor $t_3$. Otherwise, $t_2$ accepts the responsibility for processing the feedback.

Then, the impacts on transitive successors of $t_2$ need to be estimated. Active tasks which will be affected drastically have to be suspended even if they are not immediate successors of $t_2$. Other tasks will not be affected at all and may continue without disruption. Thus, feedback needs to be managed in a flexible way, relying heavily on decisions to be performed by the user. In general, it is virtually impossible to automate the processing of feedback completely, e.g., by brute-force methods such as suspending all transitive successors.

Management of feedback is illustrated in Figure 2.10 which shows a task net for the development of an assembly part. AP, SP1, SP2, and SP3 denote the assembly part and three single parts, respectively; D, P, and N stand for designs, manufacturing plans, and NC progams, respectively. Task states are represented by different symbols or fill patterns (e.g., a grey box highlights an active task). The figure shows four snapshots of the task net, corresponding to different stages of feedback processing:

1. When elaborating the manufacturing plan of SP3, it turns out that the design of the single part is not well suited for manufacturing. Therefore, feedback to the design of SP3 is raised (dashed edge labeled with 1). This results in *cascading feedback* to the design of AP because SP3 cannot be redesigned without affecting the design of other single parts contained in the assembly part.

2. The design of AP, which was already terminated, is reactivated. Since all predecessors of a terminated task must be terminated as well (conservative termination condition), all terminated successors of AP are forced into the state Created. Impact analysis is carried out to anticipate the consequences of the intended change. Since SP1 will be affected significantly, releases of outputs are revoked transitively on this branch of the task net such that the NC programming tasks at the end of the chain are suspended. On the other hand, NC programming for SP2 may continue without disruption because SP2 is supposed to remain unchanged. Releases are revoked on the SP3 branch, resulting in suspension of design and manufacturing planning. Finally, the redesign implies that another single part SP4 needs to be added. Therefore, the task net is extended with a new branch which, however, cannot be executed yet.

3. Design of the assembly part has proceeded and it has been released again to all successors. All designs of single parts are active now. In the SP1 branch, the NC programming tasks are still suspended. Even though SP2 is expected to be unaffected by the design change, the corresponding design task has to be reactivated to inspect the changes to its input. In the SP3 branch, manufacturing planning is still suspended. Finally, design and manufacturing planning have been activated in the SP4 branch.

4. All design and planning tasks have terminated; only some NC programming tasks are still active or awaiting execution. While a new version of the design of SP1 has been created, the design of SP2 has not been changed because the changes of APD only affected other components of the assembly part. Therefore, terminated successor tasks have not been reactivated at all (immediate transition from Created to Done).

## Execution Model

The execution behavior of dynamic task nets is defined by means of *cooperating state machines*. To each task, a state machine is attached, which communicates with the state machines of neighboring tasks.

The state *transition diagram* in Figure 2.11 defines in which states a task may reside and which operations affect state changes. Created serves as initial state which is also restored in case of feedback. The Defined transition indicates that the task definition has been completed. In case of feedback, Reuse skips task execution if the task is not affected. In state Waiting, the task waits for its activation condition to hold. Redefine returns to Created , while Start is used to begin execution. Execution of an Active task may be suspended, e.g., because of erroneous inputs, and resumed later on. Failed and Done indicate failing and successful termination, respectively. From both states, re-execution may be initiated by Iterate transitions.

Notably, this state transition diagram deviates in several ways from standard diagrams known from other domains (e.g., operating system processes in computer science). For example, we have introduced an intermediate state between Created and Waiting. This state is necessary because otherwise we may not separate task definition from execution. Another example is the Reuse transition which specifically takes feedback into account.

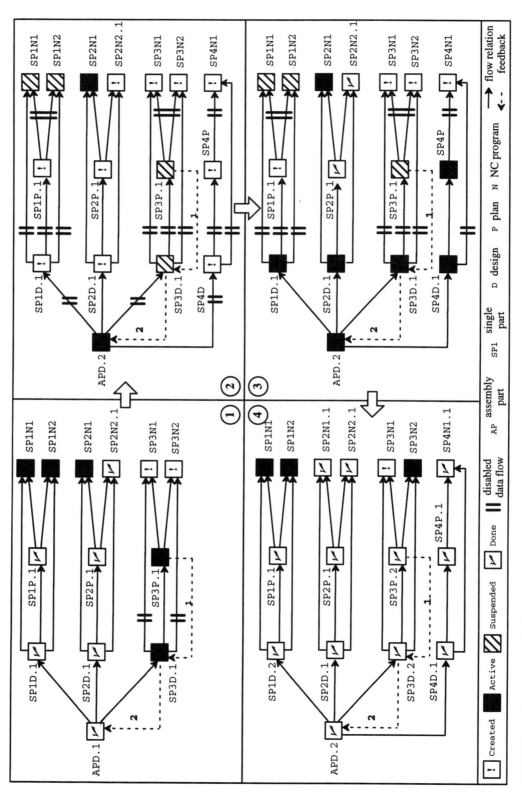

**FIGURE 2.10** Management of feedback.

**TABLE 2.1**   Compatibility Matrix for Task States (Vertical Relationships)

| supertask ▶<br>subtask ▼ | Created | Waiting | Active | Suspended | Done | Failed |
|---|---|---|---|---|---|---|
| Created | + | + | + | + | − | + |
| Waiting | + | + | + | + | − | + |
| Active | − | − | + | − | − | − |
| Suspended | − | − | + | + | − | − |
| Done | + | + | + | + | + | + |
| Failed | + | + | + | + | − | + |

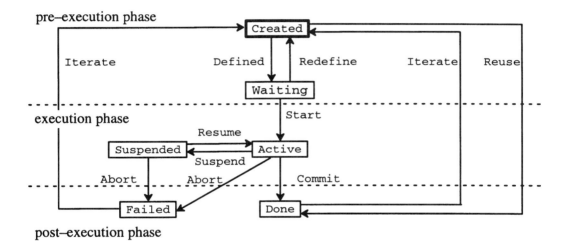

**FIGURE 2.11**   State transition diagram for tasks.

The state transition diagram considers a single task only. In addition, we have to describe the relationships between neighboring tasks. This is done by *compatibility matrices*:

- In Table 2.1, legal combinations of tasks connected by *vertical* (composition) *relationships* are specified by a *compatibility matrix*. Since Created and Waiting belong to the pre-execution phase (see Figure 2.11), subtasks residing in the execution phase (Active or Suspended) are not allowed. In case of an Active supertask, subtasks may be in any state. A Suspended supertask must not have an Active subtask: if a supertask is suspended, all subtasks are suspended as well. A supertask can only be Done if all subtasks have committed. Finally, no subtask of a Failed supertask may be Active or Suspended.

- Similarly, the compatibility matrix in Table 2.2 refers to tasks which are connected by *horizontal relationships* (data/control flows). The only constraint is that all predecessors of a Done task must be Done as well (conservative termination condition).

## Informal Cooperation

To supplement formal cooperation, the management model introduces *annotations*. An annotation may be any kind of document which is used to communicate with other employees. This includes text, graphics, audio, and video sequences. By means of annotations, informal cooperation is integrated with formal cooperation. Rather than using separate tools such as, e.g., electronic mail, employees communicate via annotations in order to get their tasks done.

**TABLE 2.2**   Compatibility Matrix for Task States (Horizontal Relationships)

| pred. task ▶<br>succ. task ▼ | Created | Waiting | Active | Suspended | Done | Failed |
|---|---|---|---|---|---|---|
| Created | + | + | + | + | + | + |
| Waiting | + | + | + | + | + | + |
| Active | + | + | + | + | + | + |
| Suspended | + | + | + | + | + | + |
| Done | − | − | − | − | + | − |
| Failed | + | + | + | + | + | + |

Annotations are characterized as follows:

- An annotation is *created* with reference to a task. The annotation is bound to the task rather than to the responsible employee, i.e., it is part of the task's workspace.
- An annotation refers to a set of *related documents*. For example, a problem report may reference a set of mutually inconsistent input documents.
- An annotation has a set of *receivers*. Like the creator, the receivers are tasks rather than employees (an employee receives an annotation only because he is working on a certain task). In many cases, annotations are received by creators and users of related documents. In other cases, the receivers are different. For example, a problem report concerning some document might be sent to the surrounding management task.
- An annotation has a *state* which is relative to the workspaces of the creator and each receiver. From the creator's point of view, an annotation is initially private until it is released to its receivers. Later on, the creator may mark it as obsolete, making it disappear from the task's workspace. Similarly, a receiver may read an annotation and finally mark it as obsolete. An annotation may be deleted only after it has been considered obsolete by all involved parties.
- Finally, an annotation may be used for both *synchronous* and *asynchronous communication*. In the first case, a conference is launched, e.g., to discuss a certain problem in a distributed work session. In the second case, an annotation is similar to an e-mail, but differs in two respects. First, any kind of tool can be used to edit and view an annotation (not just text editors like in ordinary mail tools). Second, an annotation remains under control of the management system.

An example is given in Figure 2.12. The example is taken from the development of a drill, which will be explained further in Section 2.5. Smith has designed the gear of the drill, and Clark has performed an FEM simulation on it. Since the simulation has revealed some error in the design, he creates an annotation describing the problem. For example, he may use some graphical tool to attach comments to the design. Subsequently, Clark launches a conference to discuss the problem. In addition to Smith, he invites Miller, who is responsible for the casing and will be affected by the change to the gear design.

## Resource Management Model

Since this article mainly focuses on product and process management, resource management is discussed only briefly as follows.

### Human Resources

*Resources* are required for executing development tasks. They are classified into humans and tools. For human resources, a *role model* has been developed which supports flexible assignment of employees according to the roles they play in development projects (see [8,20] for similar models). Tools are integrated in a coarse-grained manner by means of *wrappers* which prepare the data, set environment variables, and call the tools (see, e.g., [6,27] for general work on tool integration).

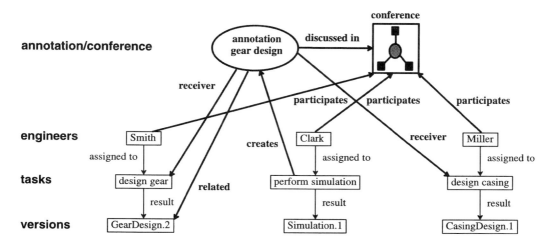

**FIGURE 2.12**   Annotations.

```
team Sample;
    employee Haseltine;    roles SinglePart;
    employee Miller;       roles SinglePart, Design;
    employee Smith;        roles Design, Plan;
    employee Clark;        roles Plan, NCProgram;
end;
```

**FIGURE 2.13**   Project team.

*Human resources* are *employees* who are organized into project teams. Employees are assigned to tasks according to the *roles* they play in a project. Each role (e.g., NC programmer) corresponds to a certain task type (e.g., creation of an NC program). In general, each team member may play any set of roles. A task may be assigned to an employee only if (s)he plays the appropriate role. For example, an employee who only plays the Designer role may not be assigned to NC programming tasks.

There is no explicit distinction between *engineers* and *managers*. Typically, engineers and managers are assigned to atomic and complex tasks, respectively. An atomic task corresponds to a *technical task* such as creation of a design or of an NC program. A complex task such as development of a single part is viewed as a *management task*. The manager who is assigned to a supertask is in charge of planning and controlling execution of its subtasks. In particular, the manager has to create subtasks and assign them to engineers of the project team (or to mangers of complex subtasks).

Figure 2.13 shows a simple example of a project team. Each role eventually corresponds to a task type. Since we have not introduced explicit identifiers of task types, the roles are instead denoted by object types. An engineer *e* playing role *r* may be assigned to any task producing an output of type *r*. For example, Haseltine plays the role SinglePart and may therefore be assigned to complex tasks for developing single parts. Thus, Haseltine may act as a manager. In addition, Miller plays the Design role and may work as an engineer as well. Smith and Clark are engineers who are both skilled in multiple working areas. Given these roles, the team members may be assigned to tasks as shown in Figure 2.7.

### Tools

Employees are supported by *tools* in executing assigned tasks. In case of atomic tasks, these are the development tools tools integrated with the management system. In case of complex tasks, the management environment is used for task execution. Unlike human resources, tools are implicitly assigned to tasks.

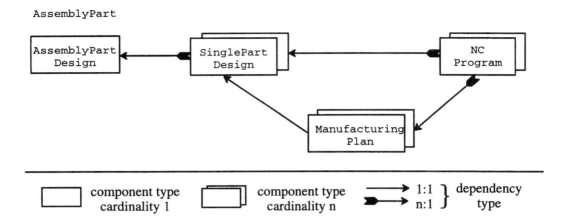

FIGURE 2.14   Specific schema for product management.

Each tool is characterized by the types of objects on which it may operate. In order to execute a task $t$ with output type $ot$ and input types $it_1 \ldots it_n$, all tools are made available which may read or write objects of type $ot$ and read objects of types $it_1 \ldots it_n$.

Task assignments are used to build up agendas. An *agenda* for an employee $e$ consists of all tasks for which $e$ is responsible. To work on a task $t$, an employee selects an item from the agenda. Then, the workspace of $t$ is presented which consists of all input, output, and auxiliary objects (see Figure 2.9). For operating on these objects, tools are made available according to the rules explained above.

## Model Adaptation

So far, the management model has been independent of a specific application domain. The domain-independent part of the model is called *generic model*. In the following, we discuss how the generic model is adapted to a specific application domain. The result of such an adaptation is called *specific model*.

In general, *adaptation* of the generic management model may involve the definition of domain-specific types of objects, relationships, and attributes, as well as domain-specific constraints and operations. Within the SUKITS project, however, adaptation has been studied only to a limited extent, mainly focusing on the definition of a domain-specific schema for product management (Figure 2.14). Due to the product-centered approach, this schema is used to adapt process management as well. For example, each component type in a configuration type corresponds to a task type. Furthermore, the resource management model is also adapted implicitly since roles determine the types of tasks which may be assigned to employees.

## Formal Specification

The management model which we have introduced informally above is fairly complex. A formal specification of the model promises the following benefits:

- It describes the model at a high level of abstraction and thereby aids in its understanding.
- The model is defined precisely and unambiguously.
- An implementation may be derived from the formal specification, which is easier than development from scratch.

The management model is formalized by means of a *programmed graph rewriting system* [23]. Graphs have already been used informally throughout this section. Formally, an attributed graph consists of typed nodes and edges. Nodes may be decorated with attributes. Operations on attributed graphs are described by graph rewrite rules which replace a graph pattern (the left-hand side) with some subgraph (the right-hand side).

Presentation of the formal specification goes beyond the scope of this article; the interested reader is referred to [13, 29].

## 2.4   Management System

This section presents the management system, which is based on the model described in the previous section. We first take a look at the management system from the user's perspective and then discuss its realization.

### Tools: Functionality and User Interface

After having presented the management model, we now turn to the tools offered by the management system. Tools are composed into environments which address different kinds of users. As explained in Section 2.2 (see Figure 2.1), the management system provides a management environment for project managers, a front end for both engineers and managers, and a parameterization environment for adapting the management system to a specific application domain. Functionality and user interface of these environments are described below in turn.

#### Management Environment

The management environment is an instance of an *IPSEN* environment [19]. IPSEN is a project that is dedicated to the development of tightly integrated, structure-oriented environments. While IPSEN has its origins in software engineering, we have applied its technology to mechanical engineering in the SUKITS project.

The most important features of IPSEN environments as perceived by their users are summarized below:

- All tools are language-based. A *language* defines the external representation of some document and may comprise both textual and graphical elements. Syntactic units of the language are denoted as *increments* (e.g., a statement in a programming language or an entity type in an ER diagram).
- (Cut-outs of) documents are displayed in *windows*. Each window has a *current increment* which can be selected by mouse click or cursor movement. Furthermore, a *menu bar* offers all commands which can be activated on the current increment.
- Commands are organized into *command groups* denoted by capital letters (e.g., EDIT or ANALYZE). Command groups affect the way in which commands are offered in the menu bar (hierarchical menu). Thus, command groups are a concept at the user interface and may or may not correspond to tools.
- A *tool* offers a set of logically related commands to the user. In general, commands in one group may be offered by different tools, and one tool may spread its commands over multiple command groups. Tools are highly integrated. Multiple tools at a time may provide commands on the current increment. Furthermore, commands from multiple tools may be activated in (almost) any order.

In particular, we have to stress the difference between tool integration in IPSEN and integration of external development tools. In IPSEN, a tool is a seamlessly integrated component of an environment. In contrast, a development tool is an application which is integrated with the management system through a wrapper. Integration is performed rather loosely at a coarse-grained level.

The management environment provides tools for managing products, processes, and resources. These are discussed in turn below, taking the examples up again which were given in Section 2.3.

*Product management* is supported through tools for editing and analyzing version and configuration graphs. Figure 2.15 shows a snapshot of a version graph for a shaft design. From the initial version 1,

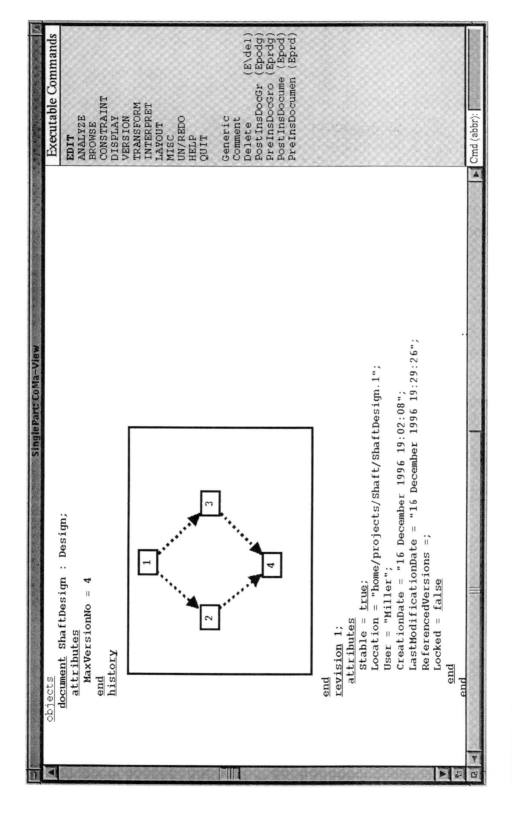

**FIGURE 2.15**  Version graph.

alternative versions 2 and 3 were derived which were merged into the common successor version 4. The text below the version graph lists the attributes of all versions in turn. Typical commands on version graphs include creation of root versions, derivation of successor versions, freezing of versions, and purging of unused versions. These commands are gathered in the command group EDIT.

Analogously, the management environment presents graphical views on configuration version graphs and configuration graphs. Commands for manipulating these graphs are again offered in the command group EDIT. In addition, several analysis commands are provided in the command group ANALYZE. For example, two configuration version graphs may be compared with the help of a Diff command which highlights components/dependencies contained in only one configuration version, as well as components occurring in different versions in both configuration versions.

*Process management* is supported through commands operating on dynamic task nets. As a consequence of the product-centered approach, task nets coincide with configuration version graphs. Therefore, structural changes of task nets are performed with the help of (redefined) edit commands inherited from product management. Commands for state transitions, workspace management, and release management are subsumed under the command group INTERPRET. Typically, only a subset of these commands are executed by project managers in order to plan a project and to account its status. For example, a project manager may execute a Defined transition to put a task onto the agenda of an engineer who will then perform the Start transition.

Figure 2.16 shows how the task net of Figure 2.7 is presented at the user interface. Note that the direction of relationships is inherited from product management, i.e., relationships have the "wrong" direction from the perspective of process management. Task attributes such as State, ReleaseState, etc. are displayed on demand in separate windows. Attributes referring to auxiliary documents, guidelines, working version, and annotations refer to the workspace of a task (see also Figure 2.9).

*Resource management* covers both humans and tools. For managing human resources, an editor is provided which is used to define team members and their roles (see Figure 2.13). Each role corresponds to a task type. A team member may only be assigned to a task if he plays the required role. Similarly, another text editor is offered for describing tools. Each tool is characterized by a set of attributes describing the types of input and output objects, the call modes (ReadOnly or ReadWrite), and the location of its wrapper.

## Front End

The front end provides a *task-centered interface* to the management system. It covers both technical and management tasks, which are represented by leaf and nonleaf nodes of the task hierarchy, respectively. Tasks are assigned via the management environment. The front end supports task execution by displaying agendas, managing workspaces, providing commands for state transitions and release management, offering queries, and activating tools. In case of a technical task, an engineer is supplied with document versions and may call development tools such as CAD systems, FEM simulators, and NC programming systems. A management task is handled in the same way as a technical task. Thus, a manager receives tasks through an agenda as well, and he may invoke the management environment on a configuration version/task net. Due to the role model, an employee may even act both as manager and engineer. In this situation, an agenda would mix technical and management tasks.

The front end was implemented with the help of a commercial tool-kit (*XVT* [3]) which was selected to ensure availability on heterogeneous platforms. Its user interface differs radically from the user interface of the IPSEN-based management environment. In particular, the front end is not language-based, i.e., the user is not aware of any textual or graphical language representing the contents of the management database. Rather, customized views are presented as tables, and mask-like windows are used to enter parameters of commands. Graphical representations were not considered because of lacking support through the tool-kit (no high-level package for developing graphical applications).

To enter the front end, an employee identifies himself by his name and his password. Figure 2.17 shows a hierarchy of windows which were opened in order to work on an output document of the task

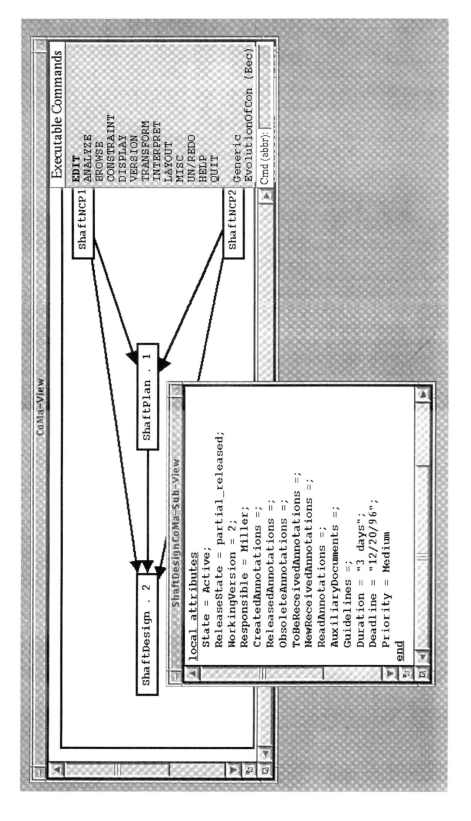

**FIGURE 2.16**    Task net.

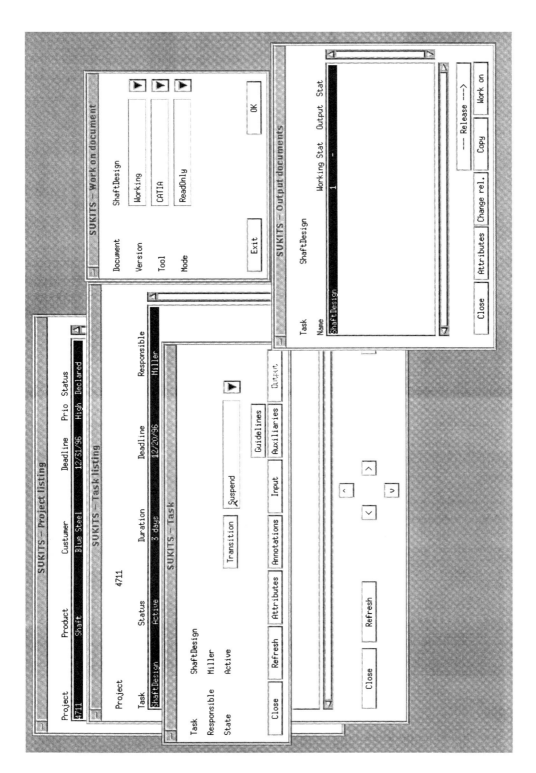

**FIGURE 2.17**  User interface of the front end.

ShaftDesign. These windows are explained in turn below:

- Project listing (topmost window on the left) lists all "projects" in which tasks need to be carried out by the employee. Here, "project" denotes an order from a customer which is described by order number, product to be developed, name of the customer, deadline, priority, and status of processing.
- After having selected an order from the main window, the employee is supplied with an agenda of tasks (window Task listing). Each task is characterized by name, state, expected duration, deadline, and responsible employee. The employee who logged into the front end may also view related taks by clicking the navigation buttons at the bottom of the task list window (for example, $<$ moves to the predecessors of a selected task).
- For a certain task selected on the agenda, the Task window offers state transitions and buttons for task attributes, annotations, input, output, and auxiliary documents, as well as guidelines.
- When the Output button is clicked, a window (Output documents) is created which shows a list of output documents. For each output, the working version is distinguished from the released version. The Stat columns denote release states ($+$, $(+)$, and $-$ correspond to the release states total, partial, and none, respectively). The buttons at the bottom are used to display version attributes, to modify release states, and to start tools. The windows for input documents, guidelines, and auxiliaries are structured similarly.
- Selection of the button Work on triggers creation of a window Work on document. Here, the employee may select either the working version or the released version, choose among available tools, and determine a call mode. After that, a wrapper is executed to start the selected tool.

Figure 2.18 demonstrates how annotations are handled in the front end. The manufacturing planner is creating a problem report concerning the design. The window Annotations on the right-hand side displays incoming and outgoing annotations (upper and lower subwindow, respectively). Both lists are currently empty. By activating the New button at the bottom, the window Annotation on the left-hand side is opened. The user inputs the name, selects a document type, and determines both related documents and receivers. The annotation is created when the OK button is clicked. After that, it will appear in the subwindow Outgoing Annotations. Work on operates in the same way as for any other document. Release "delivers" the annotation, and Conference is used to prepare a distributed work session.

**Parameterization Environment**

As already explained in Section 2.3, the generic management model has to be adapted to a specific application domain. The product management model is customized through an ER-like diagram which implicitly defines task types (process management) and roles (resource management).

Figure 2.19 shows a *schema* for the configuration type SinglePart. Boxes and arrows correspond to component and dependency types, respectively. Boxes are labeled by names of component types. Dependency types need not be named as long as there is at most one dependency type for some pair of component types. In general, attributes may be attached to entity and relationship types, but this is not shown in the screen dump.

*Cardinalities* constrain both component and dependency types. A cardinality is written as $(l1, u1) \rightarrow (l2, u2)$, where $l$ and $u$ represent lower and upper bounds, respectively. The first and the second pair constrain the numbers of outgoing and incoming relationships, respectively. Cardinalities attached to component types refer to composition relationships from configurations to components. Thus, they also refer to relationships which, however, are not represented explicitly by arrows in the graphical notation.

The schema reads as follows: A configuration of type SinglePart consists of exactly one design (cardinality $(1, 1) \rightarrow \dots$), at most one raw part design (cardinality $(0, 1) \rightarrow \dots$), and at least one NC program (cardinality $(1, N) \rightarrow \dots$). While an NC program may be used for multiple single parts (cardinality $\dots \rightarrow (1, N)$), we assume that all other documents are specific to one single part (cardinality $\dots \rightarrow (1, 1)$). If there is a raw part design, it depends on the design of the final part. Thus, the cardinality of the dependency type is $(1, 1) \rightarrow (0, 1)$. The manufacturing plan has a mandatory dependency to the design and depends on the raw part design if and only if the latter is present. Similar arguments hold for the dependencies of NC programs.

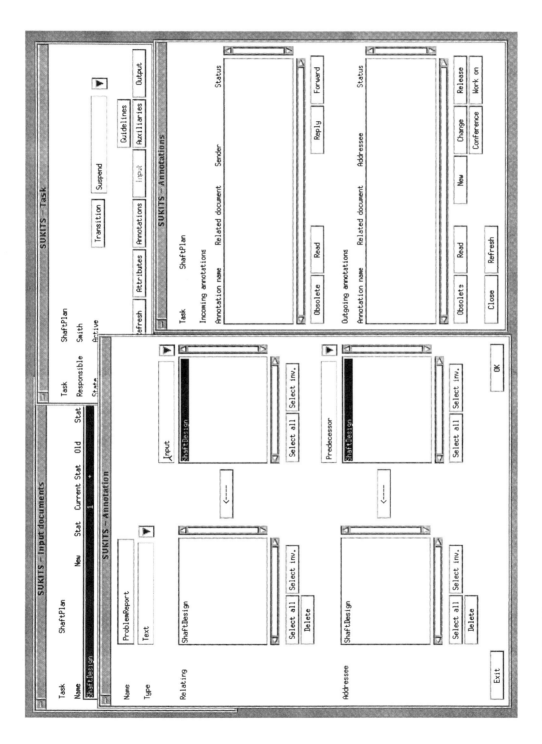

**FIGURE 2.18** Annotations.

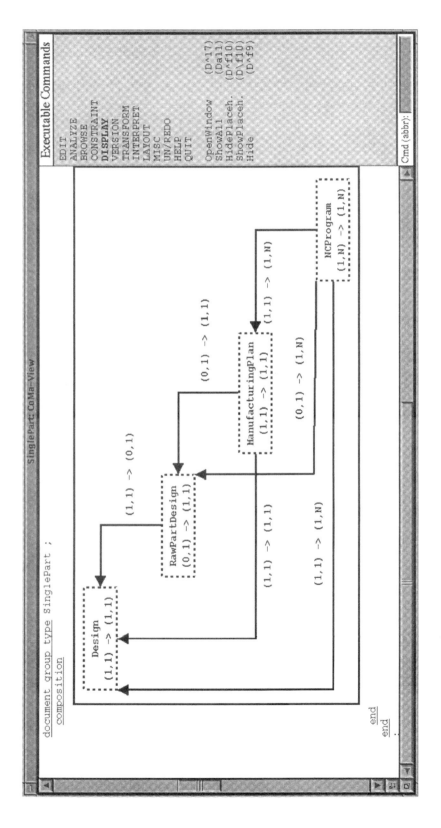

**FIGURE 2.19**   Schema diagram.

The schema is used in the following ways:

- The user receives guidance through menus generated from the schema. For example, legal object types are displayed in a menu when an object is going to be created.

- All operations violating the constraints of the schema are rejected. For example, an attempt to create two (final) designs for a single part results in a cardinality overflow.

Defining a schema is only one step in the adaptation process. In addition, tools need to be integrated with the management system. This will be discussed further below.

While the management system cannot be used without adaptation, the adaptation process may still continue at project runtime (although careful planning and coordination is required). The schema may be modified as long as changes do not introduce inconsistencies. For example, a document type may be deleted if it has not been instantiated yet, and a cardinality may be constrained from $N$ to 1 if no cardinality overflow occurs. Furthermore, descriptions of tools may be modified in any way at (almost) any time. In particular, since Perl scripts are executed in an interpretive mode in separate operating system processes, changes to Perl scripts do not require compilation or linking of any part of the management system.

## Realization

The structure of the management system was already discussed briefly earlier. Figure 2.20 is derived from Figure 2.1 by annotating each component with the system used for its realization. The following sections describe the management environment, the parameterization environment, and their underlying management database. Following this, the front end and tool integration through wrappers is described, following which communication and distribution are described.

### Management and Parameterization Environment

Both the management environment and the parameterization environment are realized with the help of *IPSEN* technology [19]. Within the IPSEN project, a standard architecture for integrated, structure-oriented development environments has been developed. We have used this architecture to design and implement management tools which can be applied in different disciplines, including both software and mechanical engineering. The architecture of the management and parameterization environment is not discussed here because this would require an intimate understanding of the internals of IPSEN environments. For an in-depth description, we recommend Chapter 4 of [19], which also includes a section on the management and parameterization environment.

Instances of IPSEN environments make use of a home-grown database management system called *GRAS* [16] which is based on attributed graphs. Types of nodes, edges, and attributes are defined in a graph schema. Derived attributes and relationships are evaluated incrementally on demand (lazy evaluation). GRAS provides primitive update operations for creating, deleting, or changing nodes, edges, or attributes. An event/trigger machine supports notification of GRAS applications and execution of event handlers (active DBMS). Furthermore, GRAS offers nested transactions with ACID properties. Finally, GRAS may be operated in a distributed environment. Within the SUKITS project, GRAS has been used for storing and accessing the management database.

### Front End

The tools used in the SUKITS scenario, e.g., CAD systems or NC programming systems, require a multitude of different operating systems, both Unix and non-Unix systems. The front end has to be easily portable across all of these platforms. We therefore selected the *XVT* tool-kit [3] to build the user interface. The main part of XVT is a library of window-related procedures. They abstract from window systems specifics and allow creation and manipulation of windows and window contents in a portable way. An editor allows to interactively define window types and window components. A generator turns these definitions into C code, which has to be compiled and linked against the platform-specific XVT base library.

Since the user interface is *event-driven*, the architecture of an XVT application consists of a set of event handlers. Each event handler processes some specific user interaction, e.g., selection of a task from an agenda.

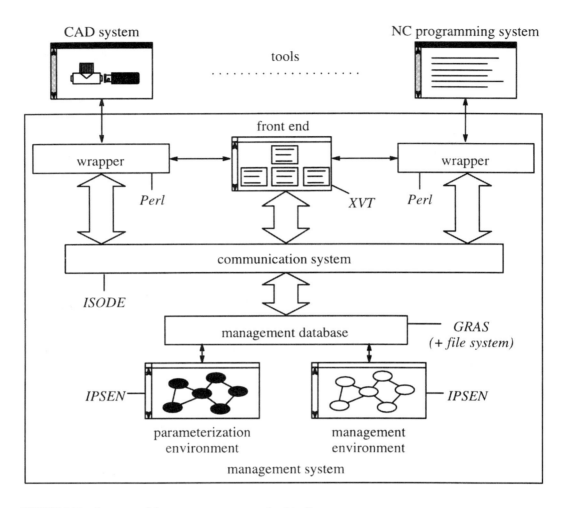

**FIGURE 2.20**    Structure of the management system (revisited).

This may involve operations against the management database, e.g., for querying task attributes, retrieving input and output documents, etc. The front end accesses the management database through a module called FrontEndView, whose interface is tailored towards the operations invoked via the front end.

An important development goal for the front end has been to make the implementation independent both from the database realization and from the details of the management model used. To achieve this, these details are not hardcoded into the interface of FrontEndView; rather, they are interpreted at run time. For example, the domain-specific schema specified via the parameterization environment defines types of documents, configurations, relationships, and attributes. This information is obtained from the management database by querying the schema; thus, the front end need not be modified when the schema is changed.

**Tool Integration**

The front end itself only works on coarse-grained management data. To modify the contents of documents, external tools are called. To start a tool, the user just has to select a document version and click at the Work on button. The front end then offers a list of those tools that are appropriate for the type of the selected document. The user is shielded from the specifics of transferring the document into a temporary session workspace and calling the environment with the necessary parameters. The front end presents to the user only a limited set of options for starting the tool. A selection between read-only and write access is usually all that is needed.

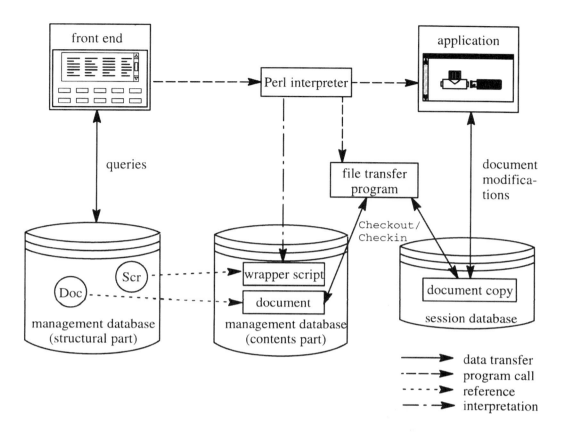

**FIGURE 2.21**   Wrappers for tool integration.

A tool-specific wrapper realizes this abstraction (Figure 2.21). The wrapper is responsible for the Checkout/Checkin operations which were illustrated in Figure 2.5. A temporary *session workspace* is populated by copying files from the contents part of a management database. Then the tool is started on the document copy. When the user quits the tool, and if the document has been modified, the changed version is transferred back into the global storage. Finally, the temporary workspace is deleted.

Some tools use a set of related files to store a document. In this case, the wrapper ensures (by packing the file set into a single archive before the file transfer, and by unpacking after a transfer) that the global database never contains mutually inconsistent document components. Similarly, the wrapper calls a converter if an input document does not have the format which is required by the respective tool.

When calling a wrapper, the front end supplies an access path that refers to the document version. It gets this path from the management database and passes it uninterpreted to the wrapper. This path contains the file transfer method used as well. The wrapper uses this information to call the right file transfer program. In this way, flexibility in the use of file transfer programs is achieved.

The management system can be used as well with tools that store their documents not in files, but in a database of their own. In this case, the access path contains a database specific path to the document. The tool has to meet several requirements in order to make this possible. As a session workspace cannot be set up by file copying, the database has to offer a programmatic interface that the wrapper can use to implement the Checkout and Checkin operations on the database, and to set the required locks to prevent modifications by other users. Furthermore, it must be possible to confine the running tool onto the part of the database that contains the checked out document version.

The wrappers have been implemented in *Perl* [26] for portability reasons. Interpreters for this language are available for a wide range of different platforms. Especially, file access and manipulation and start of external programs can be easily programmed in a way that runs unmodified on many platforms, easier than writing them in C. For debugging purposes, the script can be run manually without a front end attached. By using wrappers, new tools can be added by writing a Perl script and extending the database. The front end itself remains unchanged. In contrast, writing the wrappers in C code would require recompiling and relinking whenever a wrapper is added or modified.

**Communication and Distribution**

All components developed in the IPSEN project may be operated in a homogeneous network of workstations. Currently, these are SUN workstations which run under Unix and use the Network File System (NFS) for transparent file access. Furthermore, the GRAS database management system supports *client/server distribution* in the following way: In order to operate on a graph, an application sends an Open request and receives a handle to a (potentially remote) graph server. In this way, multiple clients may connect to a single server which encapsulates the graph and is responsible for coordinating graph accesses from the clients. Accesses are realized by remote calls to procedures offered at the GRAS interface (e.g., creation/deletion of nodes/edges).

Unfortunately, we cannot assume a homogeneous network for the overall integrated environment. In general, tools may operate on machines which run under different operating systems and mutually incompatible communication systems. In particular, neither IPSEN-based environments nor the GRAS database management system may be available on a certain machine required by some tool. Therefore, a communication system was developed which enables interoperation in a heterogeneous network.

The *SUKITS communication system* [7] is partly based on *ISODE* [22], a development environment for OSI-conforming applications. In particular, it offers services for file transfer and database access. These services are only used by the front end and the wrappers. The management and parameterization environment accesses the management database directly through the services provided by GRAS.

Figure 2.22 illustrates how the communication systems of GRAS and SUKITS are combined to access the management database. Machines C, D, and E belong to the homogeneous subnet which is covered by the GRAS communication system. The SUKITS communication system is not responsible for database access within this subnet. The management and parameterization environment (on machine E) accesses the management database via a GRAS client stub.

A front end which runs within the homogeneous subnet accesses the management configuration database through the FrontEndView, which is linked into the application. As an example, Figure 2.22 shows a front end running on machine E. The code for accessing the database is embedded into the front end down to the level of GRAS operations which are transmitted to the GRAS server.

The SUKITS communication system is used for front ends running outside the range of the GRAS communication system (machines A and B). The operations provided by FrontEndView are activated remotely via a client stub. Thus, front ends running outside the range of the GRAS communication system access the database via two inter-process communications. Note that rather complex operations are transmitted via the first communication channel, while primitive GRAS operations are sent via the second channel. Thus, the overall communication overhead is dominated by GRAS-internal communication. As a consequence, the interplay of two communication systems only induces negligible overhead.

# 2.5 Applications and Experiences

So far, we have described a model for managing product and process development and a system which realizes this model. To evaluate our work, it is essential to get feedback from applications. Throughout the SUKITS project, we have been putting great emphasis on applying our results to nontrivial (yet manageable) scenarios. Instead of doing this in "paper and pencil mode," tools were integrated with the management system, and comprehensive demo sessions were worked out. Evaluation was performed informally by the project partners themselves.

We now describe the prototypes developed in the SUKITS project (consisting of the management system and integrated tools). Following this, we present a comprehensive demo session which illustrates the use of the SUKITS prototype '96 (the final "deliverable") by a coherent example.

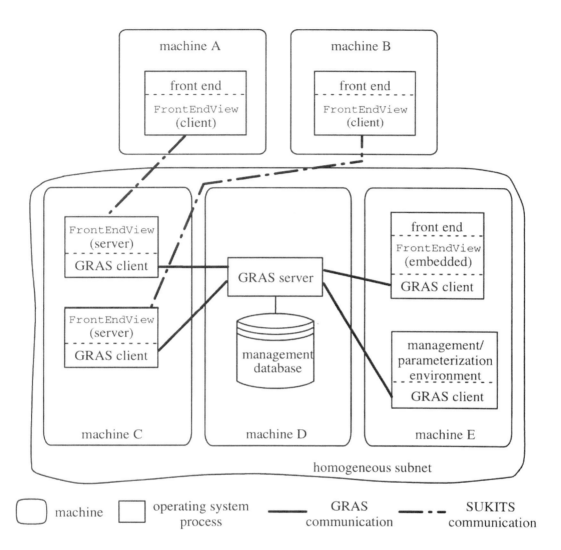

**FIGURE 2.22** Access to the management database.

## SUKITS Prototypes

Prototypes marked important milestones achieved in the SUKITS project. The first one, called *SUKITS prototype '93*, was finished in the fall of 1993. The prototype consisted of an earlier version of the management system which was integrated with heterogeneous, mainly file-based tools running under Unix, VMS, or MS-DOS. The scenario underlying the demo session covered the design of single parts. The management environment was essentially restricted to product management; process and resource management were supported in a preliminary, rather ad hoc fashion. The SUKITS prototype '93 will not be described further because all of its features are also offered by the SUKITS prototype '96 (see below).

The presentation given in this article is based on the *SUKITS prototype '96* which is distinguished from its predecessor by several extensions and improvements:

- At the level of modeling, the management model was augmented with process and resource management as described in Subsections 3.2 and 3.3, respectively.
- At the level of realization, the management environment was extended accordingly. The user interface of the front end underwent a major redesign which implied its re-implementation from scratch. Finally, integration of tools was addressed in a more systematic way.

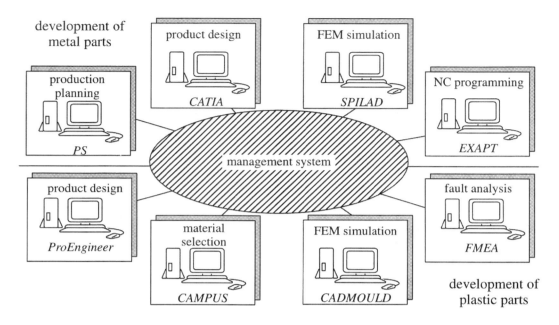

**FIGURE 2.23**   Integrated tools.

The most comprehensive (and important) demo was presented in April 1996 to the German Research Council (DFG), the funder of the SUKITS project. Considerable efforts were undertaken to prepare the demo. The tools (mostly commercial systems) which were integrated with the management system are summarized in Figure 2.23:

- The upper half depicts tools for developing metal parts. CATIA is used to design the product geometry. In the case of assembly parts, a part list is maintained using the PPC system PS. FEM simulations are performed in SPILAD, and NC programming is done in EXAPT.
- Tools for developing plastic parts are shown in the lower half. Plastic parts are designed in the CAD system ProEngineer. CAMPUS is an expert system for material selection. CADMOULD is used for simulations. If simulations reveal that the product design fails to meet the requirements (e.g., with respect to stress behaviour), FMEA supports engineers in fault detection.

To integrate the selected tools, wrappers were written as Perl scripts. Most tools are based on files. In these cases, it was a straightforward task to implement Checkout and Checkin operations against the product management database. In addition, several converters between different data formats were developed in order to support the workflow effectively. Finally, some tools use (relational or home-grown) databases which did not allow to store their documents in the product management database. In case of CAMPUS and FMEA, this restriction did not cause any problems because their databases contain expert knowledge which is used, but not modified. The PPC system PS relies on a relational database system. In this case, a converter was implemented which transforms the relational representation of a part list into an ASCII file stored in the product management database. The ASCII file is analyzed to determine the components to be developed such that the product configuration can be extended accordingly.

The hardware configuration prepared for the demo consisted of a network of workstations from different vendors as well as a file and database server (Figure 2.24). All machines ran under different variants of the Unix operating system; other operating systems (VMS, MS-DOS) were already covered by the SUKITS prototype '93.

Figure 2.24 also shows how software components are distributed over the network. The workstations and personal computers shown in the upper half host the development tools of Figure 2.23. The front

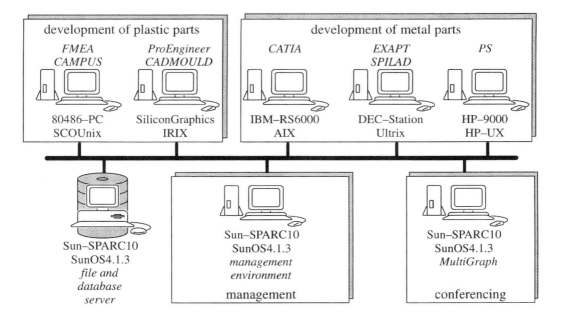

**FIGURE 2.24**  Hardware configuration and distribution of software components.

end is available on all machines of the network and was omitted from the figure. Both the management environment and MultiGraph (a conferencing tool developed in SUKITS) could only be run on the machines in the lower half because they assumed SunOS as operating system.

## Demonstration: Development of a Drill

### Overview

To demonstrate the SUKITS prototype '96, a nontrivial demo session was worked out which shows a cutout of the development of a drill. The demo focuses on the gear and the casing. In particular, it illustrates the following features:

- *Product evolution*. The task net for the gear depends on the product structure. Initially, it merely contains a few tasks which are known a priori. After the product structure has been determined, the task net is extended with tasks for developing the components of the gear.

- *Simultaneous engineering*. In order to start development of the casing, it is sufficient to know the outline of the gear. Thus, development of the gear and the casing may be overlapped. Similarly, design of the gear and creation of its part list may be performed in parallel.

- *Feedback*. For both the gear and the casing, various simulations are carried out. If they demonstrate that the requirements are not fulfilled, feedback is raised to earlier steps in the development process.

- *Integration of formal and informal cooperation*. Annotations are provided for integrating formal and informal cooperation. In the demo, annotations are used for both asynchronous communication (e.g., the manager who is reponsible for the casing is informed that development may start) and synchronous communication (e.g., designers of the gear and the casing discuss the consequences of a design change in a conference).

- *Cooperation between different companies*. The gear and the casing are developed in different companies. Thus, the demo gives a (preliminary) example of interorganizational cooperation.

Note that the demo does not cover the full range of capabilities of the SUKITS prototype. Rather, it mainly emphasizes process management. Furthermore, parameterization of the management system was performed beforehand and is not included in the demo itself.

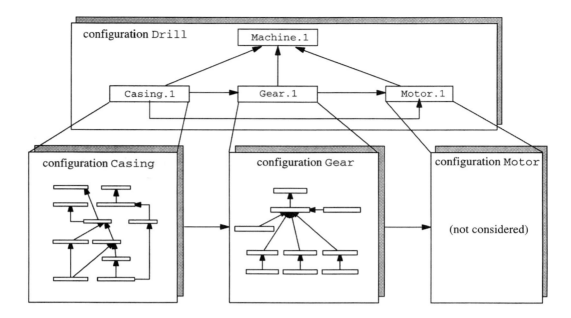

**FIGURE 2.25**   Hierarchy of configurations.

The documents for developing the drill are arranged in a hierarchy of configurations which is sketched in Figure 2.25. Recall that configurations also serve as task nets (product-centered approach). The subconfiguration Machine in the top-level configuration defines a coarse decomposition of the drill into its main parts. For each of these, there is a subconfiguration containing all documents created during development. The motor is not considered at all, the configurations for the casing and the drill are detailed below.

Figure 2.26 illustrates a (cutout of) the configuration for the gear. Bold boxes highlight the working areas covered by the demo. The normal-shaped box at the top indicates that problem analysis has been performed before. The specification list serves as input for creating a solid model of the gear; the corresponding part list is built up in parallel. For mechanical simulation, a drawing has to be derived from the solid model of the gear; the drawing serves as input to the simulator (the FE network is created implicitly at the beginning of the simulation). From the model of the complete gear, a model of its outline (mounting room) is derived which is required for developing the casing. For nonsupplied parts of the gear, drawings and NC programs need to be created. Note that the configuration depends on the product structure fixed in the part list. Thus, documents for single parts are not included in the initial configuration.

In contrast, the structure of the configuration for the casing is known beforehand (Figure 2.27). The model of the complete gear and the model of its outline are imported into the configuration and both serve as input documents for designing the casing. While the model of the outline already provides all required information (mounting room), the model of the complete gear is still needed to discuss design changes and their impacts on the casing. Another source of information is an (internal) list of requirements which are derived from the (external) specifications coming from the customer. The list of requirements is also used as input for material selection. To evaluate the design of the casing, a process simulation is carried out which depends on the selected material and an FE network. The FE network is derived from the solid model of the casing via an intermediate step (generation of a surface model). As a final step, the mold for producing the casing is designed (not included in the demo).

## Demo Steps

The steps of the demo session are summarized in Figure 2.28. The steps are arranged in a partial order which is derived partly from the task nets of Figures 2.26 and 2.27. To a great extent, the gear and the casing may be developed independently and in parallel. There are a few synchronization points where

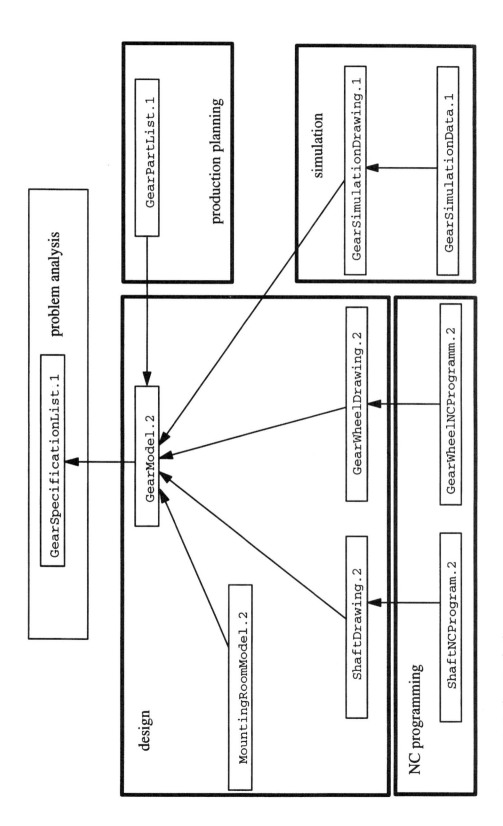

FIGURE 2.26 Configuration for the gear.

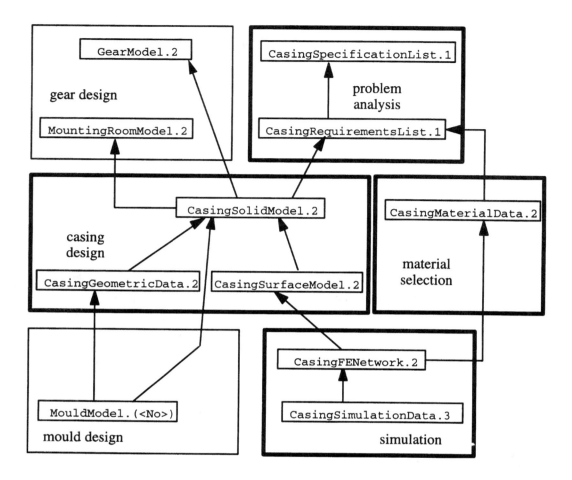

**FIGURE 2.27**　Configuration for the casing.

information is passed from gear to casing development. Furthermore, the customer's requirements are changed in the middle of development. These changes require modifications to the design of the gear; potential implications on the casing are discussed in an electronic conference. Finally, a feedback from process simulation to material selection is demonstrated at the end of the session.

From an external specification, an internal list of requirements is derived which drives the development of the casing (1). Based on these requirements (represented as a text file), the material expert selects an appropriate material from the material database CAMPUS (2). After the material has been selected, the development of the casing cannot proceed further because the models of the gear and the mounting room required for designing the casing are not available yet.

A solid model of the gear is prepared with the help of CATIA (3). The part list of the gear is built up in parallel, using the PPC system PS (4). To enable simultaneous engineering, the designer performs a partial release of an intermediate version of the gear model (Figure 2.29). Note that the gear model is not released yet to the design of the mounting room, which can be started only when sufficiently detailed and stable geometrical data are available.

When steps 3 and 4 are finished, the manager responsible for coordinating the development of the gear—called gear manager below—is notified through a maillike annotation. Subsequently, he uses the management environment to complete the configuration/task net (5). This step is automated: The part list is searched for entries corresponding to non-standard single parts. For each entry, a single part design is inserted into the configuration. For each single part design, there must be a corresponding NC program which is inserted automatically as well. Figure 2.30 shows the state after extension of the task net.

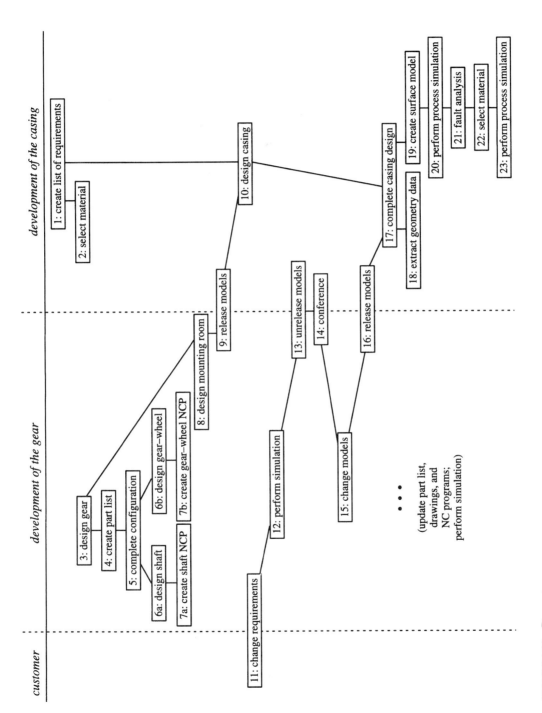

**FIGURE 2.28** Demo steps.

The manager now assigns engineers to the new development tasks (currently, this has to be done manually). Task assignment must be compatible with the roles which engineers play in this project.

Subsequently, the single parts are designed with CATIA (6), and NC programs are created with EXAPT (7). Creation of NC programs is automated to a large extent: EXAPT comes with a generator which creates NC programs from drawings.

Based on the design of the gear, the mounting room model for the casing is designed (8). When the mounting room model is completed, the gear manager is notified through an annotation. The gear manager forwards the release message to the casing manager who imports the gear model and the mounting room model into his subconfiguration (9). In this way, cooperation between different companies is supported. Afterwards, design of the casing may be started, using the CAD system ProEngineer (10).

In the middle of development, the customer changes the requirements to the drill (11). In particular, the drill needs to operate at a higher number of revolutions. Among other things, this may afffect the gear. To check this, a simulation is performed in SPILAD (12). Its results show that the gear developed so far does not withstand the increased forces caused by the higher number of revolutions.

The gear manager, who has been notified of the results of the simulation, has to estimate its consequences on ongoing development tasks (13). Here, we have to distinguish between local and global effects (within one subnet and across subnet boundaries, respectively).

Since the gear model needs to be changed, all transitively dependent tasks within the gear subnet might be affected by the change. However, it may be difficult to determine the tasks beforehand which will actually be affected. The gear manager may follow different policies, ranging from optimistic (each task may be continued without disruption) to pessimistic (all potentially affected tasks are suspended).

In the demo, the manager chooses the pessimistic policy. He makes use of a complex command offered by the management environment which transitively revokes releases of the outputs of all tasks which depend on the gear design.

In addition to local effects, global ones have to be considered as well. As in step 9, the gear manager sends an annotation to the casing manager who propagates the changes into the task net for the casing. He revokes the releases of the gear model and the mounting room model, blocking the design of the casing. Furthermore, he informs the casing designer through another annotation.

So far, annotations have been used exclusively for asynchronous communication (sending of messages). The current demo step (14) shows that synchronous communication is supported through annotations as well. The casing designer decides to discuss the probable consequences of the planned design change. To this end, he creates an annotation which refers to the mounting room model. The annotation addresses not only the designer of the mounting room model, but also the casing manager, who wants to keep track of what is going on. Based on this annotation, the casing designer initiates a conference and invites the receivers of the annotation as participants.

The conference is performed with the help of MultiGraph. A snapshot of the mounting room model is loaded into the shared whiteboard of MultiGraph, where the participants may communicate through shared cursors and may also add textual and graphical elements. Furthermore, audio and video channels are available. The contents of the shared whiteboard may be stored to document the outcome of the conference.

After the conference, the required changes are performed to the models of the gear and the mounting room. First, changes are propagated locally such that blocked tasks may be resumed (15). Consider, e.g., the creation of an NC program for a shaft, which was blocked in step 13. Later on, a new version of the shaft drawing arrives (the diameter had to be changed). Now, the NC programmer may resume his work and consume the new version.

As soon as the gear model and the mounting room model have again reached a stable, complete, and mature state, the new versions are released to the development of the casing (16). This is done in the same way as in step 9. Now, the design of the casing can be completed after it had to be suspended for some time because of the changed requirements (17).

From the solid model of the casing, characteristic geometrical data are extracted automatically by means of a converter (18). These data are used for several purposes, e.g., for designing the mold or the packing of the drill. These activities are not considered any more, i.e., the demo session ends at this point.

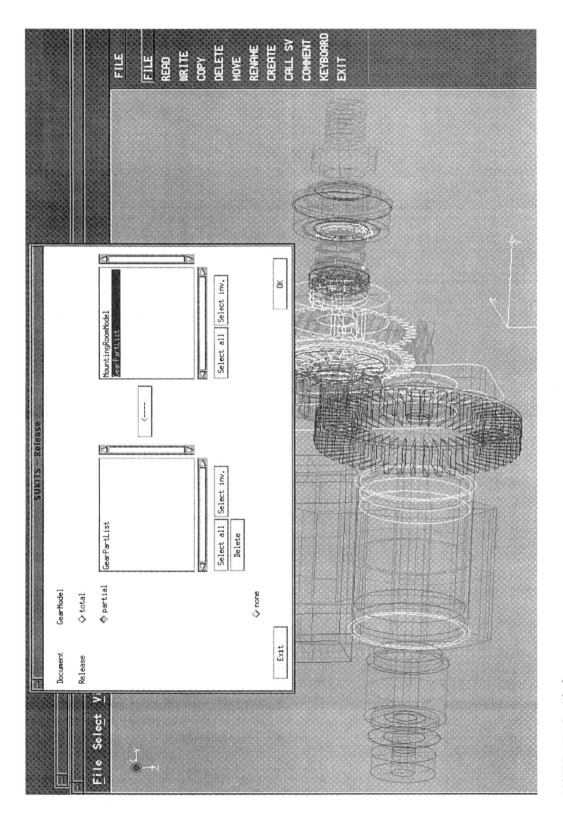

**FIGURE 2.29** Partial release.

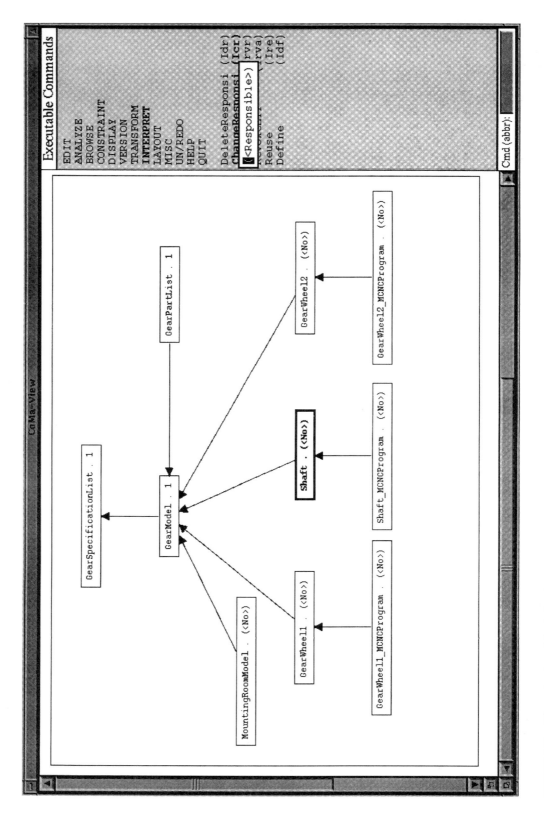

FIGURE 2.30   After extension of the task net.

Development of plastic parts may involve two kinds of simulation, namely mechanical and process simulation. Only the latter is included in the demo session. To prepare the process simulation, the solid model of the casing is converted into a surface model from which an FEM net may be generated (19). Currently, the surface model has to be created manually (in ProEngineer). From the surface model, an FE network is generated automatically (using CADMESH). Subsequently, the FE network is fed into CadMould in order to perform the process simulation (20). Its results are visualized with the help of a separately called postprocessor (SIMGRA). The simulation reveals the following problem: In order to fill the mould properly, the plastic material must be injected at a very high pressure which cannot be obtained on the machine to be used for production.

To fix this problem, feedback is raised in order to reexecute previous development steps. So far, it is by no means clear what has caused the problem and how it can be fixed. Therefore, an annotation is sent to the casing manager who has to determine how to proceed. To this end, he employs a database for fault analysis (FMEA) which is provided as a guideline of the management task (21). From the fault analysis, he concludes that the problem may be fixed by selecting another material. Therefore, he performs the Iterate transition on the respective task such that it reappears on the agenda of the material expert. All other tasks remain unaffected.

Subsequently, the material expert selects some material which presumably fixes the problem (22). The process simulation is repeated for the new material (23). Now, the pressure required to fill the mould does not exceed the limit of the machine any more, and the development of the casing may terminate successfully.

## 2.6   Conclusion

We have described techniques for integrated product and process management in engineering design which were developed in the SUKITS project. SUKITS is a interdisciplinary project carried out by computer scientists and mechanical engineers. Within this project, a management system was implemented which provides several novel features for managing products and processes.

The management model underlying the SUKITS system consists of submodels for managing products, processes, and resources, respectively:

- The product management model supports version control, configuration control, and consistency control through an integrated model based on a small number of concepts. A posteriori integration is achieved by separating the contents of documents from structures such as version and configuration graphs.
- Process management takes the dynamics of development processes into account and specifically addresses product evolution, simultaneous engineering, and feedbacks. Process management is centered around dynamic task nets, which are obtained from configuration version graphs by adding process information such as task states, release states of outputs, etc. (product-centered approach).
- Resource management covers both tools and humans. For the latter, a flexible role model has been developed which allows to assign each member of a project team any set of roles (designer, NC programmer, etc.). These roles constrain the types of tasks which may be carried out by specific employees.

Subsequently, we have described functionality and user interface of each environment provided by the management system. This includes

- a management environment which provides graphical views and tools for operating on version and configuration graphs, task nets, etc.;
- a front end which displays task agendas and work contexts for task execution and commands for activating tools such as CAD systems, NC programming systems, or simulators; and
- a parameterization environment which is used to adapt the management system to an application scenario.

Furthermore, we have also sketched the realization of the management system, which takes heterogeneity and distribution into account. The product management database is stored in a graph-based database system (GRAS). The corresponding graph contains references to the contents of documents stored in the file system or in application-specific database systems. Tools are integrated through wrappers which are realized as Perl scripts. The components of the management system are glued together by a two-tiered communication system. The implementation of the management system comprises more than 100,000 loc (Modula-2 and C).

Considerable effort has been invested in order to get feedback from usage in realistic scenarios. To this end, we have applied the management system to several scenarios involving the development of metal parts and plastic parts, respectively. In particular, we investigated the development of a drill, which consists of both metal and plastic parts (and electrical equipment which was not considered here). About a dozen tools and converters were integrated with the management system, and a nontrivial demo session was worked out to demonstrate its capabilities.

We believe that the interdisciplinary cooperation performed in SUKITS was a key to its success. Design and implementation of the management system were strongly driven by the requirements of our engineering partners. Close cooperation between computer scientists and mechanical engineers has produced a management system which supports management of complex, dynamic development processes; provides managers with detailed and accurate views; offers them sophisticated tools for planning, analyzing, and controlling development processes; maintains workspaces for executing development tasks (including document management, release control, and tool activation); and supplements formal cooperation with annotations for spontaneous, ad hoc communication, both synchronously (conferences) and asynchronously (messages).

## References

1. H.-J. Bullinger and J. Warschat, editors. *Concurrent Simultaneous Engineering Systems.* Springer-Verlag, Berlin, 1996.

2. R. Conradi and B. Westfechtel. Version models for software configuration management. *ACM Computing Surveys.* To appear.

3. Continental Graphics, Broomfield, Colorado. *XVT—Development Solution for C,* 1993.

4. B. Curtis, M. I. Kellner, and J. Over. Process modeling. *Communications of the ACM,* 35(9):75–90, 1992.

5. S. A. Dart. Parallels in computer-aided design frameworks and software development environments efforts. *IFIP Transactions A,* 16:175–189, 1992.

6. ECMA. A reference model for frameworks of computer-assisted software engineering environments. Technical Report TR/55, European Computer Manufacturers Association, 1991.

7. A. Engbrocks and O. Hermanns. Design, implementation, and evaluation of a distributed file service for collaborative engineering environments. In *Proceedings of the 3rd IEEE Workshop on Enabling Technologies—Infrastructures for Collaborative Enterprises,* pages 170–177, Morgantown, West Virginia, Apr. 1994. IEEE Computer Society Press.

8. G. Engels, M. Nagl, W. Schäfer, and B. Westfechtel. Tight integration on one document: The programming environment. In Nagl [19], pages 170–177.

9. W. Eversheim, M. Weck, W. Michaeli, M. Nagl, and O. Spaniol. The SUKITS project: An approach to a posteriori integration of CIM components. In *Proceedings GI Jahrestagung,* Informatik aktuell, pages 494–503, Karlsruhe, Germany, Oct. 1992. Springer-Verlag, Berlin.

10. A. Finkelstein, J. Kramer, and B. Nuseibeh, editors. *Software Process Modelling and Technology.* Advanced Software Development Series. Research Studies Press (John Wiley), Chichester, U.K., 1994.

11. D. Georgakopoulos, M. Hornick, and A. Sheth. An overview of workflow management: From process modeling to workflow automation infrastructure. *Distributed and Parallel Databases,* 3:119–153, 1995.

12. D. Harrison, A. Newton, R. Spickelmeier, and T. Barnes. Electronic CAD frameworks. *Proceedings of the IEEE,* 78(2):393–419, 1990.

13. P. Heimann, C.-A. Krapp, B. Westfechtel, and G. Joeris. Graph-based software process management. *International Journal of Software Engineering and Knowledge Engineering,* 7(4), 1997.

14. O. Hermanns and M. Schuba. Performance investigations of the IP multicast architecture. *Computer Networks and ISDN Systems,* 28:429–439, 1996.

15. R. H. Katz. Toward a unified framework for version modeling in engineering databases. *ACM Computing Surveys,* 22(4):375–408, 1990.

16. N. Kiesel, A. Schürr, and B. Westfechtel. GRAS, a graph-oriented software engineering database system. *Information Systems,* 20(1):21–51, 1995.

17. T. W. Malone and K. Crowston. The interdisciplinary study of coordination. *ACM Computing Surveys,* 26(1):87–119, 1994.

18. K. G. McIntosh. *Engineering Data Management—A Guide to Successful Implementation.* McGraw-Hill, Maidenhead, England, 1995.

19. M. Nagl, editor. *Building Tightly-Integrated Software Development Environments: The IPSEN Approach.* LNCS 1170. Springer-Verlag, Berlin, 1996.

20. B. Peuschel, W. Schäfer, and S. Wolf. A knowledge-based software development environment supporting cooperative work. *International Journal of Software Engineering and Knowledge Engineering,* 2(1):79–106, 1992.

21. R. Reddy et al. Computer support for concurrent engineering. *IEEE Computer,* 26(1):12–16, 1993.

22. M. T. Rose. *The Open Book —A Practical Perspective on OSI.* Prentice-Hall, Englewood Cliffs, New Jersey, 1990.

23. A. Schürr, A. Winter, and A. Zündorf. Graph grammar engineering with PROGRES. In W. Schäfer, and P. Botella, editors, *Proceedings of the European Software Engineering Conference (ESEC '95),* LNCS 989, pages 219–234, Barcelona, Spain, Sept. 1995. Springer-Verlag, Berlin.

24. A. Sheth et al. NSF workshop on workflow and process automation. *ACM SIGSOFT Software Engineering Notes,* 22(1):28–38, 1997.

25. P. van den Hamer and K. Lepoeter. Managing design data: The five dimensions of CAD frameworks, configuration management, and product data management. *Proceedings of the IEEE,* 84(1):42–56, 1996.

26. L. Wall and R. Schwartz. *Programming Perl.* O'Reilly and Associates, Sebastopol, 1991.

27. A. Wasserman. Tool integration in software engineering environments. In F. Long, editor, *Proceedings of the 2nd International Workshop on Software Engineering Environments,* LNCS 467, pages 137–149, Chinon, France, Sept. 1990. Springer-Verlag, Berlin.

28. B. Westfechtel. Engineering data and process integration in the SUKITS environment. In J. Winsor, editor, *Proceedings International Conference on Computer Integrated Manufacturing (ICCIM '95),* pages 117–124, Singapore, July 1995. World Scientific, Singapore.

29. B. Westfechtel. A graph-based system for managing configurations of engineering design documents. *International Journal of Software Engineering and Knowledge Engineering,* 6(4):549–583, 1996.

30. B. Westfechtel. Integrated product and process management for engineering design applications. *Integrated Computer-Aided Engineering,* 3(1):20–35, 1996.

# 3

# Project Selection Problems for Production-Inventory-Distribution Scheduling in Manufacturing Plants

Hochang Lee
*Kyung Hee University*

## 3.1   Introduction

Capacity planning of a manufacturing plant is the adjustment of its production capability. Especially capacity expansion is the addition of production facilities to satisfy growing demands with minimum present worth cost. The primary capacity expansion decisions typically involve the sizes of facilities to be added and the times at which they should be added. The expansion size and timing are extremely important when, for example, cost of the equipment or facilities added exhibits economies-of-scale and operating costs are significant. Often the type of capacity or location of the capacity to be added are also major concerns. Like those found in the cement and fertilizer industries, the products need to be shipped to various demand locations at substantial transportation costs. In addition to these primary decisions, there may be dependent secondary decisions involving the logistics of goods being produced and the optimal utilization of the capacity being added. On the other hand, the existing production capacity does not always exceed the growing demand at any point in time. Capacity expansion can be deferred by temporarily allowing for capacity shortages at certain shortage costs. Allowing for capacity shortages implies that either part of the demand remains temporarily unsatisfied, or that it is satisfied by temporarily importing capacity (e.g., by renting capacity or by buying the end-product from an external source). The other way of deferring capacity expansions is through the accumulation of inventory during periods in which capacity exceeds demand. The inventory is then used to satisfy demand in periods of capacity shortages. All of these concerns—size, time, type, location, flow of materials, utilization, import, and inventory—are the major operational aspects of the capacity expansion problem.

One common assumption for simplification of the analysis is that demand for the capacity in question can be determined independently of the capacity expansion decisions. Strictly speaking, it not generally true. A more complete analysis would recognize that the costs related to capacity expansion influence the prices charged for the service, which, in turn, influence the demand. This recursive relation may result in a terribly complex model which tries to solve the entire demand/expansion/cost/price problem simultaneously. We avoid this problem by assuming that a demand forecast can be made without knowledge of the expansion schedule.

We examine the problem of determining a project selection schedule and a production-distribution-inventory-import schedule for each plant so as to meet the demands of multiregional markets at minimum discounted total cost during a discrete finite planning horizon. For each market $j$, a known demand has to be met at each discrete time period $t$ over a finite planning horizon. The demand in a market can be satisfied by production and inventory from any of the plants and/or by importing from an exogenous source with a penalty cost. Whenever the excess production capacity becomes zero, the production capacity of each plant $i$ may be expanded by implementing a set of expansion projects which are chosen at the beginning of each time period. It is assumed that there is negligible time lag in implementing the projects and negligible deterioration of production facility over the planning horizon. The decision variables are selection of projects, production/import amount, inventory level, and shipment schedule for each plant $i$ at each time period $t$.

In many models the expansion size is assumed to be a continuous variable, and this seems to be instrumental in many of the algorithms that have been developed (Luss 1982). Often, however, due to political regulations and the standardization of production facilities, the set of expansion sizes is small and discrete. Typical examples can be found in public services such as electrical power, water, school, and road systems as well as in manufacturing facilities including whole factories and individual machines within a factory (Freidenfelds 1981). In these cases the facility is usually expanded by implementing capacity expansion projects.

While the project sequencing problem deals with a similar situation, it is basically different from the project selection problem. In the project sequencing problem, the capacity is expanded by implementing, at any point in time, a single project selected from a finite number of projects considered for implementation during a finite planning horizon (i.e., one must find an optimal sequence of projects that should be implemented in order to satisfy the demand). Since every candidate project should be chosen only once at a certain point in time during the planning horizon, project sequencing problems usually include vehicle routing considerations to find the best tradeoff between investment and operating costs. On the other hand, the project selection problem consists in choosing a subset of the projects which are available at the beginning of each time period (not necessarily exhausting the whole set of projects). This allows the choice of multiple projects at each discrete time point. References for the project sequencing problem include Neebe and Rao (1983, 1986), Erlenkotter (1973), and Erlenkotter and Rogers (1977). The scheduling problem for power generation in Muckstadt and Koenig (1977) is a good example of the project selection problem.

Our model is related to the Multiregion Dynamic Capacity Expansion Problem formulated in Fong and Srinivasan (1981a, 1981b, and 1986) in terms of its multifacility structure. In contrast to their work, the contribution of our model is twofold. First, we deal with the capacity expansion problem through project selection, which is common in practice. The assumption of continuous expansion size is relaxed in our model. Second, we allow accumulation of inventory during periods in which capacity exceeds demand, to delay expansion decisions at each plant. As far as we know, this is the first attempt at including inventory accumulation in the multifacility capacity expansion problem (Erlenkotter (1977) for the single facility case). Obviously, we could consider the tradeoff between savings realized by deferring expansions and costs of inventory buildup in this model. Our model could be also interpreted as an extension of the dynamic lot sizing problem, including capacity expansion and distribution over a transportation network.

We use Lagrangean relaxation to approximate the optimal solution. Through a problem reduction algorithm, the Lagrangean relaxation problem strengthened by the addition of a surrogate constraint becomes a 0-1 mixed integer knapsack problem. Its optimal solution can be obtained by solving at most two generally smaller 0-1 pure integer knapsack problems. The bound is usually very tight. To get a good primal feasible solution from a given dual solution, a Lagrangean heuristic which approximately solves constrained transportation problems is developed. During iterations of the subgradient method, both upper bound and lower bound are updated to approximate the optimal value of the original problem.

Section 3.2 briefly reviews the previous capacity planning models and their solution techniques. Section 3.3 presents the problem formulation. In Section 3.4, we introduce the Lagrangean relaxation problem, the problem reduction algorithm using network flow properties, and the strengthened problem with a surrogate constraint. In Section 3.5, we develop a Lagrangean heuristic to find a good primal feasible solution. The overall procedure is described in Section 3.6. Computational results are given in Section 3.7.

## 3.2 Previous Models and Techniques

One of the earliest and best-known works on capacity expansion problem is that of Manne (1961). The model deals with deterministic demand that grows linearly over time with a rate of $\delta$ per year. Suppose the capacity, once installed, has an infinite economic life and whenever demand reaches capacity the capacity is expanded by $x$ units. If $f(x)$ is the expansion cost of size $x$, $x/\delta$ is the time between successive expansions, and $r$ is the discount rate of money, the discounted cost, $C$, of all expansions over an infinite horizon is

$$C = \sum_{k=0}^{\infty} f(x)e^{-rkx/\delta} = f(x)/[1 - e^{-rx/\delta}].$$

The cost function $f(x)$ is usually concave, representing the economies-of-scale of large expansion sizes. For example, when $f(x) = Kx^{\alpha}, 0 < \alpha < 1$, the discounted cost, $C$, in terms of the relief interval $t = x/\delta$ is

$$C = \frac{K\delta^{\alpha}t^{\alpha}}{1 - e^{-rt}}. \tag{3.1}$$

We note that the optimal relief interval is independent of the growth rate $\delta$. Of course, the relief size is directly proportional to growth, and the present worth cost is proportional to $\delta^{\alpha}$. We can find the optimal relief interval by setting the derivative of (1) to 0, which yields

$$\frac{e^{rt} - 1}{rt} = \frac{1}{\alpha}.$$

Using the Taylor series approximation for $e^{rt}$, the optimal relief interval is approximately

$$t \approx \left(\frac{2}{r}\right)\left(\frac{1}{\alpha} - 1\right).$$

As we can expected, the optimal relief interval increases when $\alpha$ decreases (more economies-of-scale) and decreases when $r$ increases (early investments are more costly).

Instead of assuming identical capacity expansions over the planning horizon, suppose $x_0$ to be placed at time $t_0$; $x_1$ to be placed at time $t_1 = x_0/\delta$, $x_n$ to be placed at time $t_n = \sum_{i=0}^{n-1}(x_i/\delta)$. Then discounted total cost of all expansions is

$$C = f(x_0) + \sum_{n=1}^{\infty} f(x_n)e^{-(r/\delta)\sum_{i=0}^{n-1}x_i},$$

which can be rewritten

$$C = f(x_0) + \left[f(x_1) + \sum_{n=2}^{\infty} f(x_n)e^{-(r/\delta)\sum_{i=1}^{n-1}x_i}\right]e^{-(r/\delta)x_0}.$$

Designating the term that is independent of $x_0$ in brackets as $C_F$, the cost of the future,

$$C = f(x_0) + C_F e^{-(r/\delta)x_0}.$$

If the cost of future expansions $C_F$ were known, the optimal initial expansion corresponding to that future cost, $C_{opt}$, could be found by an usual backward dynamic programming formulation over a single variable,

$$C_{opt} = \min_{x \geq 0}[f(x) + C_F e^{-rx/\delta}].$$

The simplest capacity expansion model shown above is one in which demand for additional capacity is projected to grow linearly over an infinite future. Capacity must be added over time to serve that demand. Any capacity added is assumed to last forever. These simplifying assumptions make it possible to analyze the problem easily and thoroughly. This oldest model stimulated many studies until now and most of them can be classified into at least one of the following capacity expansion environments which will be briefly reviewed throughout the section.

- continuous expansion size versus discrete expansion size
- linear demand versus nonlinear demand
- infinite planning horizon versus finite planning horizon
- single facility versus multifacility
- project sequencing versus project selection

In many models, the capacity expansion size is assumed to be a continuous variable. The instrumental assumption is even valid for the discrete cases that the number of expansion sizes is very large but finite. In the application where the number of possible choices of expansion sizes is small, the feasible expansion sizes should be explicitly considered. The similar situation can be analyzed by the project-based capacity expansion problem.

The case of linear demand has often been used to examine various issues. Erlenkotter (1977) examined expansion policies in which capacity shortages and inventory accumulations are allowed through four phases. The phases are: a surplus capacity phase in which capacity exceeds demand but no inventory is built up, an inventory accumulation phase, an inventory depletion phase in which the shortages are satisfied solely by the accumulated inventory, and an import phase in which capacity shortages are satisfied by the imported capacity. A dynamic programming model was formulated for determination of optimal expansion, inventory, and import decisions under an assumption of linearly growing demand.

We still retain the assumption that demand is deterministic and known at the outset, but in place of the very restrictive assumption that it is linear over planning horizon, we can consider the nonlinear demand that introduces additional complexity to the capacity expansion problem. The problem with general nonlinear but deterministic demand can be handled readily by either backward or forward dynamic programming. Letting the cumulative demand $D(t)$ be any nonlinear function, the backward dynamic programming formulation for the capacity expansion problem with a deterministic nonlinear demand becomes

$$C(t) = min_{\tau \geq t}[f[x(t, \tau)] + C(\tau)e^{-r(\tau - t)}],$$

where $C(t)$ is the present worth cost of satisfying all additional demand starting at time $t$ with no spare capacity and $x(t, \tau) \equiv D(\tau) - D(t)$ is the additional capacity required to serve the demand arriving between $t$ and $\tau$.

We now consider the much more likely situation that we do not know at the outset just what the pattern of demand will be. The model and its solution procedure highly depend on how to specify the uncertain demand of the future. Modeling the demand as a birth-death process, Freidenfelds (1980) showed that, when capacity shortages are not allowed, the stochastic model can be reformulated as an equivalent deterministic model. The key observation was that for any birth-death process, one can generate an equivalent deterministic demand problem whose solution also solves the stochastic case. The birth-death rates are allowed to vary with the number of customers in the system resulting in nonlinear demand functions.

Many finite horizon models are useful for capacity expansion problems. In general, the time scale in these models is represented by descrete time periods $t = 1, 2, 3, \ldots, T$, where $T$ is the finite horizon. Let $r_t$ be the demand in period $t$; let $x_t$ be the expansion sizes in period $t$; and let $I_t$ be the excess capacity at the end of period $t$. The costs incurred in period $t$ include the expansion cost plus the holding cost of excess capacity. The objective is to find the expansion policy that minimizes the total discounted costs incurred over the $T$ periods provided that capacity shortages are not allowed. The simplest model with finite planning horizon is as follows:

$$min \sum_{t=1}^{T}(f_{t(x_t)} + h_t(I_t))$$
$$s.t. \quad I_t = I_{t-1} + x_t - r_t, \quad \forall t \quad\quad (3.2)$$
$$x_t, I_t \geq 0 \quad\quad \forall t$$
$$I_0 = I_T = 0,$$

where $f_t(\cdot)$ and $h_t(\cdot)$ are the expansion and holding cost functions in period $t$ respectively. This problem can be reformulated as a dynamic programming problem. Let $w_t$ be the optimal discounted cost over periods $0, 1, \ldots, t$ given $I_t = 0$, let $d_{uv}$ be the discounted cost associated with an optimal plan over periods $u + 1, \ldots, v$ given $I_u = I_v = 0$ and $I_t > 0$ for $t = u + 1, \ldots, v - 1$. These definitions leads to the following forward dynamic programming equations:

$$w_0 = 0$$
$$w_v = min_{0 \leq u < v}[w_u + d_{uv}], \quad v = 1, 2, \ldots, T.$$

In many applications the cost functions $f_t(\cdot)$ and $h_t(\cdot)$ are concave, in which case there exists an optimal solution that is an extreme point of the feasible region of (2) (Wagner and Whitin 1958, Veinott 1969). Various extensions and modification of the model was introduced. These include negative demand increments, capacity disposals, and capacity shortages. Upper bounds can also be imposed on the expansion sizes.

Multitype capacity expansion problem deals with the case where several capacity types are being used to satisfy various types of demand. Each of the capacity types is designed to satisfy a given type of demand, but it can be converted at some cost to satisfy a different type of demand. The problem of expanding the capacity of a telephone feeder cable network is an appropriate example. A basic element of the telephone feeder problem is the proper inventory of facilities in place, but the problem is complicated by interactions between several different inventories. For example, coarse gauge wire can be used to satisfy demand for a finer gauge. Also, the inventory of conduit ducts is used for the installation of cables of every gauge. Freidenfelds and McLaughlin (1979) presented a modified branch-and-bound procedure which drastically trims the amount of search by generating heuristic bounds based on analytic solutions of simpler capacity expansion problems.

Multifacility capacity expansion problems are different from multitype capacity expansion problems. In a multifacility problem capacity expansions may take place in different producing locations $i = 1, 2, \ldots, m$. Furthermore, often the products have to be shipped to geographical regions $j = 1, 2, \ldots, n$ at substantial transportation costs. These costs should be explicitly considered while making decisions regarding the timing, sizes, and locations so that the total discounted expansion and transportation costs are minimized. Let $y_{ijt}$ be the amount shipped at time $t$ from $i$ to $j$ at a price of $p_{ijt}$ and let $D_j(t)$ be the demand at location $j$ at time $t$. Obviously, one simple formulation is as follows:

$$Min \sum_{t=1}^{T}\sum_{i=1}^{m} f_{it}(x_{it}) + \sum_{t=1}^{T}\sum_{i=1}^{m}\sum_{j=1}^{n} p_{ijt}y_{ijt}$$

$$s.t. \sum_{j=1}^{n} y_{ijt} \leq \sum_{k=0}^{t} x_{ik} \qquad\qquad \forall i, t$$

$$\sum_{i=1}^{m} y_{ijt} = D_j(t), \qquad\qquad \forall j, t$$

$$x_{it} \geq 0, \quad y_{ijt} \geq 0, \qquad\qquad \forall i, j, t$$

where $x_{i0}$ is the existing capacity at $i$ prior to the first period. Holding costs for excess capacities and shortage costs incurred by importing capacity from external sources can readily be incorporated. Using a two-region capacity exchange heuristic method, Fong and Srinivasan (1981b) effectively solved the multiregion dynamic capacity expansion problem for fixed charge expansion cost functions.

Project sequencing is the problem of finding the sequencing of a finite set of expansion projects that meets a deterministic demand projection at minimum discounted cost. The problem is similar in structure to a number of job shop sequencing problems. The discrete time sequencing problem can be stated as follows: During a time horizon of $T$ periods, project $i$ from a finite set of $n$ expansion project has integer capacity $z_i$, and may be brought on stream at the start of any period $t$ at corresponding cost $c_{it}$. A non-negative integer demand projection $d_t$ is given. The total capacity available during period $t$ after any expansion must be at least equal to the demand $d_t$ in that period. The problem is to determine which projects are brought on stream, and when, to minimize total cost. Neebe and Rao (1983) formulated the discrete-time sequencing expansion problem as follows:

$$Min \sum_{t=1}^{T}\sum_{i=1}^{n} c_{it}x_{it}$$

$$s.t. \sum_{t=1}^{T} x_{it} \leq 1, \qquad\qquad \forall i$$

$$\sum_{i=1}^{n} x_{it} \leq 1, \qquad\qquad \forall t$$

$$\sum_{i=1}^{n} z_i x_{it} + y_t - y_{t+1} = d_t - d_{t-1}, \qquad \forall t$$

$$x_{it} = 0 \text{ or } 1, \qquad\qquad \forall i, t$$

$$y_t \geq 0, \qquad\qquad \forall t,$$

where $y_t$ equal to the excess capacity in period $t$ before any expansion. The problem is solved using Lagrangean relaxation. In the continuous-time sequencing expansion problem, we are given a demand function $d(t)$ which is continuous over time $t$. Since the demand function is assumed to be continuous, exactly one project will be activated whenever the excess capacity is zero. Without loss of generality, $d(t)$ is assumed to be integer. Therefore $d(t) = j$ at expansion opportunity time $t = t_j$, and period $j$ lasts from immediately after $t_{j-1}$ to $t_j$. Let $x_{ij} = 1$ if a project $i$ is activated at the start of period $j$; otherwise 0. Neebe and Rao (1986) also formulated continuous-time sequencing expansion problem as follows:

$$Min \sum_{i=1}^{m} \sum_{j=1}^{n} c_{ij} x_{ij}$$

$$s.t. \sum_{i=1}^{m} z_i x_{ij} + y_{j-1} - y_j = 1, \qquad \forall j$$

$$\sum_{j=1}^{n} x_{ij} \leq 1, \qquad \forall i$$

$$x_{ij} = 0 \quad or \quad 1, \qquad \forall i, j$$

$$y_j \geq 0, \qquad \forall j.$$

The problem is solved using a branch and bound procedure with Lagrangean relaxation, where the relaxed problem leads to the shortest path problem.

Unlike the project sequencing problem, the project selection problem is to choose multiple projects at the beginning of each discrete time points. Muckstadt and Koenig (1977) formulated the power generation scheduling problem as a project selection problem having a special structure that facilitates rapid computation. The integer variable $x_{it}$ in the model indicates whether a specified generating unit $i$ is operating during a period $t$. The continuous variable $y_{ijk}$ represents the proportion of the available capacity $M_{ik}$ that is actually used throughout period $t$, where $k = 1, 2, \ldots, k_i$. Then the total energy output from generator $i$ in period $t$ is $m_i x_{it} + \sum_{k=1}^{K_j} M_{ik} y_{ikt}$. Let $w_{it} = 1$ if unit $i$ is started in period $t$ and $w_{it} = 0$ otherwise; $z_{it} = 1$ if unit $i$ is shut down in period $t$ and $z_{it} = 0$ otherwise; $c_i$ and $d_i$ be the start-up cost and shut-down cost for unit $i$ respectively; $g_i$ be the dollar cost of operating generator $i$ at its minimum capacity for 1 hour, and $h_t$ be the number of hours in period $t$. The objective is to minimize the sum of the unit commitment and economic dispatch costs subject to demand, reserve, and generator capacity and generator schedule constraints. The mixed integer model for the basic power scheduling problem is

$$Min \sum_{i=1}^{I} \left[ \sum_{t=1}^{T} \left( c_i w_{it} + d_i z_{it} + h_i g_i x_{it} + \sum_{k=1}^{K_i} M_{ik} h_t g_{ik} y_{ikt} \right) \right]$$

$$s.t. \sum_{i=1}^{I} \left( m_i x_{it} + \sum_{k=1}^{K_i} M_{ik} y_{ikt} \right) \geq D_t, \qquad \forall t$$

$$\sum_{i=1}^{I} M_i x_{it} \geq R_t, \qquad \forall t$$

$$y_{ikt} \leq x_{it}, \qquad \forall i, t$$

$$x_{it} - x_{i,t-1} \leq w_{it}, \qquad \forall i, t$$

$$x_{i,t-1} - x_{it} \leq z_{it}, \qquad \forall i, t$$

$$w_{it}, z_{it}, x_{it} = 0 \quad or \quad 1, \quad y_{ikt} \geq 0 \qquad \forall i, k, t,$$

where $m_i$ and $M_i$ respectively represent minimum and maximum operating capacities of generating unit $i$, measured in mega-watt. A branch-and-bound algorithm was proposed using a Lagrangean method to decompose the problem into single generator problems.

## 3.3   Statement of the Problem

The basic problem for each plant is to choose a subset of capacity expansion projects at each time period to satisfy the demands during a discrete finite planning horizon. The multiregional problem setting in a transportation network adds a shipment schedule to the basic problem. The allowance of inventory accumulation at each plant makes the problem even more complicated (otherwise, the problem decomposes into one period decision problem over the planning horizon), while giving flexibility to the capacity expansion decisions because the use of inventory may delay or cancel the implementation of certain projects. We assume an economies-of-scale effect on the implementation cost of the capacity expansion project and an exponential discount factor for all unit costs.

### Notation

Given an optimization problem (P), $OV(P)$ and $OS(P)$ are an optimal value and a set of optimal solution of (P) respectively. $\chi^*$ is an optimal value of the decision variable X. $(P_i)$ denotes the $i^{\text{th}}$ subproblem of (P). $A \leftarrow B$ means assign B to A. In the development we will make use of the following notation.

$I$  number of plants ($P = \{1, \dots, I\}$)
$J$  number of markets ($M = \{1, \dots, J\}$)
$T$  number of planning periods ($H = \{1, \dots, T\}$)
$S(i, t)$ number of projects available at plant $i$ during time period $t$
$x_{it}$  amount produced at plant $i$ during time period $t$
$p_{it}$  unit cost of production at plant $i$ during time period $t$
$y_{ijt}$  amount shipped from plant $i$ to market $j$ during time period $t$
$c_{ijt}$  unit cost of shipment from plant $i$ to market $j$ during time period $t$
$I_{it}$  inventory level at plant $i$ during time period $t$
$h_{it}$  unit cost of inventory at plant $i$ during time period $t$
$z_{ist}$  1 if expansion project $s$ is chosen at plant $i$ during $t$. Otherwise, 0.
$r_{ist}$  cost of implementing the expansion project $s$ at plant $i$ during time period $t$
$e_{ist}$  size of the expansion project $s$ at plant $i$ during time period $t$
$q_i$  initial production capacity of plant $i$
$R_{jt}$  demand in market $j$ during time period $t$

Note that $(I + 1)$ is the exogenous source from which the unsatisfied demand, $y_{I+1jt}$, of market $j$ at time period $t$ can be met with a penalty cost, $c_{I+1jt}$.

### Problem Formulation

We formulate the problem with linear costs as follows:

$$(P0) \quad Min \sum_{i=1}^{I} \sum_{t=1}^{T} p_{it}x_{it} + \sum_{i=1}^{I+1} \sum_{j=1}^{J} \sum_{t=1}^{T} c_{ijt}y_{ijt}$$

$$+ \sum_{i=1}^{I} \sum_{t=1}^{T} h_{it}I_{it} + \sum_{i=1}^{I} \sum_{t=1}^{T} \sum_{s=1}^{S(i,t)} r_{ist}z_{ist} \qquad (3.3)$$

$$\text{s.t. } x_{it} \leq q_i + \sum_{\tau=1}^{t} \sum_{s=1}^{S(i,\tau)} e_{is\tau} z_{is\tau}, \qquad \forall i \neq I+1, t \tag{3.4}$$

$$\sum_{i=1}^{I+1} y_{ijt} = R_{jt}, \qquad \forall j, t \tag{3.5}$$

$$I_{it} = I_{it-1} + x_{it} - \sum_{j=1}^{J} y_{ijt}, \qquad \forall i \neq I+1, t \tag{3.6}$$

$$I_{i0} = I_{iT} = 0, \qquad \forall i \neq I+1 \tag{3.7}$$

$$x_{it}, y_{ijt}, I_{it} \geq 0, \; z_{ist} \in \{0,1\}, \qquad \forall i, j, s, t.$$

All costs are assumed to be expressed on a present value basis. Disposal of production facility is not allowed. The objective is to minimize the total present value of the costs of production, transportation, import, inventory, and expansion incurred over the entire planning horizon. (3.4) is a production capacity constraint for plant $i$ at time $t$. (3.5) is a demand constraint for market $j$ at time $t$. (3.6) is a material flow balance equation. (3.7) is a boundary condition on inventory. Figure 3.1 shows the material flows for a three plant–three period case.

Using (3.6), we substitute for $x_{it}$ and can simplify (P0) as follows:

$$(P1) \quad Min \; \sum_{i=1}^{I+1} \sum_{j=1}^{J} \sum_{t=1}^{T} (c_{ijt} + p_{it}) y_{ijt} + \sum_{i=1}^{I} \sum_{t=1}^{T} (p_{it} + h_{it} - p_{it+1}) I_{it}$$

$$+ \sum_{i=1}^{I} \sum_{t=1}^{T} \sum_{s=1}^{S(i,t)} r_{ist} z_{ist} \tag{3.8}$$

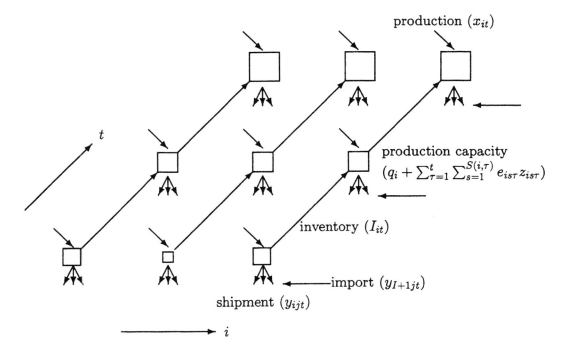

**FIGURE 3.1** Material flows for $I = 3$ and $T = 3$.

$$\text{s.t.} \sum_{j=1}^{J} y_{ijt} + I_{it} - I_{it-1} \leq q_i + \sum_{\tau=1}^{t} \sum_{s=1}^{S(i,\tau)} e_{is\tau} z_{is\tau}, \qquad \forall i \neq I+1, t \qquad (3.9)$$

$$\sum_{i=1}^{I+1} y_{ijt} = R_{jt}, \qquad \forall j, t \qquad (3.10)$$

$$\sum_{j=1}^{J} y_{ijt} + I_{it} - I_{it-1} \geq 0, \qquad \forall i \neq I+1, t \qquad (3.11)$$

$$I_{i0} = I_{iT} = 0, \qquad \forall i \neq I+1 \qquad (3.12)$$

$$y_{ijt}, I_{it} \geq 0, \quad z_{ist} \in \{0,1\}, \qquad \forall i, j, s, t \qquad (3.13)$$

where $P_{I+1t} = 0 \;\; \forall t$. (P1) is a 0-1 mixed integer linear program (MILP). One approach to solve the problem is to apply standard MILP procedures, particularly those implemented in commercially available computer codes. However, such an optimizing approach for this problem would not be efficient because of rapid growth in problem size.

We develop an efficient algorithm to approximate the optimal value of (P1). By a problem reduction algorithm, the Lagrangean relaxation problem strengthened by a surrogate constraint reduces to at most two 0-1 pure integer knapsack problems, which are relatively easy to solve. Furthermore, it is also rather easy to derive a good feasible solution from a given dual solution. During the iterations of the subgradient method, upper and lower bounds are updated to approximate the optimal value of the original problem. The strength of the overall algorithm may reduce the need for branch-and-bound in case of small duality gaps.

# 3.4 Relaxation, Reduction, and Strengthening of the Problem

## Lagrangean Relaxation Problem and Its Properties

Using Lagrangean relaxation, (P1) decomposes into subproblems which one can easily solve by exploiting their special structures. Relaxing the demand constraint (3.10) and letting $v$ be a vector of Lagrangean multipliers $v_{jt}$ associated with the constraints, the Lagrangean relaxation problem, $(LR(v))$, is decomposed into the independent subproblems, $(LR_i(v))$ for each plant $i$ and $(LR_{I+1}(v))$ for exogeneous source $(I + 1)$.

$$(LR_{I+1}(v)) \quad Min \sum_{j=1}^{J} \sum_{t=1}^{T} (c_{I+1jt} - v_{jt}) y_{I+1jt}$$
$$\text{s.t. } y_{I+1jt} \geq 0, \qquad \forall j, t.$$

If $v_{jt} > c_{I+1jt}$ for some $j$, $t$, $y_{I+1jt}$ is unbounded from above and $(LR_{I+1}(v))$ is unbounded, therefore $OV(LR(v))$ is $-\infty$ (note that $OV(LR(v)) = \sum_{i=1}^{I} OV(LR_i(v)) + OV(LR_{I+1}(v))$. If $v_{jt} \leq c_{I+1jt}$ $\forall j, t$, $y_{I+1jt}$ should be chosen as small as possible, i.e., $y_{I+1jt} = 0$ $\forall j, t$. Since the Lagrangean relaxation dual problem is to maximize $OV(LR(v))$, the solution space of $v$ is reduced as in the following lemma.

**Lemma 1** *Given that one tries to obtain the optimal value of the Lagrangean relaxation dual problem, one should only use multiplier $v$ such that $v_{jt} \leq c_{I+1jt}$ $\forall j$, $t$.*

$$(LR_i(v)) \quad Min \quad \sum_{j=1}^{J}\sum_{t=1}^{T}(c_{ijt} + p_{it} - v_{jt})y_{ijt} + \sum_{t=1}^{T}(p_{it} + h_{it} - p_{it+1})I_{it} + \sum_{t=1}^{T}\sum_{s=1}^{S(i,t)}r_{ist}z_{ist} \quad (3.14)$$

$$s.t. \quad \sum_{j=1}^{J}y_{ijt} + I_{it} - I_{it-1} \leq q_i + \sum_{\tau=1}^{t}\sum_{s=1}^{S(i,\tau)}e_{is\tau}z_{is\tau}, \qquad \forall t \qquad (3.15)$$

$$\sum_{j=1}^{J}y_{ijt} + I_{it} - I_{it-1} \geq 0, \qquad \forall t \qquad (3.16)$$

$$I_{i0} = I_{iT} = 0$$

$$y_{ijt}, I_{it} \geq 0, \quad z_{ist} \in \{0,1\}, \qquad \forall j, s, t.$$

Some insight into the structure of constraints in $(LR_i(v))$ might simplify the problem. For given $t$, all $y_{ijt}$ columns in the constraint matrix are identical, therefore no more than one $y_{ijt}$ for each $t$ need ever be positive in an optimal solution. That is, we shall set one $y_{ijt}$ with minimum coefficient in (14) for each $t$ at a positive value in the optimal solution (indeed if $y_{ij_1t}$ and $y_{ij_2t}$ were both positive in the optimal solution, one could construct a solution with equal or smaller cost by setting $y_{ij_1t} \leftarrow y_{ij_1t} + y_{ij_2t}$, $y_{ij_2t} \leftarrow 0$ if $y_{ij_1t}$ has the smaller cost).

Let $j(i,t) = argmin_{j \in M}(c_{ijt} + p_{it} - v_{jt})$ $\forall i$, $t$, then $(LR_i(v))$ reduces to $(RLR_i(v))$:

$$(RLR_i(v)) \quad Min \quad \sum_{t=1}^{T}(c_{ij(i,t)t} + p_{it} - v_{j(i,t)t})y_{ij(i,t)t}$$

$$+ \sum_{t=1}^{T}(p_{it} + h_{it} - p_{it+1})I_{it} + \sum_{t=1}^{T}\sum_{s=1}^{S(i,t)}r_{ist}z_{ist} \quad (3.17)$$

$$s.t. \quad y_{ij(i,t)t} + I_{it} - I_{it-1} \leq q_i + \sum_{\tau=1}^{t}\sum_{s=1}^{S(i,\tau)}e_{is\tau}z_{is\tau}, \qquad \forall t \qquad (3.18)$$

$$y_{ij(i,t)t} + I_{it} - I_{it-1} \geq 0, \qquad \forall t$$

$$I_{i0} = I_{iT} = 0 \qquad (3.19)$$

$$y_{ij(i,t)t}, I_{it} \geq 0, \quad z_{ist} \in \{0,1\}, \qquad \forall s, t.$$

Figure 3.2 shows the network structure of $(RLR_i(v))$ which gives some intuition for reducing the problem. $x_{it}$ $(= y_{ij(i,t)t} + I_{it} - I_{it-1})$ represents the production amount of period $t$ and must be between 0 and the production capacity $(q_i + \sum_{\tau=1}^{t}\sum_{s=1}^{S(i,\tau)}e_{ist}z_{ist})$. The objective of $(RLR_i(v))$ is to determine network flows, $I_{it}$ and $y_{ij(i,t)t}$ $\forall t$, and project selections, $z_{ist}$ $\forall s, t$, to minimize the total cost (3.17). The optimization procedure of $(RLR_i(v))$ is divided into two steps. First, the optimal flows of $I_{it}$ and $y_{ij(i,t)t}$ are expressed in terms of $z_{ist}$ through the problem reduction algorithm and a reduced pure integer program which includes only $z_{ist}$ is obtained by plugging the optimal flows, $I_{it}^*$ and $y_{ij(i,t)t}^*$, into $(RLR_i(v))$. Second, $z_{ist}^*$ is determined by solving the reduced pure integer program and, consequently, $I_{it}^*$ and $y_{ij(i,t)t}^*$ are determined.

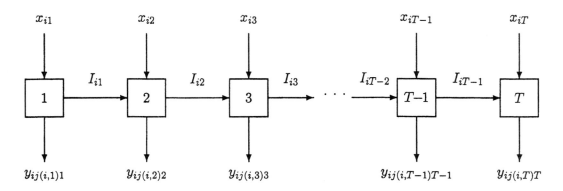

**FIGURE 3.2**   Network structure of $(RLR_i(v))$.

Exploiting the network structure, we derive the following solution properties of $(RLR_i(v))$, which later will result in the problem reduction algorithm. For notational simplicity, we define

$$CAP_i(t) = q_i + \sum_{\tau=1}^{t} \sum_{s=1}^{S(i,\tau)} e_{is\tau} z_{is\tau} \quad \forall i, t$$

$$K_i(t'',t) = \left\{ c_{ij(i,t)t} + p_{it} - v_{j(i,t)t} + \sum_{\tau=t''}^{t-1} (p_{i\tau} + h_{i\tau} - p_{i\tau+1}) \right\} \quad \forall i, t'', t\,(t \geq t'')$$

$$K_i^{(k)}(t'',t') = \min_{t \geq t''}^{(k)} \{K_i(t'',t)\} \quad \forall i, t''$$

where $t''$ is any time period such that $I_{it''-1}^* = 0$. $min^{(k)}\{b_i\}$ is the $k^{th}$ smallest $b_i$ (arbitrary tie breaking). The operators $max^{(k)}$, $argmin^{(k)}$, and $argmax^{(k)}$ are defined in an analogous manner. $CAP_i^*(t)$ $(= q_i + \sum_{\tau=1}^{t} \sum_{s=1}^{S(i,\tau)} e_{is\tau} z_{is\tau}^*)$ is the optimal production capacity of plant $i$ at time period $t$. $K_i(t'', t)$ is the sum of unit inventory holding costs during $[t'',\ t-1]$ and unit distribution cost at $t$. $t'$ is the $k^{th}$ cheapest shipping time period for the subperiod starting at $t''$.

$(RLR_i(v))$ is a problem of determining production, inventory and shipping amounts for plant $i$ at each time period such that the total cost is minimized over the planning horizon. The only constraint sets are the ones limiting the production to be less than or equal to the capacity limit $CAP_i(t)$, constraints requiring the starting and ending inventory to be equal to zero and nonnegativity constraints. If the total cost $K_i^{(1)}(t'',t')$ is positive it is optimal to produce the amount zero and since the starting inventory is zero, nothing will be stored or shipped in period $t''$. However if there exist time periods $t''$, $t'$ such that $K_i^{(1)}(t'',t') < 0$, then it is clear that one wants to produce as much as possible (i.e., up to capacity limit) in period $t''$, store it until period $t'$, and then in period $t'$ ship the whole amount. The following lemmas are just natural consequences of this reasoning.

**Lemma 2** *There exists at least one* $(y^*,\ I^*,\ z^*) \in OS(RLR_i(v))$ *such that* $y_{ij(i,t)t}^* = I_{it}^* - I_{it-1}^* = CAP_i^*(t)$ *or* $\forall t$.

**Proof**   Assume $0 < y_{ij(i,t)t}^* + I_{it}^* - I_{it-1}^* (= x_{it}^*) < CAP_i^*(t)$ for $t = t''$. We consider the following mutually exclusive and collectively exhaustive cases.

[CASE 1]: If there existed a $t^0 (>t'')$ such that $K_i(t'',t^0) < 0$, then we could improve $OV(RLR_i(v))$ by rebalancing the current optimal flows until $x_{it''}^*$ reaches $CAP_i^*(t'')$. That is, increasing $I_{it}^*\ \forall\ t'' \leq t \leq t^0 - 1$ and $y_{ij(i,t^0)t^0}^*$ each by $\Delta I = CAP_i^*(t'') - x_{it''}^*$, we could decrease $OV(RLR_i(v))$ by $K_i(t'',\ t^0) \cdot \Delta I$.
[CASE 2]: If there were no such $t^0$, we would have two possibilities,

$$\text{(i) } y_{ij(i,t'')t''}^* \leq I_{it''-1}^*, \qquad \text{or} \qquad \text{(ii) } y_{ij(i,t'')t''}^* > I_{it''-1}^*.$$

For (i), we could improve $OV(RLR_i(v))$ by decreasing $I^*_{it''}$ and rebalancing the related flows until $x^*_{it''}$ reaches 0. For (ii), we would have two possibilities.

$$\text{(ii-a)} \quad c_{ij(i,t'')t''} + p_{it''} - v_{j(i,t'')t''} < 0, \qquad \text{or} \qquad \text{(ii-b)} \quad c_{ij(i,t'')t''} + p_{it''} - v_{j(i,t'')t''} \geq 0.$$

In case (ii-a), we could improve $OV(RLR_i(v))$ by increasing only $y^*_{ij(i,t'')t''}$ until $x^*_{it''}$ reaches $CAP^*_i(t'')$. In case (ii-b), we could improve $OV(RLR_i(v))$ by decreasing $y^*_{ij(i,t'')t''}$ and $I^*_{it''}$ and rebalancing the related flows until $x^*_{it''}$ reaches 0. Thus we could improve the optimal solution under the initial assumption. This completes the proof by contradiction.

**Lemma 3** *There exists at least one* $(y^*, I^*, z^*) \in OS(RLR_i(v))$ *such that* $y^*_{ij(i,t)t} \cdot I^*_{it} = 0$ $\forall t$.

**Proof** Assume that the optimal solution violates $y^*_{ij(i,t)t} \cdot I^*_{it} = 0$ for $t = t''$. If there existed a $t^0(>t'')$ such that $K_i(t'', t^0) < c_{ij(i,t'')t''} + p_{it''} - v_{j(i,t'')t''}$, we could improve $OV(RLR_i(v))$ by increasing $I^*_{it''}$ and rebalancing the related flows until $y^*_{ij(i,t'')t''}$ becomes 0. Otherwise, we could improve $OV(RLR_i(v))$ by increasing $y^*_{ij(i,t'')t''}$ and rebalancing the related flows until $I^*_{it''}$ becomes 0. Thus we could improve the optimal solution under the assumption. This completes the proof by contradiction. $\square$

**Lemma 4** *If* $I^*_{it-1} > 0$, *then either* $y^*_{ij(i,t)t} > 0$ *or* $I^*_{it} > 0$.

**Proof** The proof follows from (3.19) and Lemma 3. $\square$

**Lemma 5** *If* $K^{(1)}_i(t'', t') < 0$, *then at the optimal solution* $y^*_{ij(i,t)t} = 0$ $\forall t'' \leq t \leq t' - 1$, $I^*_{it''} = CAP^*_i(t'')$ *and* $y^*_{ij(i,t')t'} \geq I^*_{it''}$. *Otherwise,* $y^*_{ij(i,t'')t''} = 0$ *and* $I^*_{it''} = 0$.

**Proof** If $K^{(1)}_i(t'', t') < 0$, then it is optimal to produce up to the production capacity at $t''$ and ship this amount at $t'$. Therefore $I^*_{it''} = CAP^*_i(t'')$ and $y^*_{ij(i,t')t'} \geq I^*_{it''}$. Since $K^{(1)}_i(t'', t') \leq K_i(t'', t')$ $\forall t \geq t''$, for any $t(t'' \leq t \leq t')$, $K_i(t, t') \leq K_i(t, k)$ $\forall k \geq t$. It means that $t'$ is the cheapest shipping time for the products produced at any time period during $[t'', t']$. Therefore $y^*_{ij(i,t)t} = 0$ $\forall t'' \leq t \leq t' - 1$. If $K^{(1)}_i(t'', t') \geq 0$, it is optimal to produce nothing at $t''$. So $y^*_{ij(i,t'')t''} = 0$ and $I^*_{it''} = 0$. This completes the proof. $\square$

## Problem Reduction Algorithm

By exploiting the above solution properties, we develop the problem reduction algorithm for $(RLR_i(v))$ as simple as follows: First, set $t'' = 1$ and isolate the subperiod $[t'', t']$ from the planning horizon $[t'', T]$ by calculating $K^{(1)}_i(t'', t')$. Second, for the subperiod $[t'', t']$, express $I^*_{it}$ and $y^*_{ij(i,t)t}$ in terms of $z_{ist}$. Third, for the remaining period $[t' + 1, T]$, repeat the same procedure until the entire planning horizon is covered. Finally, substitute for $I^*_{it}$ and $y^*_{ij(i,t)t}$ in $(RLR_i(v))$ and obtain a pure 0-1 integer program which includes only $z_{ist}$.

From the definition of $K^{(1)}_i(t'', t')$, we know that $t'$ is the cheapest shipping time for the products produced (if any) at $t''$. If $K^{(1)}_i(t'', t')$ is negative, Lemma 2 and Lemma 5 say that it is optimal to produce up to the production capacity, $CAP_i(t'')$, at $t''$, hold entire amount until $t'$ and ship it at $t'$, i.e., $y^*_{ij(i,t')t'} = I^*_{it} = CAP_i(t'')$ $\forall t'' \leq t \leq t' - 1$ and $y^*_{ij(i,t'')t''} = 0$. If $K^{(1)}_i(t'', t')$ is positive, it is optimal to produce nothing at $t''$, i.e., $y^*_{ij(i,t'')t''} = 0$ and $I^*_{it''} = 0$. We also know that $t'$ is also the cheapest shipping time for the products produced at any time period during the subperiod $[t'' + 1, t']$ (see the proof of Lemma 5). Therefore we only need to determine the production amount at each time period during the subperiod $[t'' + 1, t']$. The production amount at $t$ $(t'' + 1 \leq t \leq t')$ depends on the sign of $K_i(t, t')$. If $K_i(t, t')$ $\forall t'' + 1 \leq t \leq t' - 1$ is negative, the production amount at $t$ equals the production capacity at $t$, i.e., $I^*_{it} = I^*_{it-1} + CAP_i(t)$, $y^*_{ij(i,t)t} = 0$ and $y^*_{ij(i,t')t'} \leftarrow y^*_{ij(i,t')t'} + CAP_i(t)$. If $K_i(t, t')$ $\forall t'' + 1 \leq t \leq t' - 1$ is positive, $I^*_{it} = I^*_{it-1}$, $y^*_{ij(i,t)t} = 0$ and $y^*_{ij(i,t')t'} \leftarrow y^*_{ij(i,t')t'}$. If $K_i(t', t')$ is negative, $y^*_{ij(i,t')t'} = I^*_{it'-1} + CAP_i(t')$ and $I^*_{it'} = 0$. If $K_i(t', t')$ is positive, $y^*_{ij(i,t')t'} = I^*_{it'-1}$ and $I^*_{it'} = 0$.

Once all the optimal flows, $y^*_{ij(i,t)t}$ and $I^*_{it}$, during the subperiod $[t'', t']$ are expressed in terms of $z_{ist}$, a next subperiod beginning at $t' + 1$ is isolated by setting $t'' \leftarrow t' + 1$ and calculating new $K^{(1)}_i(t'', t')$.

For the next subperiod, the optimal flows are also expressed in terms of $z_{ist}$ by repeating the same procedure. The problem reduction algorithm terminates if $t'$ reaches $T$. Note that the whole planning horizon is sequentially partitioned by the time period with zero inventory (Lemma 3).

By doing this, we could assign the optimal amounts of inventory and shipment in terms of production capacities for all time periods. Following the algorithm below, we reduce $(RLR_i(v))$ by expressing the continuous variables, $I_{it}$ and $y_{ij(i,t)t}$, in terms of $z_{ist}$.

**Step 0:** (Initialization)
Set $t = 1$ and $I_{it} = y_{ij(i,t)t} = 0 \;\; \forall t$

**Step 1:** (Partition)
If $K_i^{(1)}(t'', t') \geq 0$, then go to Step 2.
If $t' = t''$, then go to Step 3.
If $t' > t''$, then go to Step 4.

**Step 2:** (No substitution)
If $t'' < T$, then $t'' \leftarrow t'' + 1$ and go to Step 1.
Otherwise, stop.

**Step 3:** (Substitution 1)
$y_{ij(i,t'')t''} \leftarrow CAP_i(t'')$
If $t'' < T$, then $t'' \leftarrow t'' + 1$ and go to Step 1.
Otherwise, stop.

**Step 4:** (Substitution 2)
$I_{it''} \leftarrow CAP_i(t'')$
$y_{ij(i,t')t'} \leftarrow CAP_i(t'')$
DO $k = t'' + 1$ TO $t'$
    If $K_i(k, t') < 0$, then
        $y_{ij(i,t')t'} \leftarrow y_{ij(i,t')t'} + CAP_i(k)$
        If $k < t'$, $I_{ik} \leftarrow I_{ik-1} + CAP_i(k)$
    Otherwise,
        $I_{ik} \leftarrow I_{ik-1}$
CONTINUE
If $t' < T$, then $t'' \leftarrow t'' + 1$ and go to Step 1.
Otherwise, stop.

It is easy to see that the computational complexity of the algorithm is bounded by $O(T^2)$. The worst case, that is $t' = t''$ at every Step 1, requires $T(T+1)/2$ comparisons during the entire planning horizon.

## Strengthening of the Problem with a Surrogate Constraint

Through the problem reduction algorithm, $y_{ij(i,t)t}$ and $I_{it}$ are expressed in terms of $z_{ist}$. Now it is easy to see that $(RLR_i(v))$ can be reduced to $(MR_i(v))$ which does not include continuous variables:

$$(MR_i(v)) \;\; Min \; \sum_{t=1}^{T} \sum_{s=1}^{S(i,t)} D_{ist} z_{ist}$$

$$s.t. \; z_{ist} \in \{0, 1\}, \;\; \forall s, t$$

where $D_{ist}$ is determined through the problem reduction algorithm. Note that $(LR(v)) = \sum_{i=1}^{I}(MR_i(v)) + (LR_{I+1}(v))$. The following result is trivial:

**Lemma 6** $(LR(v))$ *has the integrality property.*

A surrogate constraint, consisting of an aggregation of the constraints in $(P1)$, may be added to $(LR(v))$ to tighten the Lagrangean bound, which is no better than the LP relaxation bound since the relaxed problem, $(LR(v))$, has the integrality property. During the entire planning horizon, the sum of the total production capacity for all plants and the total import from an exogeneous source should be greater than or equal to the total demand for all markets. The following inequality constraint

$$\sum_{t=1}^{T}\sum_{i=1}^{I}\left(q_i + \sum_{\tau=1}^{t}\sum_{s=1}^{S(i,\tau)} e_{is\tau}z_{is\tau}\right) + \sum_{j=1}^{J}\sum_{t=1}^{T} y_{I+1jt} \geq \sum_{j=1}^{J}\sum_{t=1}^{T} R_{jt} \tag{3.20}$$

is redundant for $(P1)$, but not for $(LR(v))$. By adding it to $(LR(v))$, we could improve the lower bound on $(P1)$ for a given $v$. Furthermore, the Lagrangean relaxation bound from the strengthened problem may be stronger than the LP relaxation bound because (3.20) eliminates the integrality property of $(LR(v))$ (Geoffrion 1974).

With the surrogate constraint (3.20), $(LR(v))$ becomes a mixed integer single knapsack problem $(LRS(v))$ which does not have the integrality property:

$$(LRS(v))\quad Min \sum_{i=1}^{I}\sum_{t=1}^{T}\sum_{s=1}^{S(i,t)} D_{ist}z_{ist} + \sum_{j=1}^{J}\sum_{t=1}^{T}(c_{I+1jt} - v_{jt})y_{I+1jt}$$

$$s.t. \sum_{i=1}^{I}\sum_{t=1}^{T}\sum_{s=1}^{S(i,t)} (T-t+1)e_{ist}z_{ist} + \sum_{j=1}^{J}\sum_{t=1}^{T} y_{I+1jt} \geq L0$$

$$y_{I+1jt} \geq 0, \quad \forall j,\ t$$

$$z_{ist} \in \{0,1\}, \quad \forall i,\ s,\ t$$

where $L0 = \Sigma_{j=1}^{J}\Sigma_{t=1}^{T}R_{jt} - T\Sigma_{i=1}^{I}q_i$. For notational simplicity, define

$$\bar{jt} = argmin_{(j,t)\in M\otimes H}(c_{I+1jt} - v_{jt})$$
$$U = P\otimes G(i,t)\otimes H$$
$$U^0 = \left\{(i,s,t)\in U \Big| 0 < \frac{D_{ist}}{(T-t+1)e_{ist}} < (c_{I+1\bar{jt}} - v_{\bar{jt}})\right\}$$
$$U^+ = \left\{(i,s,t)\in U \Big| \frac{D_{ist}}{(T-t+1)e_{ist}} \geq (c_{I+1\bar{jt}} - v_{\bar{jt}})\right\}$$
$$U^- = \left\{(i,s,t)\in U \Big| \frac{D_{ist}}{(T-t+1)e_{ist}} \leq 0\right\}$$

where $P = \{1,\ldots,I\}, M = \{1,\ldots,J\}, G(i,t) = \{1,\ldots,S(i,t)\}$ and $H = \{1,\ldots,T\}$. The following result is trivial:

**Lemma 7** $(z_{ist}^*, y_{I+1jt}^*) \in OS(LRS(v))$ *satisfies* $z_{ist}^*0 = 0$; $(i,s,t)\in U^+$, $z_{1st}^* = 1$ $\forall(i,s,t)\in U^-$ *and* $y_{I+1jt}^* = 0$ $\forall(j,t) \neq \bar{jt}$.

By Lemma 7, $(LRS(v))$ reduces to $(LRS1(v))$:

$$(LRS1(v)) \quad Min \sum_{(i,s,t) \in U^0} D_{ist}z_{ist} + (c_{I+1\bar{j}t} - v_{\bar{j}t})y_{I+1\bar{j}t}$$

$$s.t. \sum_{(i,s,t) \in U^0} (T - t + 1)e_{ist}z_{ist} + y_{I+1\bar{j}t} \geq L$$

$$y_{I+1\bar{j}t} \geq 0$$

$$z_{ist} \in \{0, 1\}, \quad \forall (i, s, t) \in U^0$$

where $L = L0 - \Sigma_{(i, s, t) \in U^-}(T - t + 1)e_{ist}$.
[example of Lemma 7]

$$(LRS(v)) \quad Min \; 10z_1 + 5z_2 - 8z_3 + 2z_4 + 3y_1 + 5y_2 + 7y_3$$

$$s.t. \; 3z_1 + 9z_2 + 7z_3 + 2z_4 + y_1 + y_2 + 2y_3 \geq 17$$

$$z_1, z_2, z_3, z_4 \in \{0, 1\}$$

$$y_1, y_2, y_3 \geq 0.$$

Obviously $z_3^* = 1$ and $y_1^* = y_2^* = y_3^* = 0$ because $y_1$ dominates $z_1, y_2, y_3$. $(LRS(v))$ reduces to $(LRS1(v))$ with a continuous variable, $y_1$.

$$(LRS1(v)) \quad Min \; 5z_2 + 2z_4 + 3y_1$$

$$s.t. \; 9z_2 + 2z_4 + y_1 \geq 17 - 7$$

$$z_2, z_4 \in \{0, 1\}$$

$$y_1 \geq 0$$

**Lemma 8** *$OS(LRS1(v))$ is obtained by solving at most two 0-1 pure integer knapsack problems.*

**Proof** For fixed $z$, $(LRS1(v))$ reduces to the following linear programming problem $(LP(v))$:

$$(LP(v)) \quad Min(c_{I+1\bar{j}t} - v_{\bar{j}t})y_{I+1\bar{j}t}$$

$$s.t. \; y_{I+1\bar{j}t} \geq L - \sum_{(i, s, t) \in U^0} (T - t + 1)e_{ist}z_{ist}$$

$$y_{I+1\bar{j}t} \geq 0$$

whose optimal solution is $y_{I+1\bar{j}t}^* = max \{0, L - \Sigma_{(i, s, t) \in U^0} (T - t + 1)e_{ist}z_{ist}\}$.
$(LRS1(v))$ is equivalent to

$$(LRS2(v)) \quad Min \left[ \sum_{(i,s,t) \in U^0} D_{ist}z_{ist} \right.$$

$$\left. + \; max \left\{0, (c_{I+1\bar{j}t} - v_{\bar{j}t})\left(L - \sum_{(i,s,t) \in U^0} (T - t + 1)e_{ist}z_{ist}\right)\right\}\right]$$

$$s.t. \; z_{ist} \in \{0, 1\}, \quad \forall (i, s, t) \in U^0$$

and $OV(LRS2(v)) = min \{OV(LRS21(v)), OV(LRS22(v)) + L(c_{I+1\bar{j}t} - v_{\bar{j}t})\}$, where

$$(LRS21(v)) \quad Min \sum_{(i,s,t) \in U^0} D_{ist} z_{ist}$$

$$s.t. \sum_{(i,s,t) \in U^0} (T - t + 1) e_{ist} z_{ist} \geq L$$

$$z_{ist} \in \{0, 1\}, \quad \forall(i, s, t) \in U^0$$

$$(LRS22(v)) \quad Min \sum_{(i,s,t) \in U^0} \{D_{ist} - (c_{I+1\bar{j}t} - v_{\bar{j}t})(T - t + 1) e_{ist}\} z_{ist}$$

$$s.t. \sum_{(i,s,t) \in U^0} (T - t + 1) e_{ist} z_{ist} \leq L$$

$$z_{ist} \in \{0, 1\}, \quad \forall(i, s, t) \in U^0.$$

Alternatively, it is easily seen that the Benders' reformulation of $(LRS1(v))$ by using two extreme points of the dual of $(LP(v))$ is equivalent to $(LRS21(v))$ and $(LRS22(v))$. $\square$

[example of Lemma 8]
For $(LRS1(v))$ in the above example, we have two 0-1 pure integer knapsack problems, $(LRS21(v))$ and $(LRS22(v))$: $OV(LRS1(v)) = min \{OV(LRS21(v)), OV(LRS22(v))\} = 7$.

$$(LRS21(v)) \quad Min \ 5z_2 + 2z_4$$

$$s.t. \ 9z_2 + 2z_4 \geq 10$$

$$z_2, z_4 \in \{0, 1\}$$

$$(LRS22(v)) \quad Min \ -22z_2 - 4z_4 + 30$$

$$s.t. \ 9z_2 + 2z_4 \leq 0$$

$$z_2, z_4 \in \{0, 1\}$$

$OS(LRS1(v))$ can be obtained by solving at most two 0-1 pure integer knapsack problems, $(LRS21(v))$ and $(LRS22(v))$. If the solution of either $(LRS21(v))$ or $(LRS22(v))$ satisfies the corresponding knapsack constraint as an equality, there is no need to solve the other knapsack problem. If $OV(LRS21(v)) > OV(LRS22(v)) + L(c_{I+1\bar{j}t} - v_{\bar{j}t})$, $y^*_{I+1\bar{j}t} = L - \Sigma_{(i,s,t) \in U^0}(T - t + 1) e_{ist} z^*_{1st}$. Otherwise, $y^*_{I+1\bar{j}t} = 0$. Obviously, adding the surrogate constraint (3.20) improves the lower bound by $OV(LRS1(v))$. When import is not allowed, the corresponding improvement is given by $OV(SUR(v))$ which is at least as large as $OV(LRS1(v))$:

$$(SUR(v)) \quad Min \sum_{(i,s,t) \in U^0 \cup U^+} D_{ist} z_{ist}$$

$$s.t. \sum_{(i,s,t) \in U^0 \cup U^+} (T - t + 1) e_{ist} z_{ist} \geq L$$

$$z_{ist} \in \{0, 1\}, \quad \forall (i, s, t) \in U^0 \cup U^+.$$

Once $(z^*_{ist}, y^*_{I+1jt}) \in OS(LRS(v))$ is determined, we can also determine the solution $y^*_{ijt}$ and $I^*_{it}$.

To summarize, by using the Lagrangean relaxation method and adding a surrogate constraint, we can easily obtain a lower bound on the original problem (P1), which may be better than the lower bound from LP relaxation.

## Numerical Example

To demonstrate the problem reduction and strengthening procedure, consider the following simple example. Let $(I, J, T) = (2, 2, 4)$, $S(i, t) = 2 \; \forall \; i, t$, $e_{ist} = e_{is} \; \forall \; t$ and $c_{I+1jt} = 6 \; \forall \; j, t$.

$$[c_{ij \mid t}] = \begin{bmatrix} 1 & 3 & 1 & 2.5 & 1 & 2.5 & 0.5 & 2.5 \\ 4 & 2 & 3.5 & 2 & 3 & 2 & 2.5 & 1 \end{bmatrix}$$

$$[r_{is \mid t}] = \begin{bmatrix} 50 & 100 & 40 & 90 & 35 & 85 & 30 & 80 \\ 40 & 70 & 40 & 70 & 40 & 60 & 35 & 50 \end{bmatrix}$$

$$[p_{it} \mid h_{it} \mid e_{is} \mid q_i] = \begin{bmatrix} 2 & 2 & 1 & 1 & 1 & 1 & 0.5 & 0.5 & 5 & 20 & 0 \\ 1 & 1 & 1 & 1 & 1 & 1 & 0.5 & 0.5 & 5 & 10 & 0 \end{bmatrix}$$

$$[R_{jt} \mid v_{jt}] = \begin{bmatrix} 15 & 20 & 30 & 40 & 4 & 3 & 2 & 4 \\ 5 & 10 & 40 & 70 & 5 & 5 & 3 & 2 \end{bmatrix}.$$

For $i = 1$, $\{j(1, t), \forall t\} = \{1, 2, 1, 1\}$.

[Problem Reduction Algorithm]

**(Step 0)** $t'' = 1$, $I_{1t} = y_{1j(1,t)t} = 0 \; \forall t$

**(Step 1)** $K^{(1)}_1(1, 1) = -1 = \min\{-1, 0.5, 3, 1\}$, $t' = 1$

**(Step 3)** $y_{111} = 5z_{111} + 20z_{121}$, $t'' = 2$

**(Step 1)** $K^{(1)}_1(2, 2) = -0.5 = \min\{-0.5, 2, 0\}$, $t' = 2$

**(Step 3)** $y_{122} = 5z_{111} + 20z_{121} + 5z_{112} + 20z_{122}$, $t'' = 3$

**(Step 1)** $K^{(1)}_1(3, 4) = -2 = \min\{0, -2\}$, $t' = 4$

**(Step 4)** $I_{13} = 5z_{111} + 20z_{121} + 5z_{112} + 20z_{122} + 5z_{113} + 20z_{123}$
$\qquad y_{114} = 5z_{111} + 20z_{121} + 5z_{112} + 20z_{122} + 5z_{113} + 20z_{123}$
$\qquad K_1(4, 4) = -2.5$
$\qquad y_{114} = 10z_{111} + 40z_{121} + 10z_{112} + 40z_{122} + 10z_{113} + 40z_{123} + 5z_{114} + 20z_{124}$
$\qquad$ Stop

For $i = 2$, $\{j(2, t), \forall t\} = \{2, 2, 2, 1\}$.

[Problem Reduction Algorithm]

**(Step 0)** $t'' = 1$, $I_{2t} = y_{2j(2,t)t} = 0 \; \forall t$

**(Step 1)** $K^{(1)}_2(1, 1) = -2 = \min\{-2, -1, 2, 2\}$, $t' = 1$

**(Step 3)** $y_{221} = 5z_{211} + 10z_{221}$, $t'' = 2$

**(Step 1)** $K^{(1)}_2(2, 2) = -2 = \min\{-2, 1, 1\}$, $t' = 2$

**(Step 3)** $y_{222} = 5z_{211} + 10z_{221} + 5z_{212} + 10z_{222}$, $t'' = 3$

**(Step 1)** $K_2^{(1)}$ $(3, 3) = 0 = \min\{0, 0\}$, $\quad t' = 3$

**(Step 2)** $t'' = 4$

**(Step 1)** $K_2^{(1)}$ $(4, 4) = -0.5 = \min\{-0.5\}$, $\quad t' = 4$

**(Step 3)** $y_{214} = 5z_{211} + 10z_{221} + 5z_{212} + 10z_{222} + 5z_{213} + 10z_{223} + 5z_{214} + 10z_{224}$
　　　Stop.

Now we expressed $I_{it}$ and $y_{ijt}$ in terms of $z_{ist}$ through the problem reduction algorithm as follows:

$$y_{111} = 5z_{111} + 20z_{121}$$
$$y_{122} = 5z_{111} + 20z_{121} + 5z_{112} + 20z_{122}$$
$$y_{114} = 10z_{111} + 40z_{121} + 10z_{112} + 40z_{122} + 10z_{113} + 40z_{123} + 5z_{114} + 20z_{124}$$
$$y_{221} = 5z_{211} + 10z_{221}$$
$$y_{222} = 5z_{211} + 10z_{221} + 5z_{212} + 10z_{222}$$
$$y_{214} = 5z_{211} + 10z_{221} + 5z_{212} + 10z_{222} + 5z_{213} + 10z_{223} + 5z_{214} + 10z_{224}$$
$$I_{13} = 5z_{111} + 20z_{121} + 5z_{112} + 20z_{122} + 5z_{113} + 20z_{123}.$$

Plugging $I_{it}$ and $y_{ijt}$ into $(LR(v))$, we obtain the coefficients, $D_{ist}$, of the objective function of the reduced problem $(LRS(v))$ as follows:

$$[D_{is|t}] = \begin{bmatrix} 20 & -20 & 15 & -10 & 12.5 & -5 & 17.5 & 30 \\ 17.5 & 25 & 27.5 & 45 & 37.5 & 55 & 32.5 & 45 \end{bmatrix}$$

**When Import is Not Allowed**

Note that $(LR(v)) = \Sigma_{i=1}^{I}(MR_i(v))$ if import is not allowed. $(LR(v))$ is easily solved by inspecting the sign of $D_{ist}$. That is, if $D_{ist} \geq 0$, $z_{ist}^* = 0$. Otherwise, $z_{ist}^* = 1$. We get the solution of $(LR(v))$ as follows:

$$z_{121}^* = z_{122}^* = z_{123}^* = 1, \qquad OV(LR(v)) = \sum_{i=1}^{I}\sum_{t=1}^{T}\sum_{s=1}^{S(i,t)} D_{ist}z_{ist}^* + \sum_{j=1}^{J}\sum_{t=1}^{T} v_{jt}R_{jt} = 640.$$

The pure integer single knapsack problem $(LRS(v))$ strengthened by the following surrogate constraint

$$4(5z_{111} + 20z_{121} + 5z_{211} + 10z_{221}) + 3(5z_{112} + 20z_{122} + 5z_{212} + 10z_{222})$$
$$+ 2(5z_{113} + 20z_{123} + 5z_{213} + 10z_{223}) + 1(5z_{114} + 20z_{124} + 5z_{214} + 10z_{224}) \geq 230$$

reduces to

$$(SUR(v)) \quad Min \ 20z_{111} + 17.5z_{211} + 25z_{221} + 15z_{112} + 27.5z_{212} + 45z_{222} + 12.5z_{113}$$
$$+ 37.5z_{213} + 55z_{223} + 17.5z_{114} + 30z_{124} + 32.5z_{214} + 45z_{224}$$
$$s.t. \ 20z_{111} + 20z_{211} + 40z_{221} + 15z_{112} + 15z_{212} + 30z_{222} + 10z_{113}$$
$$+ 10z_{213} + 20z_{223} + 5z_{114} + 20z_{124} + 5z_{214} + 10z_{224} \geq 50$$
$$\text{all } z\text{'s} \in \{0, 1\}$$

we get the following solution:

$$z^*_{121} = z^*_{122} = z^*_{123} = z^*_{221} = z^*_{123} = 1, OV(SUR(v)) = 37.5, OV(LRS(v)) = 677.5.$$

## When Import is Allowed

Note that $(LR(v)) = \Sigma^I_{i=1}(MR_i(v)) + (LR_{I+1}(v))$ if import is allowed. By inspecting the sign of $D_{ist}$ and $(c_{I+1jt} - v_{jt})$, we get the solution of $(LR(v))$ as follows:

$$z^*_{121} = z^*_{122} = z^*_{123} = 1, \qquad OV(LR(v)) = 640.$$

The mixed integer single knapsack problem $(LRS(v))$ strengthened by the following surrogate constraint

$$4(5z_{111} + 20z_{121} + 5z_{211} + 10z_{221}) + 3(5z_{112} + 20z_{122} + 5z_{212} + 10z_{222})$$
$$+ 2(5z_{113} + 20z_{123} + 5z_{213} + 10z_{223}) + 1(5z_{114} + 20z_{124} + 5z_{214} + 10z_{224})$$
$$+ (y_{311} + y_{321} + y_{312} + y_{322} + y_{313} + y_{323} + y_{314} + y_{324}) \geq 230$$

reduces to

$$(LRS1(v)) \quad Min\ 17.5z_{211} + 25z_{221} + y_{321}$$
$$s.t.\ 20z_{211} + 40z_{221} + y_{321} \geq 50$$
$$y_{321} \geq 0$$
$$z_{211}, z_{221} \in \{0, 1\}$$

and equivalently to the following pure integer knapsack problems:

$$(LRS21(v)) \quad Min\ 17.5z_{211} + 25z_{221}$$
$$s.t.\ 20z_{211} + 40z_{221} \geq 50$$
$$z_{211}, z_{221} \in \{0, 1\}$$

$$(LRS22(v)) \quad Min\ -2.5z_{211} - 15z_{221}$$
$$s.t.\ 20z_{211} + 40z_{221} \leq 50$$
$$z_{211}, z_{221} \in \{0, 1\}.$$

Since $OV(LRS21(v)) > OV(LRS22(v)) + 50$, we get the following solution:

$$z^*_{121} = z^*_{122} = z^*_{123} = z^*_{221} = 1, \quad y^*_{321} = 10, \quad OV(LRS(v)) = 35, \quad OV(LRS(v)) = 675.$$

By adding the surrogate constraint to $(LR(v))$ for each cases, we obtain an improvement in the lower bound of 5.9% and 5.5% respectively. This does not necessarily mean that the strengthened Lagrangean relaxation problem $(LRS(v))$ always yields the stronger bound because the sequences of $v$ generated during the iterations of the subgradient method for $(LRS(v))$ and $(LR(v))$ may be different. But we could expect with high probability to get a stronger bound from $(LRS(v))$.

## 3.5   Determination of a Primal Feasible Solution

### Initial Primal Feasible Solution

A good initial feasible solution is needed to get a good (not too large) initial step size which might result in rapid convergence of the subgradient method. Assuming $I_{it} = 0 \; \forall \; i, t$ in (P1), we easily see that for a given production capacity, $CAP_i(t)$, of each plant at each time period, (P1) reduces to a transportation problem for each time period $t$. Since it is assumed that the import, $y_{I+1jt}$, from the exogeneous source is unlimited, each transportation problem is feasible and we can obtain an initial feasible solution by solving $T$ transportation problems separately.

In order to predetermine the production capacity, $CAP_i(t)$, we applied a greedy heuristic which expands the cheapest available project for all plants at time period $t$ until the total supply exceeds the demand at $t$. If the production capacity exceeds the demand at period $t$, we proceed to the next period $t + 1$. If not, i.e., in case there is no available project at period $t$, we have two options: (i) the import from an exogeneous source is used for the unsatisfied demand at period $t$ with a penalty cost; and (ii) another cheapest project is chosen among the projects available for all plants at period $t - 1$ (we repeat the same procedure until the capacity exceeds the demand at period $t$). The option with lower cost is chosen. Once the production capacities are fixed through the heuristic for all plants during the planning horizon, we obtain an initial feasible solution by solving the resulting transportation problems.

### Lagrangean Heuristic for a Primal Feasible Solution

The Lagrangean solutions obtained during the subgradient method can be used to construct primal feasible solution with some judicious tinkering (Fisher 1981). A clever adjustment of the Lagrangean solution might satisfy the relaxed constraint (3.10) and yield a good primal feasible solution. In this sense, $y_{ijt}$ is a natural choice for adjustment in the Lagrangean heuristic. Retaining part of the Lagrangean solution, $I_{it}^*$ and $z_{ist}^*$ in (P1), we easily see that the resulting primal problem $(P1_t(I^*, z^*))$ reduces to a constrained transportation problems $(P1_t(I^*, z^*))$ for each time period $t$:

$$(P1_t(I^*, z^*)) \quad Min \quad \sum_{i=1}^{I+1} \sum_{j=1}^{J} (c_{ijt} + p_{it}) y_{ijt} \tag{3.21}$$

$$s.t. \quad \sum_{j=1}^{J} y_{ijt} \le \left( q_i + \sum_{\tau=1}^{t} \sum_{s=1}^{S(i,\tau)} e_{is\tau} z_{is\tau}^* \right) + I_{it-1}^* - I_{it}^*, \forall i \neq I + 1 \tag{3.22}$$

$$\sum_{i=1}^{I+1} y_{ijt} = R_{jt}, \qquad \forall j \tag{3.23}$$

$$\sum_{j=1}^{J} y_{ijt} \ge I_{it-1}^* - I_{it}^*, \qquad \forall i \neq I + 1 \tag{3.24}$$

$$y_{ijt} \ge 0, \qquad \forall i, j.$$

$(P1_t(I^*, z^*))$ is feasible if $\sum_{j=1}^{J} R_{jt} \ge \sum_{i=1}^{I} \max\{0, (I_{it-1}^* - I_{it}^*)\} \forall \; t$. When $(I_{it}^*, z_{ist}^*)$ does not satisfy the feasibility condition of $(P1_t(I^*, z^*))$, we could adjust, in most cases, the inventory level, $I_{it}^*$, until the condition is satisfied. Even in case that $(I_{it}^*, z_{ist}^*)$ satisfies the feasibility condition, we can improve the primal feasible solution by adjusting $z_{ist}^*$. That is, if $\sum_{i=1}^{I} \{(q_i + \sum_{\tau=1}^{t} \sum_{s=1}^{S(i,\tau)} e_{ist} z_{ist}^*) + I_{it-1}^* - I_{it}^*\}$

is sufficiently larger than $\Sigma_{j=1}^{J} R_{jt}$ for subperiod $[t, T]$, there may exist a surplus capacity which can be eliminated by cancelling the implementation of certain projects of plant $i$ at time period $t$.

Once $I_{it}^{*}$ and $z_{ist}^{*}$ satisfy the feasibility condition of $(P1_t(I^{*}, z^{*}))$, we can obtain a primal feasible solution to $(P1)$ by solving $(P1_t(I^{*}, z^{*}))$ $\forall t$. Since $OV(P1_t(I^{*}, z^{*}))$ is not always better than the current best upper bound of $OV(P1)$ at every iteration of the subgradient method, we should not be obsessed with obtaining an exact solution of $(P1(I^{*}, z^{*}))$ to save computation time. In this sense, an efficient approximation algorithm which often yields an optimal solution is more desirable. We develop a heuristic algorithm for $(P1_t(I^{*}, z^{*}))$ which terminates in a finite number of steps as follows:

**Step 0:** (Initialization)
   Solve $(P1_t(I^{*}, z^{*}))$ ignoring (3.24).

**Step 1:** (Initial feasibility test)
   If the solution satisfies (3.24), it is optimal to $(P1_t(I^{*}, z^{*}))$.

**Step 2:** (Partitioning of the set of plants)

$$E_{it} = (I_{it-1}^{*} - I_{it}^{*}) - \sum_{j=1}^{J} y_{ijt}^{*} \quad \forall i$$
$$P' = \{i \mid E_{it} > 0 \; \forall i\}$$
$$P'' = \{i \mid E_{it} < 0 \; \forall i\}.$$

**Step 3:** (Finding minimum marginal cost)

$$(i_0', i_0'', j_0) = argmin_{i' \in P', \, i'' \in (P'' \cup \{I+1\}, j \in M)}\{(c_{i'jt} + p_{i't}) - (c_{i''jt} + p_{i''t})|y_{i''jt}^{*} > 0 \}.$$

**Step 4:** (Adjusting amount of shipment)
$$\delta = min\{-E_{i_0''t}, y_{i_0''j_0t}^{*}, E_{i_0't}\}$$
$$y_{i_0'j_0t}^{*} = y_{i_0'j_0t}^{*} + \delta$$
$$y_{i_0''j_0t}^{*} = y_{i_0''j_0t}^{*} - \delta$$
$$E_{i_0't} = E_{i_0't} - \delta$$
$$E_{i_0''t} = E_{i_0''t} + \delta.$$

**Step 5:** (Feasibility test)
   If $E_{i_0't} = 0$, then $P' = P' \backslash i_0'$
   If $E_{i_0''t} = 0$, then $P'' = P'' \backslash i_0''$
   If $P' = \emptyset$, then stop.
   Otherwise, go to Step 3.

## 3.6   Whole Approximation Procedure

In Section 3.4 we obtain a lower bound for $OV(P1)$ by solving at most two pure integer 0-1 knapsack problems, $(LRS21(\nu))$ and $(LRS22(\nu))$, for given $\nu$. If the Lagrangean solution $(I^{*}, z^{*})$ satisfies the feasibility condition of $(P1(I^{*}, z^{*}))$, the corresponding primal feasible solution obtained from the given Lagrangean solution in Section 3.5 yields an upper bound for $OV(P1)$. During iterations of the subgradient method, the current best upper bound, $UB$, and the current best lower bound, $LB$, are updated to approximate the optimal solution of the original problem until one of several terminating criteria becomes satisfied. The outline of the overall approximation procedure incorporating the subgradient method (Fisher 1981, Shapiro 1979) is given below:

**Step 0:** (Initialization)

$$v_{jt} = min \{c_{I+1jt}, min_{i \in P}(c_{ijt} + p_{it})\}, \forall j, t.$$

**Step 1:** (Updating LB and UB)

$$LB \leftarrow max \{LB, OV(LRS(v))\}$$
$$UB \leftarrow min \{UB, OV(P1(I^*, z^*))\}.$$

**Step 2:** (Termination Checking)

If F(LB, UB, k, θ, V) ≤ ε, stop.

$k$: number of iterations

θ: stepsize defined by $\theta = \lambda\{UB - OV(LRS(v))\}/\|V\|^2$

V: subgradient defined by $V_{jt} = R_{jt} - \Sigma_{i=1}^{I+1} y^*_{ijt}$ $\forall j, t$

F(·): prespecified terminating criteria.

**Step 3:** (Updating Lagrangean multiplier)

$$v_{jt} \leftarrow min (\{v_{jt} + \theta \cdot V_{jt}, c_{I+1jt}\}, \forall j, t)$$

Go to Step 1.

In choosing a value for λ we followed the approach of Held, Wolfe, and Crowder (1974) in letting λ = 2 and halving it whenever $OV(LRS(v))$ has failed to increase within 5 consecutive iterations.

## 3.7 Computational Experience

The approximation procedure described in the previous sections was programmed in FORTRAN 77 and run on a VAX 6400. The whole program consists of five main parts: (i) random problem generating subroutine; (ii) main program which comprises a problem reduction algorithm and a subgradient method; (iii) primal feasible solution finding subroutine; (iv) transportation problem solving subroutine; and (v) single knapsack problem solving subroutine (Fayard and Plateau 1982).

The computational performance of the approximation procedure was evaluated on randomly generated problem sets. Each problem set is characterized by the relative size of unit capacity expansion cost and problem size, (I, J, S(i, t), T). Table 3.1 is a summary of these problem sets.

As the relative size of unit expansion costs becomes larger, the proportion of the expansion cost in the total cost grows. This means that more emphasis is given to the determination of the discrete variable, $z_{ist}$, than the continuous variables. We assumed the same exponential discount rate of 3% for all unit costs and a randomly nondecreasing demand ($R_{jt+1} = R_{jt} + \delta_{jt}$, $\delta_{jt} \geq 0$). To include an economies-of-scale effect on expansion costs, a linearly decreasing unit expansion cost was assumed as the project size grows.

**TABLE 3.1** Description of Test Problem Set

| Type | Description |
|------|-------------|
| 1 | Max. expansion cost per unit is 20; the other unit costs are [1, 10] |
| 2 | Max. expansion cost per unit is 50; the other unit costs are [1, 10] |
| 3 | Max. expansion cost per unit is 90; the other unit costs are [1, 10] |
| A | (I, J, S, T) is (10, 10, 5, 5); 250 integer variables |
| B | (I, J, S, T) is (10, 10, 10, 20); 2000 integer variables |
| C | (I, J, S, T) is (20, 20, 10, 10); 2000 integer variables |
| D | (I, J, S, T) is (30, 20, 10, 5); 1500 integer variables |
| E | (I, J, S, T) is (30, 20, 10, 10); 3000 integer variables |
| F | (I, J, S, T) is (30, 20, 10, 15); 4500 integer variables |
| G | (I, J, S, T) is (30, 20, 10, 20); 6000 integer variables |

**TABLE 3.2**  Computational Results ($\rho = 1.2$)

| Problem Set | No. of Var.[1] | | Comp.[2] (%) | Average Bound (%)[3] | | Average CPU (sec)[4] | |
|---|---|---|---|---|---|---|---|
| | 0-1 | conti. | | with[5] imp(no-imp)[7] | w/o[6] | with imp(no-imp) | w/o |
| 1-A | 250 | 550 | 33.49 | 98.79(98.87) | 98.21 | 7.93(7.94) | 7.27 |
| 1-B | 2,000 | 2,200 | 15.22 | 98.44(98.45) | 98.65 | 40.45(40.50) | 34.24 |
| 1-C | 2,000 | 4,200 | 22.57 | 98.01(98.10) | 97.37 | 130.27(135.02) | 122.01 |
| 1-D | 1,500 | 3,150 | 25.04 | 98.65(98.77) | 98.23 | 140.18(145.11) | 135.19 |
| 1-E | 3,000 | 6,300 | 23.05 | 97.85(97.89) | 98.01 | 260.85(280.40) | 242.36 |
| 1-F | 4,500 | 9,450 | 20.64 | 98.03(98.03) | 98.02 | 400.50(437.87) | 336.96 |
| 1-G | 6,000 | 12,600 | 17.11 | 98.09(98.12) | 97.41 | 502.42(547.53) | 453.55 |
| 2-A | 250 | 550 | 53.21 | 96.98(97.12) | 92.34 | 7.99(7.97) | 6.20 |
| 2-B | 2,000 | 2,200 | 35.36 | 96.24(96.25) | 93.12 | 39.20(40.02) | 32.72 |
| 2-C | 2,000 | 4,200 | 48.77 | 96.55(96.64) | 91.44 | 128.25(132.67) | 100.31 |
| 2-D | 1,500 | 3,150 | 50.37 | 94.22(94.30) | 88.12 | 135.79(140.28) | 99.33 |
| 2-E | 3,000 | 6,300 | 45.92 | 96.12(96.24) | 90.08 | 247.50(276.22) | 195.64 |
| 2-F | 4,500 | 9,450 | 41.30 | 97.81(97.92) | 93.66 | 324.51(386.82) | 286.96 |
| 2-G | 6,000 | 12,600 | 38.25 | 97.72(97.80) | 92.34 | 499.01(558.53) | 397.93 |
| 3-A | 250 | 550 | 69.27 | 94.57(94.71) | 79.34 | 6.73(6.73) | 4.38 |
| 3-B | 2,000 | 2,200 | 52.76 | 95.27(95.29) | 89.06 | 35.77(40.10) | 30.36 |
| 3-C | 2,000 | 4,200 | 61.39 | 94.39(94.96) | 80.45 | 114.18(122.45) | 75.37 |
| 3-D | 1,500 | 3,150 | 63.92 | 93.40(93.41) | 71.43 | 99.59(122.18) | 68.84 |
| 3-E | 3,000 | 6,300 | 59.80 | 94.12(94.20) | 82.08 | 218.53(273.32) | 154.12 |
| 3-F | 4,500 | 9,450 | 52.74 | 95.71(95.84) | 87.39 | 362.40(410.48) | 232.08 |
| 3-G | 6,000 | 12,600 | 50.33 | 94.44(94.58) | 89.32 | 494.90(540.12) | 367.52 |

All statistics are averages of 5 problems solved (50 iterations of the subgradient method).
[1]Number of variables (0-1: $z_{ist}$, conti.: $y_{ijt}$, $I_{it}$).
[2]Composition Rate (%) = (Expansion Cost/Total Cost) $\times$ 100 in the best feasible solution.
[3]Bound (%) = {$1 - $ (UB $-$ LB)/UB} $\times$ 100.
[4]VAX 6400.
[5]Lagrangean relaxation problem with a surrogate constraint.
[6]Lagrangean relaxation problem without surrogate constraint.
[7]Import is allowed (import is not allowed).

The final duality gap appears to be a function of the characteristics of the data, namely the project size, $e_{ist}$, versus the magnitude of demand increase, ($R_{jt} - R_{jt-1}$), during period $t$. To demonstrate such relationships, we performed our computational experiments for two cases: (i) $\rho = \min_{t \in H} \{\sum_{i=1}^{I} \sum_{s=1}^{S(i,\, t)} e_{ist} / \sum_{j=1}^{J} (R_{jt} - R_{jt-1})\} = 1.2$, i.e., the total project size is relatively small compared to the size of the demand increase, and (ii) $\rho = 5$. For each case, we compared two types of Lagrangean relaxation problems; the first one where the subproblems are strengthened by a surrogate constraint and the other one where the subproblems have not been strengthened. We also included the computational experiments for the no-import case.

For most of the test problems the approximation procedure was terminated by the iteration limit criterion (the iteration limit was 50). Solving at most two single knapsack problems at each iteration was not too burdensome; the average number of variables to be actually determined in each knapsack problem is less than 20% of the original number of variables $|U|$, i.e., $|U^0|/|U|$ is less than 0.2 (for the no-import case, a larger knapsack problem is solved, $|U^0 \cup U^+|/|U|$ is less than 0.3). Furthermore, in some cases where $\rho = 5$, the surrogate constraint is already satisfied by variables with a negative $D_{ist}$. In these cases we do not need to solve the knapsack problems.

The performance of the Lagrangean heuristic algorithm (Section 3.5) for the *constrained* transportation problem turned out to be very good, even though we do not know its worst case behavior. For about 80% of the problems we solved, the solution of the *standard* transportation problem, ignoring (3.24), also satisfied (3.24), i.e., the optimal solution of the constrained transportation problem was obtained from the heuristic algorithm (Step 1: [Initial feasibility test] in Section 3.5). For the remaining problems,

**TABLE 3.3**    Computational Results ($\rho = 5$)

| Problem Set | No. of Var.[1] | | Comp.[2] (%) | Average Bound (%)[3] | | Average CPU (sec)[4] | |
|---|---|---|---|---|---|---|---|
| | | | | with[5] | | with | |
| | 0-1 | conti. | | imp(no-imp)[7] | w/o[6] | imp(no-imp) | w/o |
| 1-A | 250 | 550 | 30.60 | 97.54(97.65) | 97.35 | 7.16(7.18) | 7.12 |
| 1-B | 2,000 | 2,200 | 10.95 | 98.12(98.16) | 98.28 | 37.22(38.49) | 34.29 |
| 1-C | 2,000 | 4,200 | 17.93 | 97.31(97.31) | 97.31 | 128.80(132.43) | 121.33 |
| 1-D | 1,500 | 3,150 | 23.82 | 97.52(97.53) | 97.15 | 138.54(141.21) | 135.07 |
| 1-E | 3,000 | 6,300 | 18.38 | 97.15(97.19) | 97.39 | 259.49(275.55) | 241.21 |
| 1-F | 4,500 | 9,450 | 13.82 | 97.65(97.67) | 97.65 | 398.18(431.63) | 337.02 |
| 1-G | 6,000 | 12,600 | 11.31 | 97.58(97.72) | 97.49 | 500.12(521.47) | 455.81 |
| 2-A | 250 | 550 | 49.99 | 91.40(91.47) | 90.92 | 7.06(7.04) | 6.27 |
| 2-B | 2,000 | 2,200 | 30.32 | 94.75(94.81) | 94.69 | 36.81(37.50) | 30.01 |
| 2-C | 2,000 | 4,200 | 42.64 | 92.89(93.01) | 91.89 | 121.20(130.08) | 95.56 |
| 2-D | 1,500 | 3,150 | 46.47 | 88.50(88.62) | 87.01 | 128.75(138.51) | 94.98 |
| 2-E | 3,000 | 6,300 | 43.31 | 92.44(92.62) | 91.32 | 246.60(270.00) | 195.99 |
| 2-F | 4,500 | 9,450 | 35.95 | 93.71(93.77) | 93.20 | 319.54(362.45) | 285.04 |
| 2-G | 6,000 | 12,600 | 30.74 | 94.46(94.47) | 93.03 | 485.36(530.77) | 372.70 |
| 3-A | 250 | 550 | 61.13 | 86.25(86.41) | 77.40 | 5.04(5.09) | 3.95 |
| 3-B | 2,000 | 2,200 | 45.48 | 93.50(93.58) | 92.03 | 35.11(38.50) | 28.25 |
| 3-C | 2,000 | 4,200 | 58.74 | 86.79(86.89) | 80.75 | 91.29(104.51) | 73.91 |
| 3-D | 1,500 | 3,150 | 56.86 | 85.20(85.22) | 70.11 | 90.48(111.40) | 65.16 |
| 3-E | 3,000 | 6,300 | 59.33 | 90.11(90.28) | 81.51 | 200.66(259.98) | 153.97 |
| 3-F | 4,500 | 9,450 | 51.29 | 90.04(90.16) | 88.69 | 303.04(355.12) | 226.10 |
| 3-G | 6,000 | 12,600 | 46.81 | 91.65(91.82) | 89.26 | 465.07(519.17) | 351.74 |

All statistics are averages of 5 problems solved (50 iterations of the subgradient method).
[1] Number of variables (0-1: $z_{ist}$, conti.: $y_{ijt}$, $I_{it}$).
[2] Composition Rate (%) = (Expansion Cost/ Total Cost) $\times$ 100 in the best feasible solution.
[3] Bound (%) = $\{1 - (UB - LB)/UB\} \times 100$.
[4] VAX 6400.
[5] Lagrangean relaxation problem with a surrogate constraint.
[6] Lagrangean relaxation problem without surrogate constraint.
[7] Import is allowed (import is not allowed).

the approximations from the heuristic algorithm were all within 2% of the optimal solution obtained from a general LP solver (we used LINDO). Since $T$ constrained transportation problems should be solved at each iteration of a subgradient method (more than 75% of the total computation time is for the Lagrangean heuristic), this fast efficient heuristic for the constrained transportation problem is essential to our whole approximation procedure for saving computation time. The computational results based on 50 iterations at the top node of an enumeration tree are summarized in Table 3.2 and Table 3.3. They indicate that the approximation procedure is effective in solving large problems to within acceptable error tolerances.

Figure 3.3 shows the typical convergence behavior of the approximation procedure for one test problem of type 1-A with a surrogate constraint ($\rho = 5$). The initial fluctuation of $LB$ in Figure 2.3 indicates the initial adaptation process of the step size in the subgradient method. The gradual improvement of $LB$ at the final stage indirectly means that the approximation procedure generates Lagrangean multipliers which are in the neighborhood of the optimal Lagrangean multiplier (actually it is very difficult to exactly find the optimal Lagrangean multiplier in case $J \times T$ is large). For this test problem, we obtained a very strong bound of 99.4% based on 200 iterations.

With a reasonable amount of additional computation time the strengthened dual problem with a surrogate constraint generally (not always) yields a better average bound than the dual problem without the surrogate constraint. As mentioned in section 4.3, the surrogate constraint for the no-import case was also empirically proven to be stronger than one for the import case. See the atypical cases, 1-B and 1-E, showing worse bound for the surrogate constraint. It may be explained by the different sequences

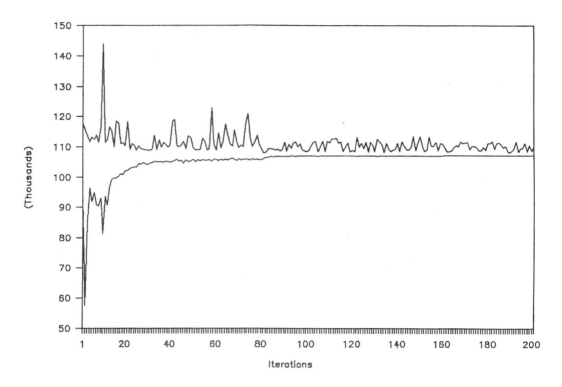

**FIGURE 3.3**    Convergence behavior of the approximation procedure (1-A, $\rho = 5$).

of Lagrangean multiplier $v$ generated during the iterations of the subgradient method for both problems. We observed several interesting results: (i) As the relative size of the unit expansion cost grows (cf. composition rates of Type 1, 2 and 3), the disparity between these two dual problem types becomes more and more striking. This is to be expected, since the Lagrangean relaxation bound without surrogate constraint approximates the LP bound where only fractions of the large fixed charges, $r_{ist}$, are incorporated in the total cost (for the strengthened dual problem, the surrogate constraint forces to incorporate entire fixed charges into the total cost). (ii) As the relative size of the unit expansion cost grows, the average bounds for both dual problem types become worse (Type 1, 2 and 3 in Tables 3.2 and 3.3). This may indicate that the duality gap is significant for large unit expansion costs. (iii) The test problems with a smaller $\rho$ yield on the average better bounds (Table 3.2 vs. Table 3.3). It is easily seen that the surrogate constraint (3.20) becomes a stronger cut for $(MR(v))$ as $\rho$ becomes smaller. Conversely, (3.20) is no longer effective when $\rho$ is sufficiently large. (iv) The computation time mainly depends on the size of the transportation problem and partly depends on the length of the planning horizon (Types B and C).

## 3.8   Conclusion

The model and the algorithm implemented here illustrate how the problem of selecting projects for capacity expansion in a transportation network can be formulated as a 0-1 mixed-integer program and how its optimal solution can be effectively approximated.

Through problem reduction, the mixed-integer program is easily reduced to a trivial pure integer program. For most problems solved we obtained significant improvements by adding a surrogate constraint whose effectiveness mainly depends on the problem structure. In order to save computation time, we only looked for approximate solutions to the constrained transportation problem while looking for feasible solutions. This heuristic algorithm turned out to be very good even though we do not know its worst-case behavior. From a computational standpoint, it may be concluded that the approximation scheme is computationally efficient and provides strong bounds even for large problems, except for problems of type 3 for large ρ.

Designing a clever branch and bound scheme for an optimal solution will be an immediate extension. The most promising area for further research is the extension of the model to multiechelon production-distribution networks, where the linking mechanism of each echelon will be a main point of the research. We also intend to study separately the *loose* case, that is, the case where unit expansion costs are large relative to other costs, and where ρ is large. Some other surrogate constraints, such as Benders inequalities, might prove more effective.

# References

ERLENKOTTER D. (1973) Sequencing Expansion Projects. *Operations Research* **21**, 542–553.

ERLENKOTTER D. (1977) Capacity Expansion with Imports and Inventories. *Management Science* **23**, 694–702.

ERLENKOTTER D. AND J. S. ROGERS (1977) Sequencing Competitive Expansion Projects. *Operations Research* **25**, 937–951.

FAYARD D. AND G. PLATEAU (1982) An Algorithm for the Solution of the 0-1 Knapsack Problem. *Computing* **28**, 269–287.

FISHER M. L. (1981) The Lagrangean Relaxation Method for Solving Integer Programming Problems. *Management Science* **27**, 1–18.

FONG C. D. AND V. SRINIVASAN (1981a) The Multiregional Capacity Expansion Problem, Part I. *Operations Research* **29**, 787–799.

FONG C. D. AND V. SRINIVASAN (1981b) The Multiregional Capacity Expansion Problem, Part II. *Operations Research* **29**, 800–816.

FONG C. D. AND V. SRINIVASAN (1986) The Multiregional Dynamic Capacity Expansion Problem: An Improved Heuristic. *Management Science* **32**, 1140–1163.

FREIDENFELDS J. (1980) Capacity Expansion when Demand Is a Birth-Death Process. *Operations Research* **28**, 712–721.

FREIDENFELDS J. (1981) *Capacity Expansion, Analysis of Simple Models with Applications.* Elsevier, New York.

FREIDENFELDS J. AND C. D. MCLAUGHLIN (1979) A Heuristic Branch-and-Bound Algorithm for Telephone Feeder Capacity Expansion. *Operations Research* **27**, 567–582.

GEOFFRION A. M. (1974) Lagrangean Relaxation for Integer Programming. *Mathematical Programming Study* 2, **82**–113.

HELD M., P. WOLFE, AND H. D. CROWDER (1974) Validation of Subgradient Optimization. *Mathematical Programming* **6**, 62–88.

LUSS H. (1982) Operations Research and Capacity Expansion Problems: A Survey. *Operations Research* **30**, 907–947.

MANNE A. S. (1961) Capacity Expansion and Probablistic Growth. *Econometrica* **29**, 632–649.

MUCKSTADT J. A. AND S. A. KOENIG (1977) An Application of Lagrangean Relaxation to Scheduling in Power Generation Systems. *Operations Research* **25**, 387–403.

NEEBE A. W. AND M. R. RAO (1983) The Discrete-Time Sequencing Expansion Problem. *Operations Research* **31**, 546–558.

NEEBE A. W. AND M. R. RAO (1986) Sequencing Capacity Expansion Projects in Continuous Time. *Management Science* **32**, 1467–1479.

SHAPIRO J. F. (1979) A Survey of Lagrangean Techniques for Discrete Optimization. *Annals of Discrete Mathematics* **5**, 113–138.

VEINOTT JR. A. F. (1969) Minimum Concave Cost Solution of Leontief Substitution Models of Multifacility Inventory Systems. *Operations Research* **17**, 262–291.

WAGNER H. M. AND T. M. WHITIN (1958) Dynamic Version of the Economic Lot Size Model. *Management Science* **5**, 89–96.

# 4

# Computer-Aided and Integrated Manufacturing Systems

Hui-Ming Wee

*Chung Yuan Christian University*

Three fundamental points regarding the differences in the manufacturing environment are made by Bertrand et al. (1990). These are

1. Different manufacturing situations require different control systems.
2. Different control systems require different information systems.
3. The production management system is the key to the competitive manufacturing operation, and its choice may influence success or failure.

The differences in the market conditions have an important bearing on which techniques in pricing and replenishment policy for inventories are to be used. The major objective of the chapter is to develop different inventory techniques that can be applied to minimize the total operation costs under different market conditions.

In this analysis, inventory control is one of the strategic factors in cost reduction. Inventory is defined by Love (1979) as: "A quantity of goods or materials in an enterprise and held for a time in a relative idle or unproductive state, awaiting its intended use or sale."

In a manufacturing system, inventory can be divided into three distinct classes. They are the raw materials, the work-in-process (WIP), and the finished goods inventory. The two key questions in inventory management for manufacturing systems are (i) When and how much should be the replenishment order of raw material? and (ii) What is the pricing policy of the finished goods?

The reasons for holding inventories are those related to economies of scale in manufacturing replenishment, to fluctuating requirements over time, to a desire for flexibility in scheduling sequential facilities, to speculation on price or cost, and to uncertainty about requirements and replenishment lead times. Almost all inventory techniques seek to find the optimum replenishment time and lot size. The objectives of the pricing policy are profit maximization and product competitiveness.

The costs considered during inventory management decisions are item cost, inventory carrying costs, ordering costs, shortage costs, and deteriorating costs.

*Item Cost:* The unit cost of the purchased item.

*Inventory Carrying Costs:* These costs include all expenses incurred by the firm for carrying the item; it includes capital costs and storage space costs.

*Ordering Costs:* Ordering costs are costs associated with placing an order.

*Shortage Costs:* Shortage costs include costs of special order and loss of goodwill when supply is less than demand.

*Deteriorating Costs:* They include costs incurred when products undergo decay, damage, obsolescence, or dissipation during storage.

In this chapter, we illustrate three replenishment techniques under different environments. Section 4.1 discusses the replenishment technique for product with increasing demand rate and decreasing cost structures. Section 4.2 discusses an inventory strategy when there is a temporary discount in the sale price. The third section develops a production lot size model for deteriorating inventory with time-varying demand.

# 4.1   Inventory Replenishment Model with Increasing Demand Rate and Decreasing Cost Structure

Classical economic order quantity technique assumes a constant demand rate and the existence of constant cost parameters. The advent of new technology is assumed to result in the emergence of a more superior product which has a better function and at the same time its cost decreases with time. As a result, demand for the new product will increase, and in the long run will substitute the existing product. In this study, a simple procedure is developed to determine the economic order policy for a new product that is replacing the existing one due to its lower cost and better functions. A numerical example is presented to illustrate the theory.

## Introduction

This study formulates an inventory replenishment policy for a product with an increasing demand rate and a decreasing cost. The change in demand rates and cost of an item are usually the result of an advent in new technology. Superior technology-fostered product is assumed to have an increasing demand rate (McLoughdin et al. 1985). Empirically the total market share of a superior technology-fostered product usually increases in the form of a logistic curve or an S-shaped curve (Leporelli and Lucertini 1980). The new product usually has a high initial price, but it will decrease after the initial launch period. Some of the products that behave in this manner are personal computers, synthetic products and other fashion goods. The model developed assumes a fixed batch size due to shipment constraints and is solved by means of finite difference method. Numerous inventory models dealing with demand variations (Silver and Meal 1969, Buchanan 1980, Mitra et al. 1984) and price changes (Buzacott 1975, Misra 1979, Aggarwal 1981, Kingsman 1986) are developed in the last two decades. Budhbhatti and Jani (1988) developed a model to discuss the influence of the mark-up prices due to damaged goods. They assumed a constant demand rate. This study considers the influence of increasing demand rate and decreasing cost simultaneously. It can be modified for decreasing demand rate and increasing cost structures. A numerical example and sensitivity analysis are carried out for the problem with finite planning horizon.

## Proposed Method

The demand rate $R(t)$ at time $t$ is assumed to be a non-linear function of time and technology advancement. Empirically it is represented as

$$R(t) = A\left(\frac{t}{\alpha} + B\right)^b \tag{4.1}$$

where $A(B)^b$ is the demand rate at the initial time ($t = 0$), and $b$ is the trend index value related to the advent of new technology, $B$ is a constant adjusting the variation in demand rate, and $\alpha$ is a constant influencing the speed of substitution by new technology. With a given planning horizon $H$, and demand rate, the total demand $D$ is

$$D = \int_0^H A\left(\frac{t}{\alpha} + B\right)^b dt$$

$$= \frac{A\alpha}{1+b}\left[\left(\frac{H}{\alpha} + B\right)^{1+b} - B^{1+b}\right]. \tag{4.2}$$

The number of replenishment during the planning horizon is

$$D = \frac{D}{Q} \tag{4.3}$$

where $Q$ is the batch size.

When no shortage is allowed, the batch size is equal to the demand in each period. The inventory system (Figure 4.1) shows a fixed lot size ordering policy for the case of increasing demand rate. The order quantity of each period is

$$Q = \int_{t_{i-1}}^{t_i} A\left(\frac{t}{\alpha} + B\right)^b dt$$

$$= \frac{A\alpha}{1+b}\left\{\left[\frac{t_i}{\alpha} + B\right]^{1+b} - \left[\frac{t_{i-1}}{\alpha} + B\right]^{1+b}\right\}. \tag{4.4}$$

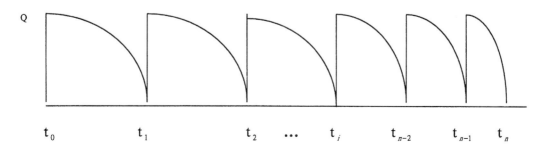

**FIGURE 4.1**    Fixed lot size inventory level for increasing demand rate.

From equations (4.2) through (4.4) and $D = nQ$, one has

$$\frac{A\alpha}{1+b}\left[\left(\frac{H}{\alpha}+B\right)^{1+b} - B^{1+b}\right] = n\frac{A\alpha}{1+b}\left[\left(\frac{t_i}{\alpha}+B\right)^{1+b} - \left(\frac{t_{i-1}}{\alpha}+B\right)^{1+\bar{b}}\right]$$

$$t_i = \alpha\left\{\frac{1}{n}\left[\left(\frac{H}{\alpha}+B\right)^{1+b} - B^{1+b}\right] + \left(\frac{t_{i-1}}{\alpha}+B\right)^{1+b}\right\}^{\frac{1}{1+b}} - \alpha B \qquad (4.5)$$

where $i = 5, \ldots, n$

With $t_0 = 0$, and $t_n = H$, the time for each replenishment can be derived. The change in inventory level with respect to time is a function of demand rate, one has

$$\frac{dI(t)}{dt} = -A\left(\frac{t}{\alpha}+B\right)^b \qquad (4.6)$$

where $t_{i-1} \le t \le t_i, i = 1, \ldots, n$.

When no shortage is allowed, $I(t_i) = 0$, the inventory at time t is

$$I(t) = \int_{t_i}^{t} -A\left(\frac{u}{\alpha}+B\right)^b du$$

$$I(t) = \frac{A\alpha}{1+b}\left\{\left(\frac{t_i}{\alpha}+B\right)^{1+b} - \left(\frac{t}{\alpha}+B\right)^{1+b}\right\}. \qquad (4.7)$$

Total amount of inventory during $i^{\text{th}}$ cycle is

$$I_i = \int_{t_{i-1}}^{t_i} I(t)dt$$

$$= \frac{A\alpha}{1+b}\left(\frac{t_i}{\alpha}+B\right)^{1+b}(t_i - t_{i-1}) + \frac{A\alpha^2}{(1+b)(2+b)}$$

$$+ \frac{A\alpha^2}{(1+b)(2+b)}\left[\left(\frac{t_{i-1}}{\alpha}+B\right)^{2+b} - \left(\frac{t_i}{\alpha}+B\right)^{2+\bar{b}}\right]. \qquad (4.8)$$

Average inventory during H is

$$I_n(n) = \frac{1}{H}\sum_{i=1}^{n} I_i$$

$$= \frac{1}{H}\left\{\frac{A\alpha}{1+b}\sum_{i=1}^{n}\left(\frac{t_i}{\alpha}+B\right)^{1+b}(t_i - t_{i-1}) + \frac{A\alpha^2}{(1+b)(2+b)}\right.$$

$$\left.\left[B^{2+b} - \left(\frac{H}{\alpha}+B\right)^{2+b}\right]\right\}\left[B^{2+b} - \left(\frac{H}{\alpha}+B\right)^{2+b}\right]. \qquad (4.9)$$

Assuming item cost decreases linearly with time due to technology advancement, the cost of the item during $i^{th}$ cycle can empirically be represented as

$$P_i = \left(1 - k\frac{t_{i-1}}{H}\right)P_0 \tag{4.10}$$

where $P_0$ is the initial cost of item and $k$ is the rate of technology advancement.
Total cost during the planning horizon is

$$\text{Total cost} = \text{replenishment cost} + \text{item cost} + \text{inventory carrying cost}$$

$$TC(n) = Sn + \sum_{i=1}^{n} D_i P_i + \sum_{i=1}^{n} I_i P_i h \tag{4.11}$$

where $S$ is the cost per replenishment and $h$ is the percentage unit cost for carrying inventory, and

$$D_i = \frac{A\alpha}{1+b}\left(\frac{t_i}{\alpha} + B\right)^{1+b} - \frac{A\alpha}{1+b}\left(\frac{t_{i-1}}{\alpha} + B\right)^{1+b}.$$

Since the number of replenishment must be an integer, the optimal number of replenishment must satisfy the following condition:

$$\Delta TC(n^* - 1) \leq 0 \leq \Delta TC(n^*)$$
$$\text{where} \quad \Delta TC(n^*) = TC(n^* + 1) - TC(n^*). \tag{4.12}$$

When the demand rate is assumed constant, $\alpha$ tends to infinity. When there is no change in demand trend, $b$ and $B$ values vanish. Equation (4.5) becomes

$$t_i - t_{i-1} = \frac{H}{n}. \tag{4.13}$$

Substituting (4.13) into (4.14) and $t_i = \frac{H}{n}i$, one has

$$I_a = \frac{AH}{2n}. \tag{4.14}$$

The result is identical to that of the periodic replenishment policy.

## Numerical Example

The proposed method is illustrated with a numerical example where the demand rate is a function of time and technology advancement. Empirically it is represented as

$$R(t) = A\left(\frac{t}{\alpha} + B\right)^{b}$$

where

| | |
|---|---|
| $A = 100$ | $B = 0.05$ |
| $b = 0.5$ | $h = 15\%$ |
| $\alpha = 0.98$ | $k = 0.8$ |
| $h = \$300$ | $H = 4$ years |
| $P_0 = \$200$ | |

Using (4.10) to (4.12), the values of $\Delta TC(n)$ and $TC(n)$ are analyzed for different $n$ values. The results are tabulated in Table 4.1 Sensitivity analyses for different values of $H$, $\alpha$, $k$, and $b$ are shown in Tables 4.2 through 4.6.

*Sensitivity analysis:*

**TABLE 4.1**   Optimal Number of Replenishment

| $H = 4$ | $\alpha = 0.98 \quad b = 0.5$ | $K = 0.9$ |
|---|---|---|
| $n$ | $\Delta TC(n)$ | $TC(n)$ |
| 1 | $-50{,}917.5$ | 149,072.1 |
| 2 | $-14{,}551.6$ | 98,154.6 |
| 3 | $-6{,}640.6$ | 83,603.0 |
| 4 | $-3{,}693.9$ | 76,962.4 |
| 5 | $-2{,}286.5$ | 73,268.5 |
| 6 | $-1{,}508.2$ | 70,981.9 |
| 7 | $-1{,}033.6$ | 69,473.8 |
| 8 | $-723.4$ | 68,440.1 |
| 9 | $-509.7$ | 67,716.7 |
| 10 | $-356.3$ | 67,207.1 |
| 11 | $-242.6$ | 66,850.7 |
| 12 | $-155.9$ | 66,608.2 |
| 13 | $-88.4$ | 66,452.2 |
| 14 | $-34.9$ | 66,363.8 |
| 15* | 8.4 | 66,328.9* |
| 16 | 43.8 | 66,337.4 |
| 17 | 73.2 | 66,381.2 |
| 18 | 97.8 | 66,454.4 |
| 19 | 118.6 | 66,552.1 |
| 20 | 136.3 | 66,670.7 |
| 21 | 151.6 | 66,807.0 |
| 22 | 164.9 | 66,958.7 |
| 23 | 176.4 | 67,123.5 |
| 24 | 186.6 | 67,299.9 |
| 25 | 195.5 | 67,486.5 |
| 26 | 203.4 | 67,682.0 |
| 27 | 210.5 | 67,885.4 |
| 28 | 216.8 | 68,095.9 |
| 29 | 222.5 | 68,312.7 |
| 30 | 227.6 | 68,535.2 |

**TABLE 4.2**   Cost Comparison of Optimal Policies for Various Values of $k$

When $H = 4 \quad \alpha = 0.98 \quad b = 0.5$,
the total number of unit ordered during $H = 548.0$

| $k$ | $n^*$ Number of Replenishment | $Q^*$ Fixed Order Quantity | $TC$ Total Cost ($) | Unit Average Cost ($) |
|---|---|---|---|---|
| 0 | 11 | 49.8 | 116,011.9 | 211.7 |
| 0.05 | 11 | 49.8 | 112,939.4 | 206.1 |
| 0.5 | 14 | 39.1 | 85,062.3 | 155.2 |
| 0.8 | 15 | 36.5 | 66,328.9 | 121.0 |

**TABLE 4.3**    Cost Comparison of Optimal Policies for Various Values of $k$

When $H = 4$  $\alpha = 0.04$  $b = 0.5$,
the total number of unit ordered during $H = 2668.6$.

| $k$ | $n^*$ Number of Replenishment | $n^*$ Fixed Order Quantity | $TC$ Total Cost (\$) | Unit Average Cost (\$) |
|---|---|---|---|---|
| 0 | 24 | 111.2 | 547,813.8 | 205.3 |
| 0.05 | 24 | 112.2 | 532,221.8 | 199.4 |
| 0.5 | 30 | 89.0 | 391,433.7 | 146.7 |
| 0.8 | 33 | 80.9 | 297,263.8 | 11.4 |

**TABLE 4.4**    Cost Comparison of Optimal Policies for Various Values of $k$

When $H = 4$  $\alpha = 0.08$  $b = 0.5$,
the total number of unit ordered during $H = 1884$

| $k$ | $n^*$ Number of Replenishment | $Q^*$ Fixed Order Quantity | $TC$ Total Cost (\$) | Unit Average Cost (\$) |
|---|---|---|---|---|
| 0 | 20 | 94.4 | 389,545.0 | 206.3 |
| 0.05 | 21 | 89.0 | 378,573.9 | 200.5 |
| 0.5 | 25 | 75.5 | 279,428.9 | 148.0 |
| 0.8 | 28 | 67.4 | 213,033.0 | 112.8 |

**TABLE 4.5**    Cost Comparison of Optimal Policies for Various Values of $k$

When $H = 8$  $\alpha = 0.08$  $b = 0.5$,
the total number of unit ordered during $H = 5337.3$

| $k$ | $n^*$ Number of Replenishment | $Q^*$ Fixed Order Quantity | $TC$ Total Cost (\$) | Unit Average Cost (\$) |
|---|---|---|---|---|
| 0 | 47 | 113.6 | 1,095,468.0 | 205.2 |
| 0.05 | 47 | 113.6 | 106,302.0 | 199.3 |
| 0.5 | 51 | 104.7 | 777,695.3 | 145.7 |
| 0.8 | 53 | 100.7 | 586,940.6 | 110.0 |

**TABLE 4.6**    Cost Comparison of Optimal Policies for Various Values of $k$

When $H = 4$  $\alpha = 0.98$  $b = 1.5$,
the total number of unit ordered during $H = 1360.1$

| $k$ | $n^*$ Number of Replenishment | $Q^*$ Fixed Order Quantity | $TC$ Total Cost (\$) | Unit Average Cost (\$) |
|---|---|---|---|---|
| 0 | 18 | 75.6 | 282,603.6 | 207.8 |
| 0.05 | 19 | 71.6 | 273,261.4 | 200.9 |
| 0.5 | 23 | 59.1 | 188,781.3 | 138.8 |
| 0.8 | 26 | 52.3 | 132,205.5 | 97.2 |

## Summary

From Table 4.1, it can be seen that (4.12) is satisfied with the total cost of \$66,328.90 when $n = 15$. Tables 4.2 to 4.6 compare the optimal policies under different rate of technology advancement. From Table 4.2, one can see that the optimal number of replenishment is 11 when $k$ value is zero. If this replenishment value is chosen for the case in the example, the total cost derived from Table 4.1 would have been \$66,850.70. The extra cost of \$521.80 (\$66,850.70 − \$66,328.90) is incurred when the rate of technology advancement of 0.8 is ignored. This shows the importance of considering the influence of technology advancement in determining the replenishment policy of a technology-fostered new product. Section 3 will consider the effect of the varying lot size and item deterioration in a production environment.

## 4.2 Deteriorating Inventory with a Temporary Price Discount Purchase

Inventory management has been widely studied in recent years. Most of the researchers assume that the price of the inventory items maintain their original value over the planning horizon. This may not be true in practice as suppliers may wish to induce greater sales by offering temporary price discounts. This is very common for items that deteriorate with time. In this section, a replenishment policy for an exponentially deteriorating item with a temporary price discount is developed. The objective of this study is to maximize the total cost savings during a temporary price discount order period. A numerical example is provided at the end of the section to illustrate the theory.

### Introduction

The classical inventory model assumes that the condition of the inventory items is unaffected by time, and the replenishment size is constant. However, it is not always true, since some items deteriorate and a special discount order may be placed to take advantage of a temporary price discount. Deterioration is the loss of potential or utility of the storage items with the passage of time. Such deterioration can be the spoilage of fruits and vegetables, the evaporation of volatile liquids such as gasoline or alcohol, or the decrease in function of radioactive substances. For example: electronic products may become obsolete as technology changes; the value of clothing tends to depreciate over time due to changes in season and fashion; batteries lose their charge as they age, and so on.

When there is a temporary price discount, a company may purchase a larger quantity of items to take advantage of the lower price. A question naturally arises as to the optimal quantity to be purchased in the situation. In this study, item deterioration is assumed to follow the exponential distribution; that is, the amount of storage item deteriorated during a given time is a constant fraction of the existing inventory.

### Background

The inventory model developed in this study incorporates two effects: (1) temporary price discount, and (2) deterioration of items. Notice that this study investigates a situation in which both effects are coexistent.

#### Temporary Price Discount

A supplier may temporarily discount the unit price of an item for a limited length of time. When the special sale occurs during a replenishment cycle, it will be beneficial for the buyer to purchase more than the regular amount. The objective of this study is to develop a model to determine the optimal purchase quantity. Tersine (1994) has developed an inventory model which incorporates the effect of a temporary price discount. Martin (1994) later discovered an oversight in Tersine for the average inventory during the special discount purchase. Both of these two authors do not consider the effects of deterioration.

#### Deteriorating Effect

Ghare and Schrader (1963) are the first researchers to consider the effect of decay on inventory items. They derived an economic order quantity model (EOQ) where inventory items deteriorate exponentially with time. Covert and Philip (1973) extended the model to consider Weibull distribution deterioration. The model was further generalized for a general distribution deterioration by Shah (1977). Hwang and Sohn (1982) developed a model for the management of deteriorating inventory under inflation with known price increases. None of these authors has considered a temporary price discount purchase.

## Model Development and Analysis

The mathematical model presented in this section has the following assumptions:

1. Demand rate is known, constant, and continuous.
2. Stock-out is not permitted.
3. Replenishment rate is infinite.
4. Deteriorated items are not replaced during cycle time.
5. Lead time is zero.
6. A constant fraction inventory deteriorates per unit time.
7. There is no constraint in space, capacity, and capital.
8. Only a single product item is considered.
9. The unit price decreased, $d$ is less than the unit purchase price, $P$.
10. Only one replenishment can be made during the temporary price discount sale period, and this occurs at the regular replenishment cycle.
11. When there is no temporary price discount purchase, regular price EOQ value is retained.

The last two assumptions are added to eliminate the oversight suggested by Martin (1994), and to strengthen the representation in Tersine (1994).

The following notation is used in this chapter:

$P$ = unit purchase cost before the discount;
$d$ = unit price decrease;
$C$ = cost per order;
$F$ = annual holding cost fraction;
$R$ = annual demand;
$\theta$ = deterioration rate;
$g$ = saving cost amount;
$T^*$ = optimal order interval for regular replenishment cycle;
$\hat{T}$ = optimal order interval for temporary price discount purchase cycle;
$TC_s$ = total cost for the temporary price discount purchase;
$TC_n$ = total cost for regular price ordering policy;
$\hat{Q}$ = optimal order quantity for temporary price discount purchase; and
$Q_0^*$ = optimal order quantity for regular price purchase;

The objective is to derive the optimal temporary price discount order quantity in order to maximize the cost saving during the cycle time $\hat{T}$. Assume that an order has been placed for an item at the temporarily reduced price of $(P-d)$ per unit, where $P$ is the per unit regular price and $d$ is the per unit price discount. The inventory system with exponential deterioration illustrated Figure 4.2 is described as follows:

The change in inventory level, $dQ$, during an infinitesimal time, $dt$, is a function of the deterioration rate $\theta$, demand rate $R$ and $Q(t)$, the inventory level at time $t$. It can be formulated as

Decrease in inventory level = deterioration + demand

$$-dQ(t) = \theta Q(t)dt + Rdt. \tag{4.15}$$

The solution of (1) after adjusting for constant of integration is

$$Q(t) = \frac{R}{\theta}[\exp(\theta(T-t)) - 1] \qquad 0 \le t \le T. \tag{4.16}$$

The initial inventory at $t = 0$ is

$$Q_0 = \frac{R}{\theta}[\exp(\theta T) - 1]. \tag{4.17}$$

Inventory

Level

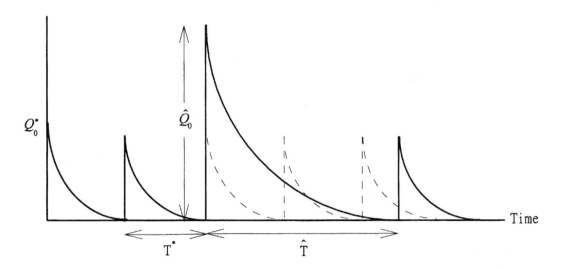

**FIGURE 4.2**   Inventory system for finite replenishment rate and decreasing demand.

For regular price purchase, the economic order quantity is

$$Q_0^* = \frac{R}{\theta}[\exp(\theta T^*) - 1].\tag{4.18}$$

The total cost, *TC*, for a regular price order interval $T^*$ consists of the item cost, the holding cost and the ordering cost. It can be formulated as:

$$
\begin{aligned}
TC &= PQ_0^* + FP\int_0^{T^*} Q(t)\,dt + C\\
&= \frac{RP}{\theta}(\exp(\theta T^*) - 1) + \frac{FPR}{\theta^2}(\exp(\theta T^*) - \theta T^* - 1) + C.
\end{aligned}\tag{4.19}
$$

Introducing Taylor's quadratic approximation $\exp(\theta T^*) \approx 1 + \theta T^* + (\theta T^*)^2/2$ for $\theta \ll 1$; setting the first derivative with respect to $T^*$ to zero, one has

$$T^* = \sqrt{\frac{2C}{RP(\theta + F)}}.$$

For the case of a temporary price discount order, (4.2) and (4.3), are modified as:

$$\hat{Q}(t) = \frac{R}{\theta}[\exp(\theta(\hat{T} - t)) - 1]\qquad 0 \le t \le \hat{T}\tag{4.20}$$

$$\hat{Q}_0 = \frac{R}{\theta}[\exp(\theta\hat{T}) - 1].\tag{4.21}$$

The total cost for a temporary price discount order cycle $\hat{T}$ consists of the item cost, the holding cost and the ordering cost. It can be formulated as:

$$
\begin{aligned}
TC_s &= (P - d)\hat{Q}_0 + (P - d)F\int_0^{\hat{T}} \hat{Q}(t)dt + C \\
&= \frac{R(P - d)}{\theta}(\exp(\theta\hat{T}) - 1) + \frac{F(P - d)R}{\theta^2}(\exp(\theta\hat{T}) - \theta\hat{T} - 1) + C.
\end{aligned}
\tag{4.22}
$$

Introducing Taylor's quadratic approximation $\exp(\theta\hat{T}) \approx 1 + \theta\hat{T} + (\theta\hat{T})^2/2$, one has

$$
TC_s = R(P - d)\left(\hat{T} + \frac{\theta\hat{T}^2}{2}\right) + \frac{(P - d)FR\hat{T}^2}{2} + C.
\tag{4.23}
$$

If no temporary price discount order is placed during $\hat{T}$, the total cost when the first order is made at $(P - d)$ per unit and all subsequent orders are made at $P$ per unit is as follows:

$$
\begin{aligned}
TC_n &= (P - d)Q_0^* + P(\hat{Q}_0 - Q_0^*) + (P - d)F\int_0^{\hat{T}} Q^*(t)dt + PF\int_0^{\hat{T}} Q^*(t)dt\left(\frac{\hat{Q}_0 - Q_0^*}{Q_0^*}\right) + C\frac{\hat{Q}_0}{Q_0^*} \\
&= (P - d)Q_0^* + P(\hat{Q}_0 - Q_0^*) + \frac{FR(P - d)}{\theta^2}(\exp(\theta T^*) - \theta T^* - 1) \\
&\quad + \frac{PRF}{\theta^2}\left(\frac{\hat{Q}_0 - Q_0^*}{Q_0^*}\right)(\exp(\theta T^*) - \theta T^* - 1) + C\frac{\hat{Q}_0}{Q_0^*}.
\end{aligned}
\tag{4.24}
$$

Substituting (4.4) and (4.8) into (4.11), one has

$$
\begin{aligned}
TC_n &= (P - d)\frac{R}{\theta}(\exp\theta T^* - 1) + P\left(\frac{R}{\theta}(\exp\theta T - \exp\theta T^*)\right) + \frac{(P - d)FR}{\theta^2}(\exp\theta T^* - \theta T^* - 1) \\
&\quad + \frac{PRF}{\theta^2}\left(\frac{(\exp\theta T - \exp\theta T^*)}{\exp(\theta T^*) - 1}\right)(\exp\theta T^* - \theta T^* - 1) + C\frac{\exp\theta T - 1}{\exp\theta T^* - 1}.
\end{aligned}
\tag{4.25}
$$

Introducing Taylor's quadratic approximation, $\exp(\theta T^*) \approx 1 + \theta T^* + (\theta T^*)^2/2$, and $\exp(\theta T^*) \approx 1 + \theta\hat{T} + (\theta\hat{T})^2/2$, one has

$$
\begin{aligned}
TC_n &= R(P - d)\left(T^* + \frac{\theta T^{*2}}{2}\right) + PR\left(T + \theta T^2 - T^* - \frac{\theta T^{*2}}{2}\right) + \frac{(P - d)FRT^{*2}}{2} \\
&\quad + \frac{PRFT^{*2}}{2}\left(\frac{T + \theta T^2/2 - T^* - \theta T^{*2}/2}{T^* + \theta T^{*2}/2}\right) + C\left(\frac{T + \theta T^2/2}{T^* + \theta T^{*2}/2}\right).
\end{aligned}
\tag{4.26}
$$

The cost savings due to a temporary price discount order is expressed by

$$
g = TC_n - TC_s
\tag{4.27}
$$

where $TC_n$ is from (4.13), and $TC_s$ is from (4.10). Setting the first derivative of $g$ with respect to $\hat{T}$ to zero, and solving for $\hat{T}$, one has

$$\hat{T} = \frac{\frac{R}{2}(P\theta - 2d - d\theta) - \frac{R}{Q_0^*}\left(\frac{PRFT^{*2}}{2} + C\right)}{R\theta\left(P + \frac{PRFT^{*2}}{2Q_0^*} + \frac{C}{Q_0^*} - \frac{F(P-d)}{\theta}\right)}. \qquad (4.28)$$

The second derivative of $g$ with respect to $\hat{T}$ is

$$\frac{d^2g}{d\hat{T}^2} = PR\theta + \frac{PRF\theta T^{*2}}{2T^* + \theta T^{*2}} + \frac{C\theta}{T^* + \theta T^{*2}/2} - (P-d)FR.$$

For $\theta \ll 1$ one can verify that the second order derivative of $g$ is negative since $d$ is less than $P$. Hence global maximum is assured.

If $\theta$ is zero, (4.15) is reduced to

$$\hat{T} = \frac{\dfrac{PRFT^{*2} + 2C}{2T^*} + Rd}{FR(P-d)} \qquad (4.29)$$

which is identical to the result obtained by Tersine as can be seen from the first row of Table 4.1.

## Numerical Example

The preceding theory can be illustrated by the following numerical example.

The parameters from Tersine are:

$P = \$10.00/\text{unit}$
$d = \$1.00/\text{unit}$
$C = \$30.00/\text{order}$
$R = 8000 \text{ units/year}$
$F = 0.30/\text{year}$

The deterioration of an item is arbitrarily assumed to follow the exponential distribution with a deteriorating rate $\theta$. The replenishment period before and after a temporary price discount order is derived from (6); the order quantity for different deterioration rates is derived from (4.4). The optimal values for $\hat{T}$ and $\hat{Q}_0$ can then be found from (4.15) and (4.8), respectively. The results are tabulated in Table 4.1.

**TABLE 4.7**    Numerical Values for Different Rates of Deterioration

| $\theta$ | $T^*$ (year) | $Q_0^*$ (units) | $T$ (year) | $\hat{Q}_0$ (units) | $g^*$ (\$) |
|---|---|---|---|---|---|
| 0 | 0.0500 | 400.0 | 0.4259 | 3407.2 | 1526.26 |
| 0.001 | 0.0499 | 399.2 | 0.4258 | 3407.1 | 1518.27 |
| 0.005 | 0.0496 | 396.8 | 0.4255 | 3408.6 | 1489.57 |
| 0.01 | 0.0492 | 392.9 | 0.4252 | 3408.8 | 1453.75 |
| 0.05 | 0.0463 | 370.8 | 0.4220 | 3411.6 | 1158.53 |

## Summary

An inventory model with a temporary price discount purchase has been developed for items that deteriorate continuously with time. Besides eliminating the oversight suggested by Martin (1994), the developed model is a more general one since it considers both Tersine's model (1994) for a temporary price discount purchase policy and Ghare and Schrader's model (1963) for exponentially decaying inventory.

An optimal temporary price discount order quantity, $\hat{Q}_0$, and the estimated optimal gain, $g^*$, are found for different rates of deterioration. When $\theta$ equals zero, the present researchers' model conforms to the model by Tersine without the deflated error suggested by Martin. When no price discount is considered, the model conforms to Ghare and Schrader. As the deterioration rate increases, the quantity order as well as the temporary price discount purchase gain tends to decrease. This is a logical result since for a higher deterioration rate, it is not advisable to keep a large stock, and the gain in the temporary price discount purchase is reduced due to item deterioration.

# 4.3 A Production Lot Size Model for Deteriorating Items with Time-Varying Demand

This section develops a production lot size model for deteriorating items with time-varying demand. Unlike most of the current research, both the replenishment cycle and the lot size in this model are variables. The classical optimization technique with the assistance of the heuristic procedures is implemented to derive at an optimal solution. The numerical example shows that this model has a significant cost saving when compared with the results of the existing literature.

## Introduction

With the rapid changes in technologies and market conditions since the industrial revolution, the demand and the value of consumer goods varies drastically with time. When there is an improvement in technologies and functions of a product, the demand will increase; when a product is being substituted by another product, its demand will decrease. For superior technology-fostered products such as semiconductors, its cost will usually decrease proportionally to the time in inventory. This decrease in cost is referred to as the obsolescence of the item during storage, very similar to the deterioration of goods such as vegetables and fruits while in storage.

Substantial research on deteriorating items has been done since Ghare and Schrader (1963) first considered the deteriorating effect of inventory in their model analysis. Misra (1975), Choudhury and Chaudhuri (1983) and Deb and Chaudhuri (1986) relaxed the instantaneous assumption to develop an optimal finite replenishment lot size model for a system with deteriorating inventory. Most of these studies (Nahmias 1978, Raafat 1991) assumed a constant demand rate. Dave and Patel (1981) and Hong et al. (1990) later considered the inventory of deteriorating items with time proportional demand. Their model assumed a finite planning horizon and a constant replenishment cycle for a no-shortage environment. Sachan (1984), Goswami and Chaudhuri (1991a, 1991b), and Wee (1995a) extended Dave and Patel's model to allow for shortages. Hollier and Mak (1983), Bahari-Kashani (1989), and Hariga and Benkherouf (1994) relaxed the fixed replenishment cycle assumption in Dave and Patel. The paper by Goswami and Chaudhuri (1991a) was recently modified by Hariga (1995) to rectify the flaws in its mathematical formulation. Xu and Wang (1992) implemented the dynamic programming approach to derive an optimal solution when both no-shortage and fixed-replenishment assumptions are relaxed. Most of the preceding studies with time-proportional demand assumed instantaneous replenishment. In 1991, Aggarwal and Bahari-Kashani developed a flexible production rate model for a fixed replenishment cycle and declining market demand which do not allow for shortages. Wee (1995b) recently extended the model to consider pricing and partial backordering under constant replenishment cycle and instantaneous replenishment. In this section, a finite production rate with variable replenishment cycle is assumed.

By means of a classical optimization technique and a heuristic procedure, a methodology is developed. A numerical example from Hollier and Mak (1983) is adopted to compare the result of this study with the results of the previous studies.

## Mathematical Modeling and Analysis

The mathematical model in this section is developed using the following assumptions:

1. A single-item inventory with a constant rate of deterioration is considered.
2. Deterioration occurs as soon as the items are received into inventory.
3. There is no replacement or repair of deteriorating items during the period under consideration.
4. The demand rate changes exponentially with time.
5. The production rate is constant; the production lot size and scheduling cycle are not constant.
6. The system operates for a prescribed period of planning horizon.
7. No shortage is allowed.
8. The order quantity, inventory level, and demand are treated as continuous variables, while the number of replenishments is treated as discrete variable.

The following notation is used:

$H$: the planning horizon;
$m$: the number of production cycles;
$P$: the production rate;
$q$: the inventory deterioration rate, $0 < \theta < 0.5$;
$c$: the unit cost in \$;
$c_1$: the inventory holding cost in \$/unit/time unit;
$c_s$: the production cost in \$/setup;
$Q(t)$: the inventory level at time $t$, $0 \leq t \leq H$;
$T_i$: the starting time for each production cycle where $i = 0, 1, \ldots, m$ (note that $T_0 = 0$ and $T_m = H$);
$x_i$: the length of each period where $x_i = T_i - T_{i-1}$ and $i = 1, 2, \ldots, m$ ($T_i > T_{i-1}$ for increasing demand rate);
$t_i$: the point of time in each cycle when production terminates, $i = 1, 2, \ldots, m$;
$TVC(m, T_i)$: the total variable cost;
$R(t)$: demand rate (unit/time unit), a function of time where

$$R(t) = \begin{cases} a \, \exp(bt) & \text{if } 0 \leq t \leq H \\ 0 & \text{otherwise} \end{cases}$$

It is noted that $a$ and $b$ are known constants where $a \geq 0$ and $\exp(bt) \geq 0$. If $b < 0$, demand rate decreases with time; if $b > 0$, demand rate increases with time.

## Model Development

Using the assumptions in Section 4.2, the inventory system depicted in Figure 4.3 can be represented by the following differential equations at any time in the cycle:

$$\frac{d}{dt}Q_{1i}(t) + \theta Q_{1i}(t) = P - R(t), \qquad T_{i-1} \leq t \leq t_i; \tag{4.30}$$

$$\frac{d}{dt}Q_{1i}(t) + \theta Q_{1i}(t) = -R(t), \qquad t_i \leq t \leq T_i; \tag{4.31}$$

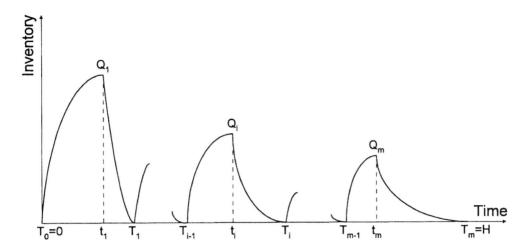

**FIGURE 4.3**    TVC values for different $T1$ values when $m = 2$.

The solutions to the differential equations (Spiegel 1960) are:

$$Q_{1i}(t) \ = \ \exp(-\theta t)\!\int_{T_{i-1}}^{t} \exp(\theta u)(P-R(u))du, \quad T_{i-1} \le t \le t_i;\tag{4.32}$$

$$Q_{2i}(t) \ = \ \exp(-\theta t)\!\int_{t}^{T_i} \exp(\theta u)R(u)du, \quad t_i \le t \le T_i;\tag{4.33}$$

From (4.32) and (4.33) the maximum inventory levels in the $i^{\text{th}}$ cycle occurring at $t = t_i$ are

$$Q_i = Q_{1i}(t_i) \ = \ \exp(-\theta t_i)\!\int_{T_{i-1}}^{t_i} \exp(\theta t)(P-R(t))dt$$

$$Q_i = Q_{2i}(t_i) \ = \ \exp(-\theta t_i)\!\int_{t_i}^{T_i} \exp(\theta t)R(t)dt;\tag{4.34}$$

From (4.34), after simplifying

$$t_i \ = \ \theta^{-1}\ell\mathrm{n}\!\left(\frac{\theta}{P}\!\int_{T_{i-1}}^{T_i} \exp(\theta t)R(t)dt + \exp(\theta T_{i-1})\right);\tag{4.35}$$

From (4.32) and (4.34), the $i^{\text{th}}$ cycle inventory is

$$I_i \ = \ \int_{T_{i-1}}^{t_i} \exp(-\theta t)\!\int_{T_{i-1}}^{t} \exp(\theta u)(P - R(t))dudt$$

$$+ \int_{t_i}^{T_i} \exp(-\theta t)\!\int_{t_i}^{T_i} \exp(\theta u)R(u)dudt;\tag{4.36}$$

Transforming (4.7) using partial integration, one has

$$I_i = \int_{T_{i-1}}^{t_i} -\theta^{-1}[\exp(\theta(t - t_i)) - 1](P - R(t))dt$$

$$+ \int_{t_i}^{T_i} \theta^{-1}[\exp(\theta(t - t_i)) - 1]R(t)dt; \qquad (4.37)$$

The production lot-size (from Figure 4.3) during the $i^{th}$ cycle is

$$L_i = P(t_i - T_{i-1}); \qquad (4.38)$$

Deterioration during the $i^{th}$ cycle is

$$D_i = \theta I_i$$

$$= L_i - \int_{T_{i-1}}^{T_i} R(t)dt; \qquad (4.39)$$

where $\int_{T_{i-1}}^{T_i} R(t)dt$ is the demand during the $i^{th}$ cycle

For a planning horizon with $m$ cycles, the total variable cost (TVC) consisting of the inventory carrying cost, the deterioration cost, and the setup cost is

$$TVC(m, T_i) = c_1 I_i + c\theta I_i + mc_s$$

$$= (c_1 + \theta c) \sum_{i=1}^{m} \left\{ \int_{T_{i-1}}^{t_i} -\theta^{-1}[\exp(\theta(t - t_i)) - 1](P - R(t))dt \right.$$

$$\left. + \int_{t_i}^{T_i} -\theta^{-1}[\exp(\theta(t - t_i)) - 1]R(t)dt \right\} + mc_s; \qquad (4.40)$$

where $m$ is a known integer and $t_i$ is given in (4.6).

Optimal solution to the problem can be found by taking the derivative of the total variable cost (*TVC*) with respect to $T_i$:

$$\frac{\partial}{\partial T_i} TVC(m, T_i) = 0, \qquad i = 1, 2, \dots, m - 1; \qquad (4.41)$$

from which one has

$$\{[(\exp(\theta(T_i - t_i)) - 1]R(T_i) + [\exp(\theta(T_i - t_{i+1})) - 1](P - R(T_i))\} = 0,$$

$$i = 1, 2, \dots, m - 1; \qquad (4.42)$$

From (4.6) and (4.13), after simplification, one has

$$[\exp(\theta(t_{i+1} - t_i)) - 1]R(T_i) = \theta\exp(-\theta T_i)\int_{T_i}^{T_{i+1}} \exp(\theta t)R(t)dt,$$

$$i = 1, 2, \dots, m - 1; \qquad (4.43)$$

When $P \to \infty$, $t_{i+1} = T_i$, (4.14) becomes

$$[\exp(\theta(T_i - T_{i-1})) - 1]R(T_i) = \theta\exp(-\theta T_i)\Big|_{T_i}^{T_{i+1}} \exp(\theta t)R(t)dt,$$

$$i = 1, 2, \dots, m - 1; \tag{4.44}$$

which is identical to the results obtained by Hariga and Benkherouf (1994).

The next stage of the analysis is to solve the above problem for all possible $m$ integer value until the following conditions are satisfied for a minimum $TVC(m^*, T_i^*)$:

$$\Delta TVC(m^* - 1, T_i^*) \le 0 \le \Delta TVC(m^*, T_i^*), \tag{4.45}$$

where

$$\Delta TVC(m^*, T_i^*) = TVC(m^* + 1, T_i^*) - TVC(m^*, T_i^*); \tag{4.46}$$

Global optimality can be proved by taking the second derivatives of the total variable cost function.

**A case study with known demand rate**

Substitute $R(t) = a\exp(bt)$ into (14) and simplify, one has

$$\left\{\begin{array}{l} \text{For the value } (\theta + b) \ne 0; \\[2mm] \exp(\theta(t_{i+1} - t_i)) = \dfrac{\theta}{\theta + b}\exp((\theta + b)(T_{i+1} - T_i)) + \dfrac{b}{\theta + b}, \\[2mm] \text{where } i = 1, 2, \dots, m - 1; \end{array}\right. \tag{4.47}$$

Substitute (4.18) into (4.6) and simplify, one has

$$t_i = \theta^{-1}\ell n\left\{\frac{\theta a}{P(\theta + b)}\exp((\theta + b)T_{i-1})[\exp((\theta + b)(T_i - T_{i-1})) - 1] + \exp(\theta T_{i-1})\right\}, \tag{4.48}$$

Equation (4.19) can be rewritten as

$$\exp(\theta t_i) - \exp(\theta T_{i-1}) = \frac{\theta a}{P(\theta + b)}[\exp((\theta + b)T_i) - \exp((\theta + b)T_{i-1})], \tag{4.49}$$

Substitute (4.20) into (4.8) and simplify, one has

$$\begin{aligned} I_i = \ &\theta^{-1}\{P([\exp(\theta(T_{i-1} - t_i)) - 1]\theta^{-1} + t_i - T_{i-1}) \\ &+ a(\theta + b)^{-1}\exp(bt_i)[1 - \exp((b + \theta)(T_{i-1} - t_i))] \\ &- ab^{-1}\exp(bt_i)[1 - \exp(b(T_{i-1} - t_i))]\} \\ &+ \theta^{-1}\{a(\theta + b)^{-1}\exp(bt_i)[\exp((\theta + b)(T_i - t_i)) - 1] \\ &- ab^{-1}\exp(bt_i)[\exp(b(T_i - t_i)) - 1]\}, \end{aligned} \tag{4.50}$$

## A Heuristic Approach to Derive the Optimal $m$

The following heuristic is developed to derive the optimal $TVC(m, T)$:

Step 1: Start with $m = 1$;
Step 2: Fix $T_{m-1} = x_m$ and substitute $T_m = H$ and $T_{m-1}$ into (4.19) to find $t_m$;
Step 3: Substitute $T_{i+1}$ and $T_i$ into (4.18) to find $t_i$; then substitute $T_i$ and $t_i$ into (4.20) to find $T_{i-1}$, where $i = m-1, m-2, \ldots, 1$;
Step 4: If $T_0 \doteq 0$, go to step 5; otherwise go to step 2;
Step 5: Calculate the $TVC(m, T)$ value;
Step 6: Continue for the next $m$ values, then go to step 2.

The value of $m$ starts at 1 and increases incrementally until $m^*$ satisfies (4.16); this $TVCm^*$ is then, the minimum total variable cost. The value of $x_m$ is found by trial and error, and equations (4.18) to (4.20) are solved by MAPLE V computer software.

## Numerical Example and Analysis

### Example 1

A numerical analysis is carried out using the data from Hollier and Mak (1983) where:

$a = 500.0,$
$b = -0.98,$
$\theta = 0.08,$
$H = 4.0 \text{(years)},$
$c = 200.0(\$/\text{unit}),$
$c_1 = 40.0(\$/\text{unit/year}), \text{ and}$
$c_s = 250.0(\$/\text{setup}).$

The production rate, $P$, is assumed to be 500.0 units/year.

Implementing the heuristic in the previous section, a program code is written to derive the optimal number of replenishment and its relevant values when the minimum total variable cost ($TVC$) is minimum. Table 4.8 shows the $TVC$ for different values of $m$.

From Table 4.8, one can see that the minimum $TVC(m, T)$ of \$4653.766 occurs at an optimal value of $m = 9$. The other relevant values for $m = 9$ production cycles are as shown in Table 4.8.

It is observed that the production time as well as the batch size decrease with time when the demand rate decreases negative-exponentially. The response will be reversed if the demand rate increases with time.

**TABLE 4.8**   Total Variable Cost
for Different Numbers of Cycles

| $m$ | $TVC(m, T^*)$ |
|---|---|
| 1 | 12,989.288 |
| 2 | 8,693.821 |
| 3 | 6,823.538 |
| 4 | 5,833.783 |
| 5 | 5,267.647 |
| 6 | 4,938.838 |
| 7 | 4,756.623 |
| 8 | 4,671.415 |
| 9* | 4,653.766* |
| 10 | 4,685.066 |
| 11 | 4,752.984 |
| 12 | 4,849.017 |
| 13 | 4,967.107 |

**TABLE 4.9**  Related Results for 9 Production Cycles

| I | $t_i - T_{i-1}$ | $T_i - t_i$ | $x_i$ | $t_i$ | $T_i$ | $Q_i$ | $L_i$ |
|---|---|---|---|---|---|---|---|
| 1 | 0.37983 | 0.09443 | 0.47426 | 0.38082 | 0.47525 | 31.166 | 189.913 |
| 2 | 0.16082 | 0.13260 | 0.29343 | 0.63607 | 0.76868 | 33.508 | 80.412 |
| 3 | 0.11917 | 0.16964 | 0.28881 | 0.88784 | 1.05748 | 32.953 | 59.585 |
| 4 | 0.09426 | 0.21072 | 0.30498 | 1.15174 | 1.36246 | 31.042 | 47.129 |
| 5 | 0.07610 | 0.26015 | 0.33625 | 1.43857 | 1.69872 | 28.319 | 38.052 |
| 6 | 0.06151 | 0.32413 | 0.38564 | 1.76023 | 2.08436 | 25.045 | 30.754 |
| 7 | 0.04903 | 0.41418 | 0.46321 | 2.13338 | 2.54757 | 21.367 | 24.513 |
| 8 | 0.03786 | 0.55649 | 0.59435 | 2.58542 | 3.14192 | 17.371 | 18.928 |
| 9 | 0.02747 | 0.83061 | 0.85808 | 3.16939 | 4.00000 | 13.097 | 13.735 |

**TABLE 4.10**  Related Results for 9 Production Cycles

| I | $t_i - T_{i-1}$ | $T_i - t_i$ | $x_i$ | $t_i$ | $T_i$ | $Q_i$ | $L_i$ |
|---|---|---|---|---|---|---|---|
| 1 | 0.94835 | 0.03696 | 0.98531 | 0.94909 | 0.98605 | 17.129 | 474.177 |
| 2 | 0.45931 | 0.04694 | 0.50625 | 1.44536 | 1.49230 | 20.909 | 229.653 |
| 3 | 0.37652 | 0.05420 | 0.43072 | 1.86882 | 1.92301 | 23.336 | 188.259 |
| 4 | 0.33094 | 0.06019 | 0.39113 | 2.25396 | 2.31415 | 25.130 | 165.471 |
| 5 | 0.30026 | 0.06544 | 0.36569 | 2.61440 | 2.67984 | 26.545 | 150.128 |
| 6 | 0.27746 | 0.07019 | 0.34765 | 2.95731 | 3.02749 | 27.700 | 138.731 |
| 7 | 0.25950 | 0.07457 | 0.33408 | 3.28700 | 3.36157 | 28.665 | 129.752 |
| 8 | 0.24478 | 0.07869 | 0.32347 | 3.60635 | 3.68504 | 29.483 | 122.392 |
| 9 | 0.23237 | 0.08259 | 0.31496 | 3.91741 | 4.00000 | 30.185 | 116.185 |

**TABLE 4.11**  Comparisons of Results with Other Studies

| | Example 1 | | Example 2 | |
|---|---|---|---|---|
| | $m^*$ | TVC | $m^*$ | TVC |
| Aggarwal and Bahari-Kashani | 7 | 2,876.02 | 5 | 2,030.79 |
| This model | 9 | 4,653.77 | 9 | 4,733.19 |
| Hollier and Mak | 13 | 6,425.64 | 28 | 13,845.90 |
| Hariga and Benkherouf | 13 | 6,430.87 | 28 | 13,874.10 |

## Example 2

The parameters in this example are similar to those of Example 1, except b is changed to $-0.08$. The minimum value of *TVC* is found to be $4733.192 when $m = 9$. Table 4.9 shows the relevant results for 9 production cycles.

Table 4.10 shows how the results of this model compared with other studies. From Table 4.10, one can see that the synchronized production policy of Aggarwal and Bahari-Kashni with varying the production rate has the least total variable cost. This can be explained by the fact that no extra cost is included in varying the production rate in each period. In this section, a fixed production rate is assumed. One can see that the result of this model is much better than that of Hollier and Mak.

Graphical analysis is demonstrated in Figure 4.4 for the case with two production cycles. The minimum *TVC* occurs when $T_1 = 1.5$ and $TVC = 8693.821$. These results are identical to the numerical results for $m = 2$ in Table 4.7. Similar but more complicated graphs can be drawn for larger values of *m*.

## Summary

In this section, a finite replenishment rate policy for a variable production cycle is developed for a time-varying demand. It is shown that when the production rate is infinite, the model conforms to the instantaneous replenishment policy. An optimal solution is found using a combination of classical and heuristic algorithms. A program code is written to facilitate the analysis. Finally, numerical examples and a

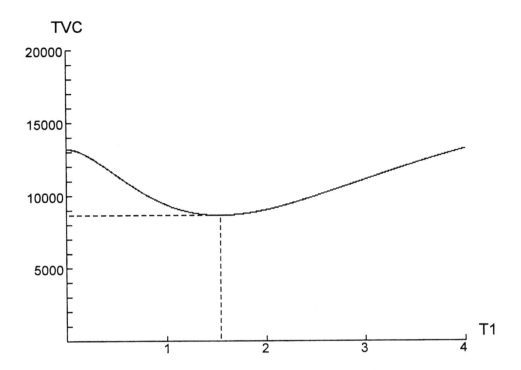

**FIGURE 4.4**    Temporary price discount purchase model for deteriorating inventory.

graphical analysis are given to elucidate the theory. It is shown that the model we developed in this section is very cost-effective.

## References

Aggarwal, S.C. (1981) Purchase-inventory decision models for inflationary condition. *Interfaces* **11**, 418–423.

Aggarwal, V. and H. Bahari-Kashani (1991) Synchronized production policies for deteriorating items in a declining market. *IIE Transactions* **23**, 185–197.

Bahari-Kashani, H. (1989) Replenishment schedule for deteriorating items with time proportional demand. *Journal of the Operational Research Society* **40**, 75–81.

Bertrand, J. W. M., J. C. Wortmann, and J. Wijngaard (1990) *Production Control: A Structure and Design Oriented Approach,* Elsevier, Amsterdam.

Buchanan, J.T. (1980) Alternative solution method for inventory replenishment problem under increasing demand. *Journal of Operational Research Society* **31**, 615–620.

Budhbhatti, H.D. and B.B. Jani (1988) Profit maximization—the ELS model under varying mark-up and different cost structures. *Journal of Operational Research Society* **39**, 351–358.

Buzacott, J.A. (1975) Economic order quantities with inflation. *Operational Research Quarterly* **26**, 553–558.

Choudhury, R. and K. S. Chaudhuri (1983) An order level inventory model for deteriorating items with finite rate of replenishment. *Opsearch,* **20**, 90–106.

Covert, R.P. and G.C. Philip (1973) An EOQ model for items with Weibull distribution deterioration. *AIIE Transactions* **5**, 323–326.

Dave, U. and L. K. Patel (1981) (T, S$_i$) policy inventory model for deteriorating items with time proportional demand. *Journal of the Operational Research Society* **40**, 137–142.

Deb, M. & K. S. Chaudhuri (1986) An EOQ model for deteriorating items with finite replenishment rate of production and variable rate of deterioration. *Opsearch* **23**, 175–181.

Ghare, P. M. and G. F. Schrader (1963) A model for exponential decaying inventory. *Journal of Industrial Engineering* **14**, 238–243.

Goswami, A. and K. S. Chaudhuri (1991a) An EOQ model for deteriorating items with shortages and a linear trend in demand. *Journal of the Operational Research Society* **42**, 1105–1110.

Goswami, A. and K. S. Chaudhuri (1991b) EOQ model for inventory with a linear trend in demand and finite replenishment. *International Journal of System Science* **22**, 181–197.

Goswami, A. and K. S. Chaudhuri (1992) Variation of order levels inventory model for deteriorating items. *International Journal of Production Economics* **27**, 111–117.

Hariga, M. A. (1995) An EOQ model for deteriorating items with shortages and a linear trend in demand. *Journal of the Operational Research Society* **46**, 398–404.

Hariga, M. A. and L. Benkherouf (1994) Optimal and heuristic inventory replenishment models for deteriorating items with exponential time-varying demand. *European Journal of Operational Research* **79**, 123–137.

Hollier, R. H. and K. L. Mak (1983) Inventory replenishment policies for deteriorating items in a declining market. *International Journal of Production Research* **21**, 813–826.

Hong, J., R. Sandrapaty, and J. Hayya (1990) On production policies for linearly increasing demand and uniform production rate. *Computers & Industrial Engineering* **18**, 119–127.

Hwang, H. and K.I. Sohn (1982) Management of deteriorating inventory under inflation. *The Engineering Economist* **28**, 191–206.

Kingsman, B.G. (1986) Purchasing raw materials with uncertain fluctuating prices. *European Journal of Operational Research* **25**, 258–372.

Leporelli, C. and M. Lucertini (1980) Substitution models for technology-fostered new production inputs. *Technological Forecasting and Social Change* **16**, 119–142.

Love, S. (1979) *Inventory Systems,* McGraw-Hill, New York.

Martin, G.E. (1994) Note on an EOQ model with a temporary sale price. *International Journal of Production Economics* **37**, 241–243.

McLoughdin, I., H. Rose, and J. Clark (1985) Managing the introduction of new technology *Omega* **13**, 251–262.

Misra, R. (1975) Optimum production lot size model for a system with deteriorating inventory. *International Journal of Production Research* **13**, 945–955.

Misra, R.B. (1979) A note on optimal inventory management under inflation. *Naval Research Logistic Quarterly* **26**, 161–165.

Mitra, A., A.F. Cox, and J.R. Jesse (1984) A note on determining order quantities with a linear trend in demand. *Journal of Operational Research Society* **35**, 141–144.

Nahmias, S. (1978) Perishable inventory theory: a review. *Operations Research* **30**, 680–708.

Raafat, F. (1991) Survey of literature on continuously deteriorating inventory models. *Journal of the Operational Research Society* **40**, 27–37.

Sachan, R. S. (1984) On (T, S$_i$) policy inventory model for deteriorating items with time proportional demand. *Journal of the Operational Research Society* **35**, 1013–1119.

Shah, Y.K. (1977) An order-level lot-size inventory model for deteriorating items. *AIIE Transactions* **9**, 108–112.

Silver, E.A. and H.C. Meal (1969) A simple modification of the EOQ for the case of a varying demand rate. *Production and Inventory Management,* 4th Quarter, 52–65.

Spiegel, M.R. (1960) *Applied Differential Equations,* Prentice-Hall, Englewood Cliffs, New Jersey.

Tersine, R.J. (1994) *Principles of Inventory and Materials Management,* Prentice-Hall, Englewood Cliffs, New Jersey.

Wee, H. M. (1995a) A deterministic lot-size inventory model for deteriorating items with shortages and a declining market. *Computers and Operations Research* **22**, 345–356.

Wee, H. M. (1995b) Joint pricing and replenishment policy for deteriorating inventory with declining market. *International Journal of Production Economics* **40**, 163–171.

Xu, H. and H. S. Wang (1992) Optimal inventory policy for perishable items with time proportional demand. *IIE Transactions* **24**, 105–110.

# 5

# Job-Shop Scheduling for Manufacturing with Random Time Operations

Dimitri Golenko-Ginzburg
*Ben-Gurion University*

Aharon Gonik
*Ben-Gurion University*

We consider a job-shop flexible manufacturing cell with $n$ jobs (orders) and $m$ machines. Each job operation $O_{i\ell}$ (the $\ell$-th operation of job $i$, $1 \le i \le n$, $1 \le \ell \le m$) has a random time duration $t_{1\ell}$ with the average value $\bar{t}_{i\ell}$ and the variance $V_{i\ell}$. Each job has its priority index $\rho_i$ and its due date $D_i$. The problem is to determine the starting time value $S_{i\ell}$ for each job-operation $O_{i\ell}$. These values are not calculated beforehand and are random values conditioned on our decisions.

Two basic concepts are imbedded in the problem's solution:

1. At each decision point, a competition among the jobs ready to be served on one and the same machine is introduced. That competition is based on the idea of pairwise comparison.
2. At each decision point, the machine is kept idle until the "bottleneck" job will be ready to be served on that machine and will enter the line. Such an approach is a combination of the pairwise comparison and the "look ahead" techniques.

Both models are modified for the case of random operations $O_{i\ell}$ and can be applied to a broad spectrum of FMS in JIT systems. For the case of a job-shop package without priorities a simplified solution is suggested. Various cost parameters have been introduced in the developed models.

Extensive experimentation is undertaken to evaluate the comparative efficiency of the job-shop scheduling models.

**Keywords:** Job-shop problem; Stochastic scheduling; Priority rule; Pairwise comparison; Tense job; Random operation; "Look Ahead" techniques.

# 5.1  Introduction

It has been well recognized in recent years that better scheduling may result in significant decrease in the life of an average workpiece in manufacturing (Narasimhan and Melnyk 1984, White 1990). Therefore, scheduling technique in various details plays a more and more essential role and tends to be concentrated in several areas, e.g., in classical job-shop scheduling problems for $n$ jobs on $m$ machines. In recent years extensive research has been undertaken in that area, e.g., by developing optimization algorithms (Coffman 1976, French 1982, Lageweg et al. 1977, Carlier and Pinson 1989) and various approximation algorithms (Balas and Zawack 1988, Spachis and King 1979, Kirkpatrick et al. 1983, van Laarhoven and Aarts 1987, van Laarhoven et al. 1992, Matsuo et al. 1988, Golenko 1973). Most optimization and approximation algorithms comprise complicated and large running time procedures, e.g., branch-and-bound methods, shifting bottleneck procedures, various simulated annealing methods, etc. Those procedures require large running times and are often not effective for high-dimension industrial scheduling problems ($m$, $n \geq$ 40–50). Such problems, especially for models of large size, can be solved by using numerous priority scheduling rules that have been developed in the last three decades (Muth and Thompson 1963, Gere 1966, Panwalker and Iskander 1977, Golenko 1973, Grabowski and Janiak 1987, Graves 1981, Blackstone et al. 1982, Barker and Graham 1985, Scully 1987, Haupt 1989, Dlell' Amico and Trubian 1993, Golenko-Ginzburg and Kats 1995). The objective is to minimize the makespan (the schedule time) according to the starting times of job operations obtained by using priority rules.

Note that unlike dynamic scheduling models (Conway et al. 1967, Eilon 1979, Ramasesh 1990), with the objective function of minimizing mean flow time rather than the makespan, minimizing the makespan is the most common criterion for static job-shop models (Coffman 1976, French 1982, Sen and Gupta 1984, Carlier and Pinson 1989, van Laarhoven et al. 1992).

In many practical cases, priority rules may provide a suboptimal solution. However, the closeness of the solution cannot be proved analytically and may deviate essentially from an optimal schedule (Wein and Chevalier 1992). Moreover, a priority rule which is effective for one job-shop may fail for another. Note that problems of this type usually deal with jobs (orders) of equal importance and delivery performance.

Most of the scheduling systems with jobs of different importance use heuristic procedures in combination with material requirements planning (MRP), capacity requirements planning (CRP), and master scheduling (New 1973, Mather 1984, 1985, White 1986). In recent years there were some good practical achievements of MRP systems comprising scheduling procedures and inventory problems. However, some MRP systems had failed (Snyder et al. 1985) since MRP combined with scheduling techniques often do not provide even a feasible solution. The latter has to be obtained by using a difficult and highly costly iterative process.

Another approach has been recently carried out by White (1986, 1990) and White and Hastings (1986). Their system JOBCODE produces a feasible solution which, in turn, results in material requirements. This approach deals more with priority scheduling technique. The load sequence is determined by starting jobs using a hierarchy of priority indices, due date, earliest start date, and job number. Two scheduling techniques are available designated Forwards (CCLSS-F) and Backwards (CCLSS-B). The first one scheduled are the operations (of the job under consideration) forward due to technological operation and capacity constraints. Backwards scheduling is carried out by scheduling all the operations backwards from the job's due date while honoring all the constraints.

In many manufacturing systems, job-operations are of random time duration even when being processed with a fixed average speed. Due to random disturbances from the environment, breakdowns of the equipment, etc., most operations are of random time duration with a random deviation from the average speed. Due to such random influences, orders are often not manufactured and not delivered to the customers in time even when using various scheduling techniques. Thus, the problem of increasing the orders' delivery performance becomes very essential. Unfortunately, the JOBCODE system does not solve problems with probability parameters. Several interesting publications (Pinedo 1982, 1983, Pinedo and Weiss 1979, 1987, Hunsucker and Shah 1994, Weber 1982, Weber et al. 1986, Huang and Weiss 1990), deal mostly with flow-shop problems or with scheduling on parallel machines. But for job-shop scheduling problems with operations of random time duration there are practically very few achievements, especially for the case of jobs with different importance (Golenko-Ginzburg et al. 1995, Golenko-Ginzburg and Gonik 1997).

It can be well recognized that each priority rule is fully determined by the way a job is chosen for the machine from a line of jobs ready to be processed on that machine. Examining classical priority rules for $m \times n$ job-shop problems (with deterministic time durations), we came to the conclusion that one of the most fruitful approaches is the idea of pairwise comparison among the jobs waiting in the line (Gere 1966, Barker and Graham 1985, Golenko 1973, Golenko-Ginzburg and Kats 1995, 1997). If at a certain moment several different job-operations are waiting to be processed on machine $M_k$, a pairwise competition between the first two competitive jobs is arranged. The winner competes with the next job in the line, etc., until only one winner will be left. The latter has to be chosen for machine $M_k$.

Note that such an approach does not take into account the so-called "tense jobs" which, in reality, cause bottlenecks in the job-shop. Moreover, in certain cases it is reasonable not to choose a job from the line waiting for a machine when it becomes available, but to keep the machine idle until the bottleneck job is ready to be served on that machine. Gere (1966) suggested the so-called "look ahead" and "insert" mechanisms for such cases, and obtained successful numerical results. However, such techniques have never been applied to job-shops with random time operations. Thus, they have to be modified for stochastic job-shop scheduling.

The goal of this chapter is to describe a variety of job-shop scheduling models based on expanding both approaches to the case of jobs (orders) with different priorities and random time operations. The competition between two jobs is based on comparing two alternative options:

*Option A.* The first job is chosen for the machine and the second job will be processed after the first job operation will be finished.

*Option B.* The second job is chosen for the machine and the first one waits until the second job operation will be processed.

The idea of such a comparison is to calculate for each alternative option the jobs' delivery performances, i.e., the probabilities for both jobs to meet their due dates on time. The option which ensures the maximal delivery performance for two competitive jobs has to be chosen.

Besides the two main approaches in applying priority rules to job-shop flexible manufacturing cells (pairwise comparison and forecasting techniques) other particular cases will be considered. For all the developed models the problem is to determine the starting time values for each job operation. For the case of random time operations, those values are not determined beforehand and are random variables conditioned on our future decisions in the course of the job-shop's manufacturing.

Note, in conclusion, that, besides various heuristic priority rules, we will not use any optimization or approximation algorithms [23–26, 31, etc.] to solve the job-shop scheduling problem with random time operations. Those algorithms are based on completely different conceptions which have not been applied as yet to stochastic scheduling.

## 5.2 The Problem's Description

We are concerned with a problem of machine scheduling known as the job-shop scheduling problem (Conway et al. 1967, Coffman 1976, French 1982, Muth and Thompson 1963). A flexible manufacturing cell consists of a chain of operations each of which needs to be processed during an uninterrupted period on a given machine. Each machine can process at most one operation at a time. Each operation is carried out under random disturbances.

Note that operations with random time duration cover a broad spectrum of production systems. For example, in producing jet engines, production begins with a motor casting unit, which goes through various machining operations on numerical control machines. In the course of each operation a team of engineers performs various quality assurance tests. Many random disturbances may occur in this multistage operation, such as repetition due to low-quality, delays in tool replacement, etc. For some operations there are continuous stochastic changes in the processing speed during the operation's time period. In metallurgy, especially in hot strip mills, continuous stochastic changes in high temperature and other factors affect production speed, too. This results in random time duration of almost each operation. Various other man-machine production systems under random disturbances, e.g., building systems, mining, fishing, forestry, agriculture, etc., comprise operations with random time durations.

Assume, further, that jobs (orders) are of different importance. This occurs in many production systems, e.g., in defence-related industries. Thus, a priority index has to be set for each job by the management, i.e., by practitioners who are responsible for the jobs-hop. This can be done by using various expert methods, e.g., the Delphi method. Note that for each job its level of significance can be practically specified by the job's delivery performance. For jobs with random time operations delivery performance is nothing else but the probability of the job to meet its deadline on time. Thus, besides the priority index, the management has to set for each job (order) two additional probability values: $p^*$ − desired probability for the job to be accomplished on time; and $p^{**} < p^*$ − least permissible probability for the job to meet its due date on time. Both values $p^*$ and $p^{**}$ also have to be set by practitioners by using expert methods. Note that for jobs with equal priority indices their confidence probabilities $p^*$ and $p^{**}$ must be equal too.

The problem is to determine starting time values for each job operation. Those values are not calculated beforehand and are random values conditioned on our decisions.

Two different models will be imbedded in the problem. The first one is based on examining delivery performance values together with priority indices. The second one deals with examining confidence probabilities $p^*$ and $p^{**}$ and does not take into account priority indices. We will examine both models via extensive simulation in order to evaluate their comparative efficiency.

For each model two different heuristic solutions will be considered. The first one is based on the idea of pairwise comparison without taking into account the bottleneck jobs. The second heuristic is a combination of pairwise comparison and "look ahead" techniques. We shall show that such a modified heuristic is more effective than the first one.

To formulate the problem let us introduce the following terms:

| | |
|---|---|
| $n$ | number of jobs $J_i$, $1 \le i \le n$; |
| $m$ | number of machines $M_k$, $1 \le k \le m$; |
| $O_{i\ell}$ | $\ell$-th operation of $J_i$, $1 \le \ell \le m_i$; |
| $m_i$ | number of operations of $J_i$, $m_i \le m$; |
| $t_{i\ell}$ | random processing time of $O_{i\ell}$; |
| $\bar{t}_{i\ell}$ | expected value of $t_{i\ell}$; |
| $V_{i\ell}$ | variance of $t_{i\ell}$; |
| $m_{i\ell}$ | index of the machine on which $O_{i\ell}$ is processed, $1 \le m_{i\ell} \le m$; |
| $\|\bar{t}_{i\ell}, V_{i\ell}, m_{i\ell}\|$ | initial data matrix; |
| $\rho_i$ | priority index of job $J_i$ to indicate the level of the job's importance (pregiven); if job $J_{i_1}$ has a higher importance than job $J_{i_2}$, relation $\rho_{i_1} > \rho_{i_2}$ holds; |
| $E_i$ | the earliest possible time moment to start processing job $J_i$ (pregiven); |
| $D_i$ | due date for job $J_i$ to be accomplished (pregiven); |
| $S_{i\ell}$ | time moment job-operation $O_{i\ell}$ starts (a random value conditioned on our decisions); |
| $F_{i\ell} = S_{i\ell} + t_{i\ell}$ | the actual moment job-operation $O_{i\ell}$ is finished; |
| $F_i$ | actual time for job $J_i$ to be accomplished; |

$p_i = P\{F_i \leq D_i\}$ — delivery performance value of $J_i$, i.e., its confidence probability to be accomplished on time;

$p_i^*$ — desired probability for job $J_i$ to be accomplished on time (pregiven);

$p_i^{**} < p_i^{**}$ — the least permissible probability for job $J_i$ to meet its due date on time (pregiven); in case $\rho_{i_1} = \rho_{i_2}$ relations $p_{i_1}^* = p_{i_2}^{**}, p_{i_1}^{**} = p_{i_2}^{**}$ hold;

$\bar{F}_{i\ell}(S_{i\ell}, \Delta)$ — expected value of $F_{i\ell}$ on condition that operation $O_{i\ell}$ has started at $S_{i\ell}$ and has been processed not less than $\Delta$;

$\bar{t}_{i\ell}(\Delta)$ — expected value of $t_{i\ell}$ on condition that $O_{i\ell}$ has been processed not less than $\Delta$

$T = \underset{i}{\text{Max}} F_i - \underset{i}{\text{Max}} S_{i1}$ — the actual time period for all jobs $J_i$ to be manufactured;

$C_i^*$ — the penalty cost for not delivering job $J_i$ on time (pregiven, to be paid once);

$C_i^{**}$ — the penalty cost *per time unit* when waiting for the job's delivery within the period $[D_i, F_i]$ (pregiven);

$C_i = C_i^* + C_i^{**} \cdot (F_i - D_i)$ — total penalty expenses for job $J_i$ (in case $F_i < D_i$).

The problem is to determine values $S_{i\ell}$, $1 \leq i \leq n$, $1 \leq \ell \leq m_i$. The following two models will be considered.

**Model I** (*with priority indices*). Determine values $S_{i\ell}$ to maximize the objective

$$\underset{S_{il}}{\text{Max}} \sum_{i=1}^{n} \rho_i \text{Pr}\{F_i \leq D_i\} \tag{5.1}$$

subject to

$$S_{i\ell} \geq E_i \tag{5.2}$$

where $\text{Pr}\{F_i \leq D_i\}$ is the job's delivery performance. Note that maximizing objective (5.1) results in the policy us follows: the management takes all measures to accomplish first jobs with higher priorities; only afterwards it takes care of other jobs.

**Model II** (*with confidence probabilites*). Determine values $S_{i\ell}$ to minimize the makespan

$$\underset{S_{i\ell}}{\text{Min}} \; T = \underset{S_{i\ell}}{\text{Min}} \left\{ \underset{i}{\text{Max}} \; F_i - \underset{i}{\text{Min}} \; S_{il} \right\} \tag{5.3}$$

subject to (5.2) and

$$\text{Pr}\{F_i \leq D_i\} \geq p_i^{**}, \qquad l \leq i \leq n. \tag{5.4}$$

Both models are stochastic optimization problems which cannot be solved in the general case; they allow only heuristic solutions.

The basic idea of the heuristic solution of both problems is as follows. Decision-making, i.e., determining values $S_{i\ell}$, is carried out at moment $E_i$ and $F_{i\ell}$, when either one of the machines is free for service or a certain job-operation ($O_{i\ell}$ or $O_{i,\ell+1}$) is ready to be processed. If job-operation $O_{i\ell}$ is ready to be processed on machines $M_k$ (free for services) and there is no line for that machine, the job is passed to the machne. Otherwise a competition is arranged based on the idea of pairwise comparison.

We will show later one that model (1–2) is simpler and easier is usage than the second model. Thus, we will mostly use Model I to schedule the job-shop.

A more effective, although more complicated, heuristic to solve problem (1–2) is as follows: *in order to choose the job to be passed on to the machine, the model considers at each decision point not only the jobs waiting for that machine in the line, but takes into account other jobs under way which will soon be in need of that machine and which may serve as a bottleneck for the job-shop.* Some basic principles of such "tense jobs" for a deterministic job-shop will be outlined below.

In the case when all the jobs are of equal priority, they can be regarded as an order package which has to be manufactured as soon as possible. The problem is to minimize the makespan for all jobs entering the package, i.e., to minimize value T. Thus, **Model III** can be formulated as follows:

Determine values $S_{i\ell}$, $1 \leq i \leq n$, $1 \leq \ell \leq m_i$, to minimize

$$\underset{S_{i\ell}}{\text{Min}} \; T = \underset{S_{i\ell}}{\text{Min}} \left[ \underset{i}{\text{Max}} \; F_i - \underset{i}{\text{Min}} \; S_{il} \right]$$

subject to (5.2) only. Note that the due dates $D_i$, $1 \leq i \leq n$, are not imbedded in the model. Model III can be solved by introducing heuristics based on the unification of the concepts of pairwise comparison and "tense" jobs in a bottleneck situation (McMahon, Florian 1975).

## 5.3    Priority Rules for Deterministic Job-Shop Scheduling

We shall outline below two most effective priority rules which, for different job-shop problems of medium and large size ($m \cdot n \geq 500$), perform better than the well-known bench-marks SPT (Shortest processing time), LRT (longest remaining time), FIFO (first in–first out) (Muth and Thompson 1963, Gere 1966). Those rules can be used in job-shop scheduling with deterministic processing times $t_{i\ell}$ of all job-operations $O_{i\ell}$, $1 \leq i \leq n$, $1 \leq \ell \leq m_i$. Note that each priority rule fully determines the way a job is chosen for the machine from a line of jobs ready to be processed on that machine. Assume that at a certain moment $t$, $q$ arbitrarily enumerated job-operations $O_{i_1\ell_1}, \dots, O_{i_q\ell_q}$ are awaiting to be processed on machine $M_k$, i.e., relationship $m_{i_1\ell_1} = m_{i_2\ell_2} = \cdots = m_{i_q\ell_q} = k$ hold.

### Rule PC (Pairwise Comparison) (Gere 1966, Golenko 1973, Barker and Graham 1985)

Arrange a pairwise comparison between the first competitive job-operations $O_{i_1\ell_1}$ and $O_{i_2\ell_2}$ as follows. If

$$t_{i_1\ell_1} + \max \left\{ \sum_{p=\ell_1+1}^{m_{i_1}} t_{i_1 p}, \; \sum_{p=\ell_2}^{m_{i_2}} t_{i_2 p} \right\}$$

$$< t_{i_2\ell_2} + \max \left\{ \sum_{p=\ell_2+1}^{m_{i_2}} t_{i_2 p}, \; \sum_{p=\ell_1}^{m_{i_1}} t_{i_1 p} \right\}, \tag{5.5}$$

job-operation $O_{i_1\ell_1}$ wins the competition. The winner competes with the next job-operation $O_{i_3\ell_3}$ in the line, etc., until only one winner will be left. The latter has to be chosen for the machine $M_k$.

Note that using relation (5.5) means that two alternatives and mutually exclusive decision-making have been examined: I. Job-operation $O_{i_1\ell_1}$ is operated first, and $O_{i_2\ell_2}$ afterwards (option A). II. Job-operation $O_{i_1\ell_1}$ is operated after $O_{i_2\ell_2}$ (option B).

If relation (5.5) holds that means that by using option A we will accomplish *both* jobs $J_{i_1}$ and $J_{i_2}$ earlier than by using option B. Since due dates $D_{i_1}$ and $D_{i_2}$ are not introduced and both jobs are of equal importance, such a decision-making seems to be reasonable. Note that (5.5) is a heuristic and is based on the assumption that job-operations $O_{i_1,\ell_1+2}, \ldots, O_{i_1,m_1}$ and $O_{i_2,\ell_2+1}, O_{i_2,\ell_2+2} \ldots O_{i_2,m_2}$ will not wait in lines.

It can be proven that the winner's determination by using PC does not depend on the order of enumerating the competitive jobs.

**Theorem:** If job $J_{i_1}$ wins the competition with job $J_{i_2}$ and job $J_{i_2}$ is the winner with $J_{i_3}$, then job $J_{i_1}$ wins the competition with $J_{i_3}$.

**Proof:** Denote, correspondingly,

$$A_{i_1} = \sum_{p=\ell_1}^{m_{i_1}} t_{i_1 p}, \qquad A_{i_2} = \sum_{p=\ell_2}^{m_{i_2}} t_{i_2 p}, \qquad A_{i_3} = \sum_{p=\ell_3}^{m_{i_3}} t_{i_3 p}.$$

Since $J_{i_1}$ is the winner in competition with $J_{i_2}$,

$$t_{i_1 \ell_1} + \text{Max}\{A_{i_1} - t_{i_1 \ell_1}, A_{i_2}\} < t_{i_2 \ell_2} + \text{Max}\{A_{i_2} - t_{i_2 \ell_2}, A_{i_1}\} \tag{5.6}$$

holds. Relation (5.6) is equivalent to

$$\text{Max}\{A_{i_1}, A_{i_2} + t_{i_1 \ell_1}\} < \text{Max}\{A_{i_2}, A_{i_1} + t_{i_2 \ell_2}\}. \tag{5.7}$$

Note that $\text{Max}\{A_{i_2}, A_{i_1} + t_{i_1 \ell_1}\} = A_{i_1} + t_{i_2 \ell_2}$, otherwise (5.7) does not hold. Since

$$\text{Max}\{A_{i_1}, A_{i_2} + t_{i_1 \ell_1}\} < A_{i_1} + t_{i_2 \ell_2},$$

relation     $$A_{i_2} + t_{i_1 \ell_1} < A_{i_1} + t_{i_2 \ell_2} \tag{5.8}$$

is an evident one. Since $J_{i_2}$ wins the competition with $J_{i_3}$

$$A_{i_3} + t_{i_2 \ell_2} < A_{i_2} + t_{i_3 \ell_3} \tag{5.9}$$

holds. Summarizing (5.8) and (5.9) we finally obtain

$$A_{i_3} + t_{i_1 \ell_1} < A_{i_1} + t_{i_3 \ell_3}. \tag{5.10}$$

Thus, due to (5.10),

$$t_{i_1 \ell_1} + \text{Max}\{A_{i_1} - t_{i_1 \ell_1}, A_{i_3}\} < t_{i_3 \ell_3} + \text{Max}\{A_{i_3} - t_{i_3 \ell_3}, A_{i_1}\} \tag{5.11}$$

holds.                                                                                              □

**Corollary:** *Choosing a job for the machine from the line of q competitive jobs is carried out by realizing q–1 consecutive pairwise comparions.*

Computational experiments show (Golenko-Ginzburg and Kats 1995) that priority rule PC performs well and is very effective, especially for cases of large *m* and *n*.

Another effective priority rule is based on determining the so-called "tense" job which is chosen from *all n* jobs and which, in reality, causes a bottleneck in the job-shop.

### Rule TSW (tense + SPT + waiting) (Gere 1966, Golenko 1973, Golenko-Ginzburg and Katz 1995)

From all jobs $J_1, \ldots, J_n$, choose, a moment $t$, job $J_\gamma$ with the longest remaining processing time, i.e.,

$$\sum_{v=d_\gamma}^{m_\gamma} t_{\gamma v} = \underset{1 \le i \le n}{\text{Max}} \sum_{v=d_i}^{m_i} t_{iv},$$

where

$$d_i = \text{Max}\{b: S_{i,b-1} + t_{i,b-1} \le t\}$$

and

$$t_{id_i} = \begin{cases} t_{id_i} & \text{if } O_{id_i} \text{ is } not \text{ processed at moment } t; \\ t_{id_i} - (t - S_{id_i}) & \text{if } O_{id_i} \text{ is processed at moment } t. \end{cases}$$

If $J_\gamma$ (we will henceforth call it the tense job) is not in the line for $M_k$ then carry out a check: has $J_\gamma$ already passed machine $M_k$? If yes, then choose the job according to rule SPT [33], i.e., choose job $J_{i_\xi}$ with

$$t_{i_\xi \ell_\xi} = \underset{1 \le r \le q}{\text{Min }} t_{i_r \ell_x}.$$

If not, then calculate $T_{\gamma k}$-the earliest moment $J_\gamma$ can be processed on machine $M_k$. If

$$\underset{1 \le r \le q}{\text{Min }} t_{i_r \ell_r} \le T_{\gamma k} - t,$$

then choose job $J_{i_\xi}$. If $t_{i_\xi \ell_\xi} > T_{\gamma k} - t$ wait until $T_{\gamma k}$ and then process $J_\gamma$ on machine $M_k$.

If $J_\gamma$ waits in the line for machine $M_k$, it has to be chosen, i.e., $S_{\gamma d_\gamma} = t$.

The operational significance of "waiting" is as follows: in certain cases, it is reasonable not to choose a job from the line waiting for a machine when it becomes available for use, but to keep the machine idle until the bottleneck job is ready to be served on that machine. Such a policy may result in decreasing the makespan. The rule is a modification of "look ahead" and "insert" mechanisms suggested by Gere (1966).

We shall show below that a unification of both rules is more effective than each of them separately.

## 5.4 Determining Starting Time Moments for Job-Operations of Random Durations (Pairwise Comparison)

Consider $q$ jobs $J_{i_1}, J_{i_2}, \ldots, J_{i_q}$ with different priority indices $\rho_{i_1}, \rho_{i_2}, \ldots, \rho_{i_q}$ which at moment $t$ are ready to be processed on machine $M_k$. Assume that random processing time durations $t_{i_1 \ell_1}, \ldots, t_{i_q \ell_q}$ have a normal distribution $N(\bar{t}_{i_r \ell_r}, V_{i_r \ell_r})$, $1 \le r \le q$, respectively.

The idea of pairwise comparison $(\bar{t}_{i_r \ell_r}, V_{i_r \ell_r})$ outlined in Section 5.3 is expanded to develop reasonable decision-marking in stochastic job-shop scheduling (Golenko-Ginzburg et al. 1995). To carry out the comparison we will henceforth use two types of forecasting:

- Short-term forecasting is used to forecast the moment a certain operation will be finished. For that purpose the average processing time $\bar{t}_{i,\ell_r}$ has to be used.
- Long-term forecasting is used to calculate the delivery performance, i.e., the probability for a certain job $J_{i_r}$ to be accomplished on time.

Such a calculation is carried out by

$$P_t(J_{i_r}) = \text{Pr}_t\{F_{i_r} \le D_{i_r}\} = \Phi\left\{\frac{D_{i_r} - t - \sum_{s=\ell_r}^{m_{i_r}} \bar{t}_{i_r s}}{\sqrt{\sum_{s=\ell_r}^{m_{i_r}} V_{i_r s}}}\right\} \tag{5.12}$$

where $\Phi(x) = (1/\sqrt{2\pi})\int_{-\infty}^{x} e^{-t^2/2}\, dt$ is a standard normal distribution.

Note that *even for the case of non-normal distributed values* $t_{i,s}$, $s \ge \ell_r$ relation (5.12) is asymptotically true since from the Central Limit Theorem the sum is $\Sigma_{s=\ell_r}^{m_{i_r}} t_{i,s}$ is asymptotically normal. Thus, to carry out the comparison via (5.12) we can consider *random processing time* $t_{i\ell}$ *with arbitrary distributions.*

Pairwise comparison is carried out as follows: for the first pair $J_{i_1}$ and $J_{i_2}$ in the line four values are calculated.

1. Probability performance $P_t(J_{i_1}) = p_1$ for job $J_{i_1}$ to be accomplished on time on condition that job $J_{i_1}$ will be passed to machine $M_k$ at moment $t$.
2. Probability performance $P_t(J_{i_2}) = p_2$ for job $J_{i_2}$ to be accomplished on time on condition that job $J_{i_2}$ will be passed to machine $M_k$ at moment $t$.
3. Probability performance $P_{t+\bar{t}_{i_1\ell_1}}(J_{i_2}) = p_3$ for job $J_{i_2}$ to be accomplished on time on condition that first job $J_{i_1}$ will be passed to the machine and later on, at the time moment $t + \bar{t}_{ij\ell_1}$ job $J_{i_2}$ will start to be processed on that machine.
4. Probability performance $P_{t+\bar{t}_{i_2\ell_2}}(J_{i_1}) = p_4$ for job $J_{i_1}$ to be accomplished on time on condition that first job $J_{i_2}$ will start to be processed at moment $t$ and later on, at moment $t + \bar{t}_{i_2\ell_2}$, job $J_{i_1}$ will be passed to the machine.

After calculating values $p_1, p_2, p_3$ and $p_4$ decision-making is carried on by analyzing those values together with either priority indices $\rho_{i_1}, \rho_{i_2}$ (Model I) or with confidence probabilities $p_{i_1}^*, p_{i_1}^{**}, p_{i_2}^*, p_{i_2}^{**}$ (Model II). Let us consider both models separately.

**Model I** (*On the basis of priority indices $\rho$*). Two alternative cases will be examined:

*Case A.* Jobs $J_{i_1}$ and $J_{i_2}$ are of one and the same importance, i.e., $\rho_{i_1} = \rho_{i_2}$. Referring to objective (5.1) if relation

$$\rho_{i_1}p_1 + \rho_{i_2}p_3 \ge \rho_{i_1}p_4 + \rho_{i_2}p_2 \tag{5.13}$$

holds, job $J_{i_1}$ wins the competition and has to be processed first on the machine. Otherwise, job $J_{i_2}$ has to be chosen for the machine. Since $\rho_{i_1} = \rho_{i_2}$ we can substitute decision making rule (13) for a simpler one:

$$\begin{cases} p_1 + p_3 \ge p_2 + p_4 \Rightarrow \text{job } J_{i_1} \text{ wins the competition;} \\ p_1 + p_3 < p_2 + p_4 \Rightarrow \text{job } J_{i_2} \text{ is the winner.} \end{cases} \tag{5.14}$$

Another decision-making rule can be recommended in case A. Namely, if relation

$$\text{Min}(p_1, p_3) > \text{Min}(p_2, p_4) \tag{5.15}$$

holds, job $J_{i_1}$ wins the competition. Relation

$$\text{Min}(p_1, p_3) < \text{Min}(p_2, p_4) \tag{5.16}$$

results in choosing first job $J_{i_2}$ for the machine. If $\text{Min}(p_1, p_3) = \text{Min}(p_2, p_4)$ we choose the job which provides the maximum of values $p_1$, $p_2$. Note that relation (5.16) is nothing else but a probabilistic analogue of relation (5.5) for the deterministic job-shop scheduling. We simply substitute the minimal time to accomplish both two jobs for their minimal probability performance. Note that although rules (5.15) and (5.16) do not correspond with objective (5.1) it is a reasonable rule since using the maximin concept for two competitive jobs results in *protecting the slowest job* on account of the faster one. The comparative efficiency of this rule will be outlined below.

*Case B.* Jobs $J_{i_1}$ and $J_{i_2}$ are of different importance, e.g., $\rho_{i_1} > \rho_{i_2}$.

Honoring objective (1) results in decision-making as follows:

$$\rho_{i_1} p_1 + \rho_{i_2} p_3 \geq \rho_{i_1} p_4 + \rho_{i_2} p_2 \Rightarrow J_{i_1} \text{ is the winner.} \tag{5.17}$$

Otherwise job $J_{i_2}$ wins the competition.
*We will henceforth call decision-makings based on*

- rules (5.14) and (5.17)
- rules (5.15–5.17)
- heuristics $I^*$ and $I^{**}$, correspondingly.

**Model II** (*on the basis of confidence probabilities $p^*$ and $p^{**}$*). In case A, if two competigive jobs $J_{i_1}$ and $J_{i_2}$ are of one and the same importance, relations $p_{i_1}^* = p_{i_2}^*$ and $p_{i_1}^{**} = p_{i_2}^{**}$ holds. For this case we recommend to determine the winner by using relations (5.15) and (5.16).

In case B, for jobs with different importance, we will not consider all possible situations, but we will limit ourselves to nontrivial cases only.
  I. In cases

$$\begin{cases} p_4 < p_{i_1}^{**} \\ p_1 > p_{i_1}^{**} \end{cases} \tag{5.18}$$

or

$$\begin{cases} p_1 > p_4 \geq p_{i_1}^{**}, \\ p_3 \geq p_{i_2}^{**}, \end{cases} \tag{5.19}$$

job $J_{i_1}$ wins the competition.
  II. In cases

$$\begin{cases} p_3 < p_{i_2}^{**}, p_2 \geq p_{i_2}^{**}, \\ p_1 > p_{i_4}^{**}, \geq p_{i_1}^{**}, \end{cases} \tag{5.20}$$

**TABLE 5.1** Initial Data Matrix (3 Jobs, 4 Machines)

| | | | |
|---|---|---|---|
| (60, 100, 3) | (40, 60, 2) | (60, 80, 4) | (80, 60, 1) |
| (50, 80, 30) | (90, 70, 2) | (100, 60, 1) | (50, 50, 4) |
| (50, 25, 1) | (80, 60, 3) | (50, 50, 4) | (25, 16, 2) |

or

$$\begin{cases} p_1 > p_4 \ge p_{i_1}^*, \\ p_{i_2}^* > p_2, \end{cases} \tag{5.21}$$

job $J_{i_2}$ is the winner. We will henceforth call decision-makings based on (5.15–5.16), (5.18–5.21) heuristic II.

It goes without saying that other policies may be suggested instead of rules (5.18–5.21). But the general approach remains unchanged. For example, if relations

$$\begin{cases} p_1 \ge p_{i_1}^*, \\ p_2 < p_{i_2}^{**}, \\ p_{i_1}^* > p_4 \ge p_{i_1}^{**}, \\ p_{i_2}^* > p_2 \ge p_{i_2}^{**} \end{cases} \tag{5.22}$$

hold, two different decision-makings may be introduced. If job $J_{i_1}$ is *essentially* more important than job $J_{i_2}$, job $J_{i_1}$ has to be processed first even taking into account relation $p_3 < p_{i_2}^{**}$. But if the difference between two competitive jobs is not so great, choosing job $J_{i_2}$ first may be considered preferable.

A numerical example is outlined below to illustrate heuristic rules $I^*$, $I^{**}$, and II. The intial data matrix $\|\bar{t}_{i\ell}, V_{i\ell}, m_{i\ell}\|$ is given in Table 5.1. Assume that each processing time $t_{i\ell}$ has a normal distribution with average $\bar{t}_{i\ell}$ and variance $V_{i\ell}$.

Some jobs are of different importance and the corresponding priority indices are $\rho_1 = \rho_2 = 10$, $\rho_3 = 6$. Due dates and the earliest time moments are $E_i = 0$, $1 \le i \le 3$, $D_1 = 300$, $D_2 = 350$, $D_3 = 320$. The desired and the least permissible probabilities $p_{i_1}^*$ and $p_{i_1}^{**}$ are:

$$\begin{aligned} P_1^* &= 0.8, & P_1^{**} &= 0.6, \\ P_2^* &= 0.8, & P_2^{**} &= 0.6, \\ P_3^* &= 0.7, & P_3^{**} &= 0.5. \end{aligned}$$

At moment $t = 0$ two jobs $J_1$ and $J_2$ with equal priority indices are ready to be processed on the machine $M_3$. Values $P_{t=0}(J_1)$, $P_{t=0}(J_2)$, $P_{t=60}(J_2)$ and $P_{t=50}(J_1)$ according to (5.12) are

$$\begin{aligned} P_1 &= 0.999, & P_2 &= 0.9999, \\ P_3 &= 0.5, & P_4 &= 0.72. \end{aligned}$$

Since $p_2 + p_4 > p_1 + p_3$(rule $I^*$) job $J_2$ is the winner. The same decision-making can be obtained by using rules $I^{**}$ and II according to (5.16). Thus $J_2$ *is* chosen for the machine and $S_{21} = 0$. Job $J_1$ proceeds waiting in the line for machine $M_3$. Job $J_3$ is passed to machine $M_1$ at $t = 0$ since at that moment there is no line for that machine; value $S_{31} = 0$. Assume that the values of random time durations for job-operations $O_{21}$ and $O_{31}$ are as follows:

$$\begin{aligned} t_{21} &= 52, & F_{21} &= 52, \\ t_{31} &= 48, & F_{31} &= 48. \end{aligned}$$

Examine the system at $t = 48$. Job $j_3$ leaves machine $M_1$ and is ready to be processed on machine $M_3$. Since $M_3$ is not free for service job, $J_3$ enters the line for that machine. At moment $t = 52$ job $J_2$ leaves machine $M_3$ and is ready to be processed on $M_2$. Since that machine is free job $J_2$ is passed to $M_2$; value $S_{22} = 52$.

At moment $t = 52$ jobs $J_1$ and $J_3$ with different priorities are ready to be processed on machine $M_3$. Values $p_1 - p_4$ are as follows:

$$p_1 = p_{52}(J_1) = 0.68,$$
$$p_2 = P_{52}(J_3) = 0.9999,$$
$$p_3 = P_{52+(60)}(J_3) = 0.9999,$$
$$p_4 = p_{52+(80)}(J_1) = 0.0001.$$

Using rules $I^*$ and $I^{**}$ results in choosing job $J_2$ (Grabowski and Janiak 1987), since relation $10 \cdot 0.68 + 6 \cdot 0.9999 > 10 \cdot 0.0001 + 6 \cdot 0.9999$ holds. Using rule II results in the same decision-making. Since job $J_1$ has a higher priority than $J_3$ values $p_1$, $p_2$, $p_3$ and $p_4$ represent case (18) outlined above. If $J_1$ is not processed first it has no chance to be accomplished on time. Thus according to all rules we choose job $J_1$ first for machine $M_3$. Value $S_{11} = 52$ and job $J_3$ remains in the line for $M_3$. Assume that the values of random time durations of $O_{11}$ and $O_{22}$ are $t_{11} = 63$, $t_{22} = 84$. This results in $F_{11} = 52 + 63 = 115$ and $F_{22} = 52 + 84 = 136$. At moment $t = 115$ job $J_1$ leaves machine $M_3$ and is ready to be processed on machine $M_2$. Since the latter is occupied job $J_1$ enters the line for $M_2$. At moment $t = 115$ job $J_3$ remains the only one for machine $M_3$ and is passed to that machine, $S_{32} = 115$. Assume that for $O_{32}$ the random processing time is 78; this results in $F_{32} = 115 + 78 + 193$. At moment $t = 136$ job $J_2$ leaves machine $M_2$ and is passed to the free machine $M_1$; value $S_{32} = 136$. Assume that the random value $t_{23} = 109$; this results in $F_{23} = 109 + 136 = 245$. Since at $t = 136$ machine $M_2$ is free job $J_1$ is passed to that machine for $O_{12}$; value $S_{12} = 136$. Simulating $t_{12} = 35$ results in $F_{12} = 136 + 35 = 17$. At moment $t = 171$ job $J_1$ leaves machine $M_2$ and is ready to be processed on $M_4$. Since the latter is free and there is no line $S_{13} = 171$. Simulating $t_{13} = 70$ results in $F_{13} = 241$. At moment $t = 193$ job $J_3$ leaves machine $M_3$ and is ready to be processed on machine $M_4$. However, that machine is occupied (by job $J_1$), and $J_3$ enters the line for $M_4$. At moment $t = 241$ job $J_1$ leaves machine $M_4$ and is ready to be processed on machine $M_1$. Since the latter is occupied until $t = 245$ job $J_1$ enters the line for $M_1$. Since $J_3$ is the only job in the line for $M_4$ it can be passed to that machine; value $S_{33} = 241$. Simulating $t_{33} = 49$ results in $F_{33} = 241 + 49 = 290$. At moment $t = 245$ job $J_2$ leaves machine $M_1$ and is ready to be processed on $M_4$. However, since $M_4$ is occupied (until $t = 290$), $J_2$ enters the line for $M_4$. Note that $J_1$ is the only job in line for $M_1$ (now free). Thus $S_{14} = 245$, and simulating $t_{14} = 81$ we obtain value $F_1 = F_{14} = 326$. That means that $J_1$ does not meet its deadline on time. At moment $t = 290$ job $J_3$ leaves $M_4$ and is ready to be processed on $M_2$. Since there is no line for that machine value $S_{34} = 290$ is determined. Simulating $t_{34} = 27$ results in obtaining $F_{34} = F_3 = 290 + 27 = 317 < D_3$. Job $J_3$ has been accomplished on time.

At moment $t = 290$ job $J_2$ is in the line for $M_4$ (now free for service). Determining $S_{24} = 290$ and simulating $t_{24} = 48$ results in $F_{24} = S_{24} = t_{24} = 338 < D_2$. Thus $J_2$ meets its deadline on time, and $J_1$ is the only job which has not been accomplished on time.

All the heuristics (based on rules I and II) are performed in real time; namely, each decision-making can be introduced only after a certain machine finishes to process a job-operation. Values $S_{i\ell}$ cannot be predetermined. However, if we want to evaluate the efficiency of the heuristic under consideration we can simulate the job-shop's work by random sampling of the actual job-operations' time durations. By simulating the job-shop's work many times, the probability of completion on time for each job can be evaluated.

We have implemented our heuristics on an IBM PC in the Borland C++ programming language. In order to evaluate the performance of each heuristic a job-shop flexible manufacturing cell has been chosen. The job-shop FMS comprises 10 jobs to be processed on 5 different machines. Priority indices together with confidence probabilities have been set by practitioners. Note that to set priority indices is a much easier task than to set confidence probabilities: practitioners usually avoid probabilistic terms since they are not trained sufficiently. But any practitioner is able to set priority indices.

The initial data matrix is given in Table 5.2. Three distributions of random processing time on the machines have been considered:

1. normal distribution with mean $\bar{t}_{i\ell}$ and variance $V_{i\ell}$,
2. uniform distribution in the interval $[\bar{t}_{i\ell} - 3\sqrt{V_{i\ell}}, \bar{t}_{i\ell} + 3\sqrt{V_{i\ell}}]$,
3. exponential distribution with value $\lambda = 1/\bar{t}_{i\ell}$.

Other job-shop parameters are given in Table 5.3. Three heuristics have been considered;

1. Heuristic based on rule $I^*$, i.e., taking into account (5.13–5.14) and (5.17);
2. Heuristic based on rule $I^{**}$, i.e., referring to (5.15–5.16);
3. Heuristic based on rule II, i.e., taking into account (5.15–5.16) and (5.18–5.21).

For each combination of heuristics and distributions 500 runs were done. For each jobs its delivery performance, i.e., the probability of meeting the due date on time, has been evaluated. The summary of results is presented in Table 5.4

The following conclusions can be drawn from the summary

1. Normal distribution results for each job in the highest delivery performance while using exponential distribution we obtain the lowest one. However, even those lower delivery performances are reliable enough for practical industrial problems. Thus, the heuristics under consideration can be used for arbitrary distributions of random processing times on the machines.
2. For identical distributions using three comparative heuristics ($I^*$, $I^{**}$, and II) results for one and the same job in performance values which are practically very close to each other. The difference between the corresponding values is not essential and depends only on random deviations.

**TABLE 5.2**  Initial Data Matrix (10 Jobs, 5 Machines)

| Job | Operation 1 | Operation 2 | Operation 3 | Operation 4 | Operation S |
|-----|-------------|-------------|-------------|-------------|-------------|
| 1 | (10, 3, 2) | (15, 16, 1) | (20, 36, 5) | (12, 16, 4) | (18, 25, 3) |
| 2 | (30, 36, 1) | (25, 16, 2) | (60,16, 5) | (50, 25, 3) | (40, 25, 4, 1) |
| 3 | (10, 9, 2) | (25, 16, 1) | (40, 80, 5) | (30, 80, 3) | (40, 100, 4) |
| 4 | (50, 16, 1) | (100, 144, 5) | (90, 400, 2) | (100, 400, 4) | (75, 300, 3) |
| 5 | (100, 100, 1) | (80, 144, 5) | (100, 100, 4) | (60, 100, 3) | (100, 150, 2) |
| 6 | (100, 900, 1) | (60, 200, 2) | (400, 900, 3) | (100, 900, 5) | (60, 400, 4) |
| 7 | (90, 900, 5) | (400, 1600, 2) | (30, 100, 1) | (90, 400, 3) | (100, 900, 4) |
| 8 | (120, 900, 4) | (70, 900, 3) | (80, 625, 5) | (50, 100, 2) | (40, 80, 1) |
| 9 | (140, 1000, 5) | (100, 400, 3) | (100, 900, 2) | (50, 100, 1) | (60, 400, 4) |
| 10 | (400, 1000, 5) | (100, 900, 4) | (100, 900, 2) | (70, 400, 3) | (100, 900, 1) |

**TABLE 5.3**  Job-Shop Parameters

| Job | Due Date | Priority Index $\rho$ | Probability $p^*$ | Probability $p^{**}$ |
|-----|----------|------------------------|--------------------|----------------------|
| 1 | 1300 | 10 | 0.7 | 0.6 |
| 2 | 1140 | 20 | 0.8 | 0.7 |
| 3 | 1180 | 10 | 0.7 | 0.6 |
| 4 | 1000 | 30 | 0.9 | 0.8 |
| 5 | 1440 | 25 | 0.85 | 0.75 |
| 6 | 935 | 25 | 0.85 | 0.75 |
| 7 | 1620 | 20 | 0.8 | 0.7 |
| 8 | 950 | 25 | Q85 | 0.75 |
| 9 | 1625 | 25 | 0.85 | 0.75 |
| 10 | 1000 | 30 | 0.9 | 0.8 |

**TABLE 5.4**    Summary of Results

| | Exponential | | | Uniform | | | Normal | | |
|---|---|---|---|---|---|---|---|---|---|
| Job | Rule $I^*$ | Rule $I^{**}$ | Rule II | Rule $I^*$ | Rule $I^{**}$ | Rule II | Rule $I^*$ | Rule $I^{**}$ | Rule II |
| 1 | 0.562 | 0.602 | 0.519 | 0.653 | 0.779 | 0.702 | 0.724 | 0.875 | 0.906 |
| 2 | 0.458 | 0.457 | 0.434 | 0.824 | 0.851 | 0.799 | 0.896 | 0.892 | 0.927 |
| 3 | 0.472 | 0.531 | 0.487 | 0.793 | 0.829 | 0.865 | 0.901 | 0.906 | 0.787 |
| 4 | 0.515 | 0.495 | 0.572 | 0.778 | 0.806 | 0.859 | 0.905 | 0.934 | 0.990 |
| 5 | 0.495 | 0.522 | 0.612 | 0.659 | 0.752 | 0.812 | 0.748 | 0.915 | 0.986 |
| 6 | 0.420 | 0.472 | 0.465 | 0.698 | 0.726 | 0.660 | 0.911 | 0.885 | 0.877 |
| 7 | 0.568 | 0.571 | 0.483 | 0.757 | 0.801 | 0.637 | 918 | 0.899 | 0.776 |
| 8 | 0.565 | 0.560 | 0.631 | 0.775 | 0.762 | 0.728 | 0.873 | 0.892 | 0.955 |
| 9 | 0.672 | 0.571 | 0.631 | 0.721 | 0.657 | 0.702 | 0.871 | 0.784 | 0.797 |
| 10 | 0.532 | 0.549 | 0.603 | 0.793 | 0.805 | 0.839 | 0.899 | 0.915 | 0.992 |

3. Delivery performance values do not depend on the scale of priority indices. Table 5.5 presents comparative computational results obtained by using priority indices set on three different scales as follows (rules $I^*$ and $I^{**}$):

  (a) intial scale,
  (b) scale obtained by dividing the initial scale by 10,
  (c) scale obtained by multiplying the initial scale by 10.

It can be well-recognized that for three different scales the results are practically the same. Thus, delivery performances depend only on the ratios of priority indices but not ontheir values. This is an additional reason in favour of using heuristics $I^*$ and $I^{**}$ since practitioners when setting priority indices usually take less notice of the scale than of the ratios of the indices.

## 5.5  Determining Starting Time Moments by Using Combined Approaches

Unlike the decision-making model based on pairwisecomparison and outlined above, the model described below is based on the concept of tense jobs. The idea of the model is as follows: Assume that at moment $t$, $q \geq 1$ job-operations, $O_{i_1\ell_1}, \ldots, O_{i_q\ell_q}$ are ready to be processed on machine $M_k$. Decision-making is carried out in several steps (Golenko-Ginzburg and Gonik 1997):

**Step 1** is similar to the decision-making suggested in Section 5.4 and results in determining the winner *among the jobs that are waiting in the line*. To realize the pairwise comparison we will henceforth use only Model I, namely rule $I^{**}$ as being a very efficient one. Decision-making is based on using (5.14) and (5.17), while calculating probabilities $p_1 \div p_4$ has been outlined in the previous section. Let the winner be job $J_{i_\xi}$, $1 \leq \xi \leq q$, which needs machine $M_k$ to process job-operation $O_{i_\xi\ell_\xi}$. But the winner, as yet, is not passed on to the machine.

**Step 2** results in forecasting the moment operation on the winner finishes on the machine, if its job-operation starts at moment $t$. For that purpose, the average processing time $\bar{t}_{i_\xi\ell_\xi}$ is used. Thus, the forecasting time for job-operation $O_{i_\xi\ell_\xi}$ to be accomplished is $t + \bar{t}_{i_\xi\ell_\xi}$.

**Step 3** A check has to be carried out: do jobs exist which are being processed at moment $t$ and will need machine $M_k$ for their next operation? If there are no such jobs, the winner, i.e., job $J_{i_\xi}$ is passed on to the machine. Otherwise we apply the next step.

**Step 4** Denote the jobs which have been singled out in Step 3, by $J_{\eta_1}, J_{\eta_2}, \ldots, J_{\eta_r}, r \geq 1$. Denote, further, their job-operations to be processed at moment $t$, by $O_{\eta_1 k_1}, \ldots, O_{\eta_r k_r}$, correspondingly. Note that since their next operations have to be processed on machine $M_k$, relations $M_{\eta_v k_v + 1} = k$, $1 \leq v \leq r$, hold.

Step 4 forecasts the moments that operations on jobs $J_{\eta_v k_v}$ finish and, thus, the jobs are ready to be processed on machine $M_k$. Such a forecasting is carried out for each job independently.

**TABLE 5.5**  Computational Results for Different Scales of Priority Indices

| Job | Exponential | | | | | | Uniform | | | | | | Normal | | | | |
| | $I^*$ | | | $I^{**}$ | | | $I^*$ | | | $I^{**}$ | | | $I^*$ | | | $I^{**}$ | |
| | $p$ | $p \times 10$ | $p{:}10$ | $p$ | $p \times 10$ | $p{:}10$ | $p$ | $p \times 10$ | $p{:}10$ | $p$ | $p \times 10$ | $p{:}10$ | $p$ | $p \times 10$ | $p{:}10$ | $p$ | $p \times 10$ |
|---|---|---|---|---|---|---|---|---|---|---|---|---|---|---|---|---|---|
| 1 | 0.562 | 0.585 | 0.597 | 0.602 | 0.584 | 0.606 | 0.653 | 0.618 | 0.676 | 0.779 | 0.774 | 0.797 | 0.724 | 0.699 | 0.676 | 0.875 | 0.862 |
| 2 | 0.458 | 0.429 | 0.484 | 0.457 | 0.433 | 0.4M | 0.824 | 0.822 | 0.833 | 0.851 | 0.847 | 0.834 | 0.896 | 0.903 | 0.907 | 0.892 | 0.951 |
| 3 | 0.472 | 0.476 | 0.501 | 0.531 | 0.487 | 0.506 | 0.793 | 0.793 | 0.819 | 0.829 | 0.868 | 0.863 | 0.901 | 0.901 | 0.925 | 0.906 | 0.929 |
| 4 | 0.515 | 0.503 | 0.510 | 0.495 | 0.473 | 0.473 | 0.778 | 0.765 | 0.807 | 0.806 | 0.802 | 0.796 | 0.905 | 0.943 | 0.951 | 0.934 | 0.956 |
| 5 | 0.495 | 0.503 | 0.501 | 0.522 | 0.559 | 0.549 | 0.659 | 0.598 | 0.626 | 0.752 | 0.772 | 0.770 | 0.748 | 0.718 | 0.738 | 0.915 | 0.991 |
| 6 | 0.420 | 0.465 | 0.471 | 0.472 | 0.447 | 0.483 | 0.698 | 0.698 | 0.710 | 0.726 | 0.720 | 0.724 | 0.911 | 0.833 | 0.875 | 0.885 | 0.909 |
| 7 | 0.568 | 0.581 | 0.588 | 0.571 | 0.554 | 0.537 | 0.757 | 0.782 | 0.751 | 0.801 | 0.758 | 0.754 | 0.918 | 0.892 | 0.902 | 0.899 | 0.929 |
| 8 | 0.565 | 0.587 | 0.569 | 0.560 | 0.560 | 0.586 | 0.775 | 0.767 | 0.724 | 0.762 | 0.751 | 0.746 | 0.873 | 0.904 | 0.889 | 0.892 | 0.889 |
| 9 | 0.672 | 0.627 | 0.649 | 0.571 | 0.619 | 0.594 | 0.721 | 0.713 | 0.724 | 0.657 | 0.679 | 0.713 | 0.871 | 0.885 | 0.871 | 0.784 | 0.787 |
| 10 | 0.532 | 0.528 | 0.567 | 0.549 | 0.543 | 0.557 | 0.793 | 0.760 | 0.780 | 0.805 | 0.797 | 0.806 | 0.899 | 0.918 | 0.907 | 0.915 | 0.911 |

Consider, for example, job $J_{\eta_\nu}$, $1 \leq \nu \leq r$. The job has started to be processed at moment $S_{\eta_\nu k_\nu}$ and at moment $t$, it is still under way. *Thus, the problem is to calculate the mean value of the processing time, on condition that the latter exceeds value* $\Delta_{\eta_\nu, k_\nu} = t - S_{\eta_\nu, k_\nu}$. The expected value of $F_{\eta_\nu, k_\nu}$ is as follows

$$\bar{F}_{\eta_\nu k_\nu}(S_{\eta_\nu k_\nu}, \Delta_{\eta_\nu k_\nu}) = S_{\eta_\nu k_\nu} + \bar{t}_{\eta_\nu k_\nu}(\Delta_{\eta_\nu k_\nu}). \tag{5.23}$$

To solve the problem in the general case, one has to determine for a random value $\alpha$, a conditional average value $E(\alpha/\alpha \geq \Delta)$. It can be well recognized that if $f_\alpha(x)$ is the probability density function of value $\alpha$, relation

$$E(\alpha/\alpha \geq \Delta) = \frac{1}{1 - F_\alpha(\Delta)} \cdot \int_\Delta^\infty x \cdot f_\alpha(x)\,dx \tag{5.24}$$

holds. Here $F_\alpha(x)$ is the cumulative probability function

$$F_\alpha(x) = \int_\infty^x f_\alpha(y)\,dy. \tag{5.25}$$

We have determined values $E(\alpha/\alpha \geq \Delta)$ analytically for three distributions of random processing time on the machines:

1. normal distribution with mean $\bar{t}_{i\ell}$ and variance $V_{i\ell}$;
2. uniform distribution in the interval $[\bar{t}_{i\ell} - 3\sqrt{V_{i\ell}}, t_{i\ell} + 3\sqrt{V_{i\ell}}]$;
3. exponential distribution with value $\gamma = \frac{1}{\bar{t}_{i\ell}}$.

For the exponential distribution with the p.d.f. $f_\alpha(x) = \lambda \ell^{-\lambda x}$ using (5.24) results in $E(\alpha/\alpha \geq \Delta) = \Delta + \frac{1}{\lambda}$.

In the case of a uniform distribution in $[a, b]$, i.e., for p.d.f. $f_\alpha(x) = \frac{1}{b-a}$, $E(\alpha/\alpha \geq \Delta) = \frac{(b+\Delta)}{2}$.

For the normal distribution with parameters $(a, \sigma^2)$, i.e., for p.d.f. $f_\alpha(x) = \frac{1}{\sqrt{2\pi}}e^{-(x-a)^2/2\sigma^2}$, we obtain

$$E(\alpha/\alpha \geq \Delta) = a + \frac{\sigma \cdot e^{-\frac{(\Delta - a)^2}{2\sigma^2}}}{\sqrt{2\pi\left[1 - \phi\left(\frac{\Delta - a}{\sigma}\right)\right]}}$$

where

$$\phi(x) = \frac{1}{\sqrt{2\pi}}\int_{-\infty}^x e^{-\frac{y^2}{2}}\,dy.$$

The analytical determination of values $E(\alpha/\alpha \geq \Delta)$ is outlined below, in the Appendix.

To simplify the notations, we shall henceforth denote values $\bar{F}_{\eta_\nu k_\nu}(S_{\eta_\nu k_\nu}, \Delta_{\eta_\nu k_\nu})$ by $\bar{F}_{\eta_\nu k_\nu}$. After calculating those values we apply the next step.

**Step 5** singles out all jobs $J_{\eta_\nu}$ with

$$\bar{F}_{\eta_\nu k_\nu} \in [t, t + \bar{t}_{i_\xi \ell_\xi}], \tag{5.26}$$

i.e., all the jobs which are forecasted to be finished in the interval $[t, t + \bar{t}_{i_\xi \ell_\xi}]$.

If there are no such jobs, job $J_{i_\xi \ell_\xi}$ is passed on to the machine, i.e., $S_{i_\xi \ell_\xi} = t$. Otherwise go to the next step.

**Step 6** is the central part of the model, where decision-making is actually carried out. A competition is arranged between the winner $J_{i_\xi}$ and each of jobs $J_{\eta_\nu}$, $1 \leq \nu \leq r$, that have been singled out at step 5. The competition is, in essence, a combination of the pairwise comparison techniques and a modification of the "look ahead" mechanisms outlined in Section 5.3.

Let job $J_{i_\xi}$ compete at moment $t$ with the first unfinished job $J_{\eta_1}$ satisfying (26). Two alternative decision-makings are imbedded in the model:

1. Job $J_{i_\xi}$ is passed on to the machine at moment $t$, i.e., is operated first, and job $J_{\eta_v}$ afterwards, at moment $t + \bar{t}_{i_\xi \ell_\xi}$;

2. Machine $M_k$ remains idle until the moment $\bar{F}_{\eta_1 k_1}$ when job $J_{\eta_1}$ arrives and is ready to be passed on to the machine. Note that moment $\bar{F}_{\eta_1 k_1}$ is an *average value that is forecasted and calculated* by decision-makers at decision point $t$, i.e., it is not the moment job $J_{\eta_1}$ will *actually arrive*. The latter moment is a random value which has to be simulated. At moment $\bar{F}_{\eta_1 k_1}$, job $J_{\eta_1}$ is passed on to the machine and is operated first, while job $J_{i_\xi}$ starts afterwards at moment $\bar{F}_{\eta_1 k_1} + \bar{t}_{\eta_1, k_1 + 1}$.

Four modified probability values are calculated by (5.12), namely

$$p_1 = P\{J_{i_\xi}/t\}, \tag{5.27}$$

$$p_2 = P\{J_{\eta_1}/\bar{F}_{\eta_1 k_1}\}, \tag{5.28}$$

$$p_3 = P\{J_{\eta_1}/t + \bar{t}_{i_\xi \ell_\xi}\}, \tag{5.29}$$

$$p_4 = P\{J_{i_\xi}/\bar{F}_{\eta_1 k_1} + t_{\eta_1, k_1 + 1}\}. \tag{5.30}$$

It can be well-recognized that values (5.27–5.30) are nothing else but modified delivery performances values based on the operational significance of "waiting". Namely, values $p_2$ and $p_4$ are based on the decision not to pass job $J_{i_\xi}$ (ready to be processed) on to the machine, but to keep the machine idle until the tense job $J_{\eta_1}$ is ready to be served and is passed on to that machine. Only after operation on job $J_{\eta_1}$ finishes, the competitive job $J_{i_\xi}$ is passed on to the machine.

The decision-making is as follows: if relation

$$\rho_{i_\xi} \cdot p_1 + \rho_{\eta_1} \cdot p_3 \geq \rho_{i_\xi} \cdot p_4 + \rho_{\eta_1} \cdot p_2 \tag{5.31}$$

holds, job $J_{i_\xi}$ wins the competition and has to be processed first on the machine, i.e., without waiting. Otherwise the machine has to remain idle, until the arrival of the "tense job" $J_{\eta_1}$. Such a competition has to be carried out independently between job $J_{i_\xi}$ and all jobs $J_{\eta_v}$, $1 \leq v \leq r$, that, according to the short-term forecasting, are ready to be served on machine $M_k$ within the interval $[t, t + \bar{t}_{1_\xi \ell_\xi}]$.

**Step 7** *If job $J_{i_\xi}$ wins the competition with all the jobs $J_{\eta_v}$, $1 \leq v \leq r$, it is passed on the machine $M_k$ at moment $t$. Otherwise, i.e., if job $J_{i_\xi}$ fails in the course of even one pairwise comparison, the machine remains idle until the next decision moment $t^*$, when a new job actually enters the line and is ready to be served on the machine. At that moment we apply Step 1 of the model, i.e., the process of choosing the job to be passed on the machine begins anew.*

Note, in conclusion, that even if a new job, that enters the line at moment $t^*$, is just the very "overall winner" to be awaited from moment $t$, we cannot pass the job on to the machine straightforwardly, without making a new decision. This is because the next decision moment $t^*$.

$$t^* = \underset{1 \leq v \leq r}{\text{Min}} \{S_{\eta_v k_v} + t_{\eta_v k_v}\} \tag{5.32}$$

is a *random value*, while decision-making at the previous moment to is based on forecasted values $\bar{F}_{\eta_v k_v}$, which are *calculated conditional mean values*. Thus, in practice values $t^*$ and $\bar{F}_{\eta_v k_v}$ never coincide, and the decision-making to be undertaken deals with a new situation. That is why the heuristic decision-making model has to be applied anew.

In order to compare the efficiency of the decision model based on "look ahead" techniques versus the previous one based on pairwise comparison only, experimentation has been carried out. A flexible manufacturing cell of medium size was chosen.

The job-shop FMS comprises 16 jobs on 8 different machines. Priority indices have been set by practitioners. The initial data matrix and other job-shop parameters are given in Tables 5.6 and 5.7. Three distributions of random processing time on the machines have been considered:

1. normal distribution with mean $\bar{t}_{i\ell}$ and variance $V_{i\ell}$,
2. uniform distribution in the interval $[\bar{t}_{i\ell} - 3\sqrt{V_{i\ell}}, t_{i\ell} + 3\sqrt{V_{i\ell}}]$,
3. exponential distribution with value $\lambda = 1/\bar{t}_{i\ell}$.

Both decision models under comparison have been implemented on an IBM/PC 486 in the Turbo C++ 3.0 programming language. In order to compare the efficiency of those models, we have evaluated the delivery performances $p_i$, $1 \leq i \leq 16$, for all jobs entering the flexible manufacturing cell, by using heuristic (5.12). Two models are compared, namely:

1. Heuristic $I^{**}$ based only on pairwise comparisons, i.e., not comprising the "look ahead" mechanisms for tense jobs; we will henceforth call it Model A.
2. The model based on a combination of pairwise comparisons and "look ahead" forecasting (Model B).

For each model, a sample of 500 simulation runs has been obtained. Each simulation run results in simulating the job-shop's work by random sampling of the actual durations of the job-operations and by using decision-making at each decision point. For each job, its delivery performance, i.e., the probability of meeting the due date on time, has been evaluated. The statistical outcome data of the sample is presented in Table 5.8.

It can be well recognized that such a sample does not provide sufficient statistical data to undertake the comparison: each of the two competitive samples includes only 16 units. In order to obtain representative statistics to carry out decision-making by means of the theory of statistical hypotheses, 10 samples have been simulated. Thus, for each distribution and each model under comparison, 5000 simulation runs have been realized to obtain an enlarged sample comprising 160 units. The statistical parameters of such a sample are presented in Table 5.9.

By examining Table 5.9, it can be well-recognized that most of the delivery performance values for model B, including the average value, are higher than those for model A. Thus, a hypothesis can be introduced that model B is superior to model A. To check this hypothesis, a one-sided test of hypotheses is introduced (Walpole and Myers 1978) as follows:

Hypothesis $H_0$: The delivery performance of model B is *not* essentially better than that of model A;
Hypothesis $H_1$: Model B provides an essentially better delivery performance than model A.

In order to undertake decision-making, a level of significance $\alpha$ has to be chosen. We have chosen $\alpha = 0.01$ which guarantees in practice that hypothesis $H_1$ is not chosen by mistake. Assume the sample under comparison to be approximately normal, with equal variances. Let the first sample include $n_1$ items with an average of $\bar{x}_1$ and a standard deviation of $s_1$, while the corresponding parameters of the second sample are $n_2$, $\bar{x}_2$ and $s_2$. The alternative hypotheses are as follows:

$H_0$: $\mu_1 = \mu_2$, i.e., model B is *not* better than model A.
$H_1$: $\mu_2 = \mu_2$, i.e., model B *is* better than model B.

The classical one-sided test comprises the following steps (Walpole and Myers 1978):

**Step 1** Calculate the value

$$S_p = \sqrt{\frac{(n_1 - 1)s_1^2 + (n_2 - 1)s_2^2}{n_1 + n_2 - 2}};$$

**TABLE 5.6** Initial Data Matrix of the (16 × 8) Job-Shop Problem

| Operations Orders | Op.#1 | Op.#2 | Op.#3 | Op.#4 | Op.#5 | Op.#6 | Op.#7 | Op.#8 |
|---|---|---|---|---|---|---|---|---|
| $J_1$ | (140,100,3) | (120,100,4) | (160,100,1) | (400,400,8) | (300,225,6) | (250,225,5) | (120,100,2) | (190,100,7) |
| $J_2$ | (180,100,4) | (160,100,1) | (500,400,8) | (300,225,6) | (250,100,3) | (180,100,2) | (170,100,7) | (400,225,5) |
| $J_3$ | (160,100,2) | (190,225,5) | (600,400,1) | (100,100,4) | (140,100,3) | (150,100,7) | (600,400,8) | (400,225,6) |
| $J_4$ | (180,100,1) | (200,100,2) | (140,100,4) | (400,400,8) | (190,225,5) | (320,225,6) | (170,100,7) | (160,100,3) |
| $J_5$ | (200,225,5) | (140,100,4) | (180,100,1) | (540,400,8) | (280,225,6) | (280,100,3) | (100,100,2) | (200,100,7) |
| $J_6$ | (170,100,1) | (180,100,2) | (480,400,8) | (380,225,5) | (150,100,3,) | (130,100,4) | (180,100,7) | (290,225,6) |
| $J_7$ | (400,225,5) | (150,100,4) | (580,400,1) | (600,400,8) | (160,100,7) | (300,225,6) | (200,100,3) | (190,100,2) |
| $J_8$ | (110,100,7) | (500,100,8) | (600,400,1) | (190,100,2) | (400,225,5) | (100,100,4) | (300,225,6) | (180,100,3) |
| $J_9$ | (100,100,1) | (500,400,2) | (250,100,3) | (300,225,5) | (210,225,6) | (120,100,4) | (650,400,8) | (90,100,7) |
| $J_{10}$ | (100,100,3) | (400,400,8) | (100,100,4) | (400,400,1) | (180,100,2) | (220,225,5) | (310,225,6) | (130,100,7) |
| $J_{11}$ | (240,225,5) | (140,225,7) | (400,225,8) | (500,400,1) | (190,225,2) | (160,100,3) | (170,225,4) | (250,225,6) |
| $J_{12}$ | (140,100,4) | (400,400,1) | (450,400,8) | (170,225,7) | (150,100,6) | (180,225,5) | (170,100,2) | (100,100,3) |
| $J_{13}$ | (120,100,3) | (400,400,2) | (190,225,5) | (80,100,1) | (450,400,8) | (90,100,7) | (400,400,6) | (160,225,4) |
| $J_{14}$ | (250,225,6) | (110,100,3) | (400,400,8) | (260,225,5) | (180,100,7) | (400,400,2) | (180,225,4) | (500,400,1) |
| $J_{15}$ | (400,400,8) | (100,100,1) | (400,400,7) | (110,100,2) | (280,225,5) | (110,100,6) | (280,225,4) | (380,400,3) |
| $J_{16}$ | (140,100,6) | (280,225,5) | (380,400,1) | (100,100,2) | (400,400,8) | (160,225,7) | (180,100,3) | (500,400,4) |

**TABLE 5.7**   Parameters of the Job-shop

| Order Number $i$ | Prirority Index $\rho_i$ | Earliest Date $E_i$ | Due Date $D_i$ |
|---|---|---|---|
| 1 | 30 | 0 | 2450 |
| 2 | 20 | 0 | 9000 |
| 3 | 30 | 0 | 5400 |
| 4 | 25 | 0 | 6500 |
| 5 | 30 | 0 | 3900 |
| 6 | 20 | 0 | 10200 |
| 7 | 30 | 0 | 3600 |
| 8 | 25 | 0 | 4700 |
| 9 | 20 | 0 | 7250 |
| 10 | 25 | 0 | 7100 |
| 11 | 20 | 0 | 8200 |
| 12 | 20 | 0 | 7150 |
| 13 | 20 | 0 | 8800 |
| 14 | 25 | 0 | 5600 |
| 15 | 30 | 0 | 3300 |
| 16 | 30 | 0 | 5450 |

**TABLE 5.8**   Computational Results of the Jobs' Delivery Performances: Model A versus Model B (an Example of a Sample Based on 500 Simulation Runs)

| Jobs | Exponential Model A | Exponential Model B | Uniform Model A | Uniform Model B | Normal Model A | Normal Model B |
|---|---|---|---|---|---|---|
| 1 | 0.632 | 0.634 | 0.839 | 0.826 | 0.861 | 0.803 |
| 2 | 0.780 | 0.854 | 0.803 | 0.845 | 0.789 | 0.892 |
| 3 | 0.664 | 0.700 | 0.906 | 0.911 | 0.888 | 0.906 |
| 4 | 0.691 | 0.709 | 0.843 | 0.906 | 0.892 | 0.915 |
| 5 | 0.596 | 0.587 | 0.798 | 0.836 | 0.767 | 0.817 |
| 6 | 0.839 | 0.812 | 0.874 | 0.915 | 0.870 | 0.906 |
| 7 | 0.614 | 0.624 | 0.919 | 0.887 | 0.901 | 0.906 |
| 8 | 0.538 | 0.596 | 0.834 | 0.878 | 0.789 | 0.859 |
| 9 | 0.789 | 0.864 | 0.946 | 0.977 | 0.978 | 0.995 |
| 10 | 0.623 | 0.624 | 0.780 | 0.831 | 0.857 | 0.864 |
| 11 | 0.659 | 0.643 | 0.901 | 0.873 | 0.897 | 0.915 |
| 12 | 0.704 | 0.681 | 0.865 | 0.859 | 0.870 | 0.859 |
| 13 | 0.632 | 0.667 | 0.704 | 0.775 | 0.722 | 0.793 |
| 14 | 0.484 | 0.474 | 0.753 | 0.793 | 0.949 | 0.793 |
| 15 | 0.466 | 0.437 | 0.879 | 0.864 | 0.874 | 0.869 |
| 16 | 0.623 | 0.638 | 0.937 | 0.920 | 0.879 | 0.920 |

**Step 2** Calculate $t$

$$t = \frac{\bar{x}_2 - \bar{x}_1}{s_p \sqrt{\dfrac{1}{n_1} + \dfrac{1}{n_2}}} \;;$$

**Step 3** Determine the upper bound $T$ of the critical region on the basis of the $t$-distribution [45, p. 514] with $v = 160 + 160 - 2 = 318$ degrees of freedom. For $\alpha = 0.01$ we obtain $T = 2.326$.

**Step 4** if $t > T$ accepts hypothesis $H_1$. Otherwise, a decision can be drawn that $H_1$ is likely to be true but model B does not differ sufficiently from model A.

The computations are as follows (see Table 5.9):

## I. Operations Duration $t_{i\ell}$ Have an Exponential Distribution

$$\bar{x}_1 = 0.632; \qquad s_1 = 0.14; \qquad n_1 = 160;$$

$$\bar{x}_2 = 0.639; \qquad s_2 = 0.14; \qquad n_2 = 160;$$

Value $s_p = 0.14$ and $t = \dfrac{0.007}{0.14\sqrt{0.0125}} = 0.44 < 2.326$.

Thus, hypothesis $H_1$ has to be rejected in favor of the alternative hypothesis $H_0$. That means that model B does not refine the delivery performance of model A.

## II. Values $t_{i\ell}$ Are Distributed Uniformly

$$\bar{x}_1 = 0.837; \qquad s_1 = 0.07; \qquad n_1 = 160;$$

$$\bar{x}_2 = 0.856; \qquad s_2 = 0.07; \qquad n_2 = 160;$$

Value $s_p = 0.07$ and $t = \dfrac{0.019}{0.07\sqrt{0.0125}} = 2.43 > 2.326$.

Since $t$ exceeds the critical value $T$ with the level of significance $\alpha = 0.01$ hypothesis $H_0$ has to be rejected in favor of hypothesis $H_1$. Thus, model B is taken to be essentially better than model A.

## III. Values $t_{i\ell}$ Have a Normal Distribution

$$\bar{x}_1 = 0.842; \qquad s_1 = 0.07; \qquad n_1 = 160;$$

$$\bar{x}_2 = 0.863; \qquad s_2 = 0.067; \qquad n_2 = 160;$$

Value $s_p = 0.069$ and $t = \dfrac{0.021}{0.069\sqrt{0.0125}} = 2.72 > 2.326$.

Hypothesis $H_1$ is accepted with a very high level of reliability, and model B can be seen to refine model A essentially.

The following conclusions can be drawn from the experimentation:

1. *For the case of uniform and, especially, normal distributions,* using model B results in higher delivery performance than model A.
2. *For the case of the exponential distribution,* model B is practically no better than model A. This phenomenon has an explanation: model B is based on the conception of a "tense job." But for the case of random operations, a "tense job" is usually the result of a random "jump" to the right tail, i.e., when the simulated random duration is close to the upper bound of the distribution. For an exponential distribution, such events occur very seldom, since the distribution is asymmetric and 2/3 of the random values are concentrated to the left of the average.
3. *For symmetrical distributions and especially for asymmetrical ones which are concentrated to the right tail,* model B is essentially more effective than model A.
4. *Substituting model A for model B does not result in any considerable increase of the computational time:* realizing 500 simulation runs on PC 486 (model A) takes approximately twenty seconds, while using model B increases that time by ten seconds.
5. We previously assumed that the samples are approximately normal, i.e., that the delivery performance values $p_i$ have a normal distribution. But since test hypotheses can also be used with non-normal populations when $n \geq 30$ (Walpole and Myers 1978, p. 251), we have the right to apply decision theory to the enlarged sample in Table 5.9, independently of the general distribution.

**TABLE 5.9**    Statistical Parameters of the Enlarged Sample (160 units)

| Parameters | Distribution | | | | | |
|---|---|---|---|---|---|---|
| | Exponential | | Uniform | | Normal | |
| | Model A | Model B | Model A | Model B | Model A | Model B |
| Average | 0.632 | 0.639 | 0.837 | 0.856 | 0.842 | 0.863 |
| Variance | 0.019 | 0.019 | 0.005 | 0.005 | 0.005 | 0.0045 |
| Number of units for which the delivery performance values of model B are higher than those of model A | 81 | | 104 | | 111 | |
| Number of units for which they are lower | 75 | | 51 | | 43 | |
| Number of units for which they are equal | 4 | | 5 | | 6 | |

## 5.6 An Order Package with Random Operations and without Due Dates

For certain job-shop FMS an order package has to be manufactured as soon as possible, but the corresponding due dates are not pregiven. The problem is to determine the starting times $S_{i\ell}$ of each job-operation in order to minimize the average value of the schedule time, i.e., the average value of the time needed to manufacture all the jobs (orders) entering the package. All jobs have common values $E_i$ which for the sake of simplicity are taken to be zero. Thus, the problem is as follows: determine values $S_{i\ell}$ to minimize the schedule time (the makespan)

$$\underset{S_{i\ell}}{\text{Min}}\left[\underset{i}{\text{Max}}F_i - \underset{i}{\text{Min}}S_{i1}\right] \tag{5.33}$$

subject to

$$S_{i\ell} \geq 0, \quad 1 \leq i \leq n, \quad 1 \leq \ell \leq m_i, \tag{5.34}$$

$$S_{i,\ell+1} \geq S_{i\ell} + t_{i\ell}, \quad 1 \leq \ell \leq m_i - 1. \tag{5.35}$$

We suggest a simple but effective scheduling procedure based on the results of Golenko-Ginzburg and Kats (1995, 1997). By selecting several of the most effective and best-known priority rules in job-shop scheduling and examining their relative improvement rates for different job-shop problems with deterministic durations $t_{i\ell}$, a conclusion can be drawn (Golenko-Ginzburg and Kats 1995) as follows:

1. In case $m > 3.3n - 10$ the most effective rule is TSW outlined in Section 5.3.
2. In case $0.75n - 15 < n \leq 3.3n - 10$ using rule PC outlined above results in minimizing the makespan relative to all other priority rules.
3. In case $m < 0.75n - 15$ the most effective priority rule (TS) is as follows:
   If the tense job $J_y$ (see rule TSW) is not waiting in line for machine $M_k$, choose job $J_\xi$ according to the priority rule SPT (see Section 5.3). If the tense job is in the line it has to be chosen.

We shall use those results in stochastic job-shop scheduling to arrange the competition among several jobs waiting for one and the same machine. If at a certain moment $t$, several job operations $O_{i_r, \ell_r}, 1 \leq r \leq q$

with random $t_{i\ell}$ are waiting to be processed on machine $M_k$, the competition among them as to be arranged as follows:

1. Determine all the job-operations that at moment $t$ have not yet started to be operated. Substitute their random values $t_{i\ell}$ for the corresponding average values $\bar{t}_{i\ell}$:
2. For job-operations $O_{i\ell}$ entering the job-shop FMS *and being under operation at moment t*, calculate their *remaining average* durations $\bar{F}_{i\ell} - t$ according to the techniques outlined in Section 5.5 and the Appendix. For those operations take $t_{i\ell} = \bar{F}_{i\ell} - S_{i\ell}$.
3. For all other job-operations with $F_{i\ell} < t$ take their actual time realizations $F_{i\ell} - S_{i\ell} = t_{i\ell}$. Thus, in the course of realizing 1–3, all job-operations $O_{i\ell}$ entering the package obtain deterministic values $t_{i\ell}$.
4. For the $(n - m)$-job-shop FMS determine the best priority rule (PC, TSW or TS) for the corresponding area.
5. Apply that rule to choose the job from the line of $q$ competitive job-operations $O_{i,\ell_r}, 1 \le r \le q$, ready to be processed on the machine $M_k$.
6. Apply steps 1–5 for all decision moments in the course of scheduling the job-shop.

In order to verify the suggested procedure experimentation has been carried out. The initial data of the job-shop problem compris 16 jobs and 8 machines and is presented in Table 5.6. Since the job-shop FMS refers to the second (n · m)-area (see above), priority rule PC has to be applied.

The suggested heuristic to schedule a job-shop package in order to minimize the total schedule time value (33) has been compared with four job-shop scheduling heuristics as follows:

1. Model $I^*$ described in Section 5.4 (model $A^*$);
2. Model $I^{**}$ described in Section 5.4 (model $A^*$);
3. Combined model $B^*$ comprising scheduling model $I^*$ and the "look ahead" techniques (see Section 5.5);
4. Combined model $B^{**}$ comprising model $I^{**}$ and the "look ahead" techniques (see Section 5.5).

For each competitive heuristic three distribution laws have been considered, namely, the exponential, the uniform and the normal density distributions. The distributions are similar to those outlined in Sections 5.4 and 5.5. For each distribution and each heuristic under comparison 500 simulation runs have been realized in order to obtain the average makespan value (5.33). The results are presented in Table 5.10.

It can be well-recognized that for all types of distributions the suggested heuristic (we will henceforth call it model C) provides essentially smaller makespan values than heuristics $A^*$, $A^{**}$, $B^*$, and $B^*$. Thus, we recommend to use the latter only for job-shop packages comprising jobs with different due dates and different priorities. For the case of a package with equal job priorities and the makespan to be minimized heuristic C has to be used.

Note that, similar to all heuristics outlined above, using exponential distribution results in obtaining worse results than by using uniform or normal distributions. The two latter distributions are practically of similar efficiency.

**TABLE 5.10** Summary of Results for a Job-Shop Package without Priorities and Due Dates

| | Distribution | | |
|---|---|---|---|
| | Exponential | Uniform | Normal |
| Heuristic | Average Makespan Values | | |
| Model $A^*$ | 9229 | 8190 | 8192 |
| Model $A^*$ | 9172 | 8172 | 8151 |
| Model $B^*$ | 9185 | 8135 | 8121 |
| Model $B^{**}$ | 9139 | 8141 | 8113 |
| Model C | 8551 | 7659 | 7657 |

## 5.7 Job-Shop Problems with Random Operations and Cost Objectives (Coffman, Hofri, and Weiss 1989)

In various job-shop problems the objective is to minimize the total penalty expenses for not delivering jobs on time, i.e., in case $F_i > D_i$. The corresponding penalty expenses for each job $J_i$ can be subdivided into two parts, namely:

$C_i^*$    the penalty cost for not delivering job $J_i$ on time, i.e., for the case $F_i > D_i$;

$C_i^{**}$   the penalty cost *per time unit* within the waiting interval between the due date $D_i$ and the moment $F_i$ when the job will be delivered to the customer.

Thus, the total penalty expenses for the job-shop flexible manufacturing cell are as follows:

$$C = \sum_{i=1}^{n} C_i = \sum_{i-1}^{n} \{[C_i^* + C_i^{**}(F_i - D_i)] \cdot \delta(F_i - D_i)\} \qquad (5.36)$$

where

$$\delta(x) = \begin{cases} 0 & \text{if } x \leq 0 \\ 1 & \text{otherwise.} \end{cases}$$

The problem is to determine values $S_{i\ell}$ in order to minimize the total penalty expenses

$$\underset{S_{i\ell}}{\text{Min}} \left\{ \sum_{i-1}^{n} [C_i^* + C_i^{**}(F_i - D_i)] \cdot \delta(F_i - D_i) \right\} \qquad (5.37)$$

subject to (5.34–5.35).

We will use and later on compare four heuristics, as follows:

1. Heuristic $I^*$ based on pairwise comparison PC (model $A^*$).
2. Heuristic $I^{**}$ based on PC (model $A^{**}$).
3. Combined heuristic based on decision rule $I^{**}$ and the "look ahead" techniques outlined in Sections 5.5 (model $B^*$).
4. Combined heuristic based on decision rule $I^{**}$ and the "look ahead" techniques (model $B^{**}$).

In order to compare the efficiency of those models, we have calculated the average total penalty expenses for all jobs entering the flexible manufacturing cell, as described in Tables 5.6–5.7. For each model, a sample of 500 simulation runs has been obtained. Each simulation run, like in Sections 5.4 and 5.5, results in simulating the job-shop's work by random sampling of the actual durations of the job-operations and by using decision-making at each decision point. For each job, two kinds of penalty expenses have been considered, as follows:

1. For a job $J_i$ with a highest priority index $\rho_i = 30$:

$$C_i^* = 1000; \qquad C_i^{**} = 100; 150;$$

2. For a job $J_i$ with a medium priority index $\rho_i = 25$:

$$C_i^* = 800; \qquad C_i^{**} = 80; 120;$$

3. For a job $J_i$ with the lowest priority index $\rho_i = 20$:

$$C_i^* = 600; \qquad C_i^{**} = 60; 100;$$

TABLE 5.11 Summary of Results for a Job-Shop Package without Priorities and Due Dates

| Model | Set | Distribution | | |
| --- | --- | --- | --- | --- |
| | | Exponential | Uniform | Normal |
| $A^*$ | 1 | 756,617 | 88,844 | 94,851 |
| | 2 | 1,157,318 | 138,140 | 146,005 |
| $A^{**}$ | 1 | 736,972 | 96,646 | 92,316 |
| | 2 | 1,127,242 | 150,217 | 142,610 |
| $B^*$ | 1 | 756,397 | 95,680 | 85,396 |
| | 2 | 1,155,872 | 147,811 | 131,550 |
| $B^{**}$ | 1 | 775,294 | 80,466 | 84,159 |
| | 2 | 1,186,823 | 124,928 | 130,957 |

Three distributions of random processing time on the machine, similar to those described in Sections 5.4 and 5.5, have been considered. The computer program to simulate the job-shop is written in the Turbo C++ 3.0 programming language and has been implemented on an IBM/PC 486. Thus, we have checked four models ($A^*$, $A^{**}$, $B^*$, $B^{**}$), each of them by using three distributions laws (exponential, uniform and normal) and two sets of penalty expenses ($C_1$ and $C_2$) as follows:

Set 1   $\rho_i = 20; C_i^* = 600; C_i^{**} = 60$     $\rho_i = 25; C_i^* = 800; C_i^{**} = 80$     $\rho_i = 30; C_i^* = 1000; C_i^{**} = 100$
Set 2   $\rho_i = 20; C_i^* = 60; C_i^{**} = 100$     $\rho_i = 25; C_i^* = 800; C_i^{**} = 120$    $\rho_i = 30; C_i^* = 1000; C_i^{**} = 150$

Thus a total of 24 combinations ($4 \times 3 \times 2$) has been considered. For each combination, 500 simulations runs were performed. The number of simulation runs, like similar cases described in Section 5.5, was determined by applying classical estimation theory (Walpole and Myers 1978). The summary of results is presented in Table 5.11.

The following conclusions can be drawn from the summary:

1. For the case of exponential distribution, the total penalty expenses for a job-shop are essentially higher than for other distributions. Thus, a job-shop flexible manufacturing cell with operations of exponential random distribution is the most costly one.
2. For the case of the exponential distribution, the use of models $B^*$ and $B^{**}$ does not result in decreased total penalty expenses, in comparison with models $A^*$ and $A^{**}$. Thus, model $B^*$ and $B^{**}$ are practically no better than models $A^*$ and $A^{**}$. This fully coincides with the conclusions outlined in Section 5.5.
3. For the case of unifor and normal distributions, the use of improved models B results in decreased total penalty expenses by 9–10%. Thus, models $B^*$ and $B^{**}$ with cost objectives are, essentially, more effective for symmetrical distributions than models $A^*$ and $A^{**}$.
4. When comparing the improved models $B^*$ and $B^{**}$, it can be concluded that model $B^{**}$ is more effective than model $B^*$. Note that model $B^{**}$ has been used successfully for the case of non-cost objectives, too (see Section 5.5).

## 5.8   Conclusions

1. The decision models based on the idea of pairwise comparison amongst competing jobs are effective and easy in usage. The pairwise approach has been used successfully in other areas of operations management, e.g., the analysis of PERT networks. Simulation results show that the pairwise approach is a very useful procedure also in job-shop scheduling with random processing times.
2. Two models based on the idea of pairwise comparison are suggested. Model I has the same efficiency as Model II, but is simpler and easier to be used by practitioners. Thus Model I (i.e., rules $I^*$ and $I^{**}$) can be recommended for industrial job-shop scheduling.

3. Beside the concept of pairwise comparison, there is another fruitful approach for choosing jobs from the line for the machine, namely, by determining the so-called tense job, which practically serves as a bottleneck for the job-shop. Thus, in some cases it is reasonable not to forward jobs to the machine (even in the case of the nonempty line for that machine) but to wait for the moment the tense job will enter the line. Such a "look ahead" approach may be introduced in stochastic scheduling too. A unification of both approaches is from 2.5% to 3.5% more effective than each of them separately.

4. The developed unified decision model can be widely used in various job-shop FMS as a support tool. One has only to undertake decision-making at each decision point. The model can be applied to JIT FMS with random operations. For those systems the number of publications remains very scanty.

5. Both decision models outlined in Sections 5.4 and 5.5 are easy to handle; they can be implemented on a PC. The models can be used in real-time projects. Realizing decision-making takes very little computational time.

6. Besides the principal models outlined in Sections 5.4 and 5.5, other effective heuristics can be suggested for specific cases in job-shop scheduling, e.g., for the case of a unified order package without due dates and job priorities. To minimize the average makespan's duration, one can use simplified priority rules based on the idea of pairwise comparison.

7. Various cost parameters can be introduced in the developed model, e.g., penalty expenses for manufacturing jobs on time. A conclusion can be drawn that using the unified decision model comprising both the pairwise comparison and the "look ahead" techniques results in an essential reduction of total penalty expenses relative to noncombined decision-making rules.

8. Future research can be undertaken in solving optimization problems of maximizing the FMS's net profit gained by performing the job-shop orders.

## Acknowledgments

This research has been partially supported by the Paul Ivanier Center of Robotics and Production Management, Ben-Gurion University of the Negev. The authors are most thankful to Professor Z. Landsman for very helpful comments.

The authors are very grateful to Elsevier Science-NL, Saraburgerhartstraat 25, 1055 KV Amsterdam, The Netherlands, for their kind permission to reprint material from publications (Golenko-Ginzberg et al. 1995, Golenko-Ginzberg and Gonik 1997).

## References

Balas, A. E. and D. Zawack, 1988. The shifting bottleneck procedure for job-shop scheduling. *Mgmt. Sci.,* 34: 391–401.

Barker, J. R. and B. M. Graham, 1985. Scheduling the general job-shop. *Mgmt. Sci.,* 31: 594–598.

Blackstone, J. H. Jr., D. T. Phillips, and D. L. Hogg, 1982. A state-of-the-art survey of dispatching rules for manufacturing job-shop operations. *Int. J. Prod. Res.,* 20: 27–45.

Carlier, J. and E. Pinson, 1989. An algorithm for solving the job-shop problem. *Mgmt. Sci.,* 35: 164–176.

Coffman, E. G., 1976. *Computer and Job-shop Scheduling Theory,* Wiley, New York.

Coffman, E. G., Jr., M. Hofri, and G. Weiss, 1989. Scheduling stochastic jobs with a two room distribution on two parallel machines. *Probab. Eng. Informat. Sci.,* 3: 89–116.

Conway, R. W., W. L. Maxwell, and L. W. Miller, 1967. *Theory of Scheduling,* Addison-Wesley, Reading, Massachusetts.

D'ell' Amico, M. and M. Trubian, 1993. Applying tabu-search to the job-shop scheduling problem. *Ann. Oper. Res.,* 41: 231–252.

Eilon, S., 1979. Production Scheduling, in: K. B. Haley (Ed.), *OR '78,* North-Holland, Amsterdam, pp. 237–266.

French, S., 1982. *Sequencing and Scheduling,* Wiley, New York.

Gere, W. S., 1966. Heuristics in job-shop scheduling. *Mgmt. Sci.,* 13: 167–190.

Golenko (Ginzburg), D. I., 1973. *Statistical Models in Production Control.* Statistika, Moscow (in Russian).

Golenko-Ginzburg, D., Sh. Kesler, and Z. Landsman, 1995. Industrial job-shop scheduling with random operations and different priorities. *Int. J. Prod. Econ.,* 40: 185–195.

Golenko-Ginzburg, D. and V. Kats, 1995. Priority rules in job-shop scheduling. Proceedings of the International Workshop "Intelligent Scheduling of Robots and Flexible Manufacturing Systems," Holon, July 2, Israel, CTEN Press.

Golenko-Ginzburg, D. and V. Kats, 1997. Comparative efficiency of priority rules in industrial job-shop scheduling (submitted paper).

Golenko-Ginzburg, D. and A. Gonik, 1997. Using "look-ahead" techniques in job-shop scheduling with random operations. *Int. J. Prod. Econ.,* 50: 13–22.

Grabowski, J. and A. Janiak, 1987. Job-shop scheduling with resource time models of operations. *Eur. J. Oper. Res.,* 28: 58–73.

Graves, S. C., 1981. A review of production scheduling. *Oper. Res.,* 29: 646–675.

Harrison, A., 1992. *Just-in-Time Manufacturing in Perspective, The Manufacturing Practitioner Series,* Prentice-Hall, Englewood Cliffs, New Jersey.

Haupt, R., 1989. A survey of priority rule-based scheduling. *OR-Spektrum,* 11: 3–16.

Huang, C. C. and G. Weiss, 1990. On the optimal order of M machines in tandem. *Oper. Res. Lett.,* 9: 299–303.

Hunsucker, J. L. and J. R. Shah, 1994. Comparative performance analysis of priority rules in a constrained flow-shop with multiple processors environment. *Eur. J. Oper. Res.,* 72: 102–114.

Kirkpatrick, S., C. D. Gelatt, Jr., and M. P. Vecchi, 1983. Optimization by simulated annealing. *Mgmt. Sci.,* 220: 671–680.

van Laarhoven, P. J. M and E. H. L. Aarts, 1987. *Simulated Annealing: Theory and Applications.* Reidel, Dordrecht.

van Laarhoven, P. J. M., E. H. L. Aarts, and J. K. Lenstra, 1992. Job-shop scheduling by simulated annealing. *Oper. Res.,* 40: 113–125.

Lageweg, B. J. et al., 1977. Job-shop scheduling by implicit enumeration. *Mgmt. Sci.,* 24: 441–450.

McMahon, G. B. and M. Florian, 1975. On scheduling with ready times and due dates to minimize maximum lateness. *Oper. Res.,* 22: 475–482.

Mather, H. 1984. *How to Really Manage Inventories.* McGraw-Hill, New York.

Mather, H. 1985. MRP II won't schedule the factory of the future. *Perspectives,* 57: 1–5.

Matsuo, M., C. J. Suh, and R. S. Sullivan, 1988. A controlled search simulated annealing method for the general job-shop scheduling problem. Working Paper 03-04-88, Department of management, The University of Texas at Austin.

Muth, E. M. and G. L. Thompson (Eds.), 1963. *Industrial Scheduling.* Prentice-Hall, Englewood Cliffs, New Jersey.

Narasimhan, R. and S. Melnyk, 1984. Assessing the transient impact of lot sizing rules following MRP implementation. *Int. J. Prod. Res.,* 22: 759–772.

New, C. C., 1973. *Requirements Planning.* Gower Press, Hampshire.

Panwalker, R. and W. Iskander, 1977. A survey of scheduling rules. *Oper. Res.,* 25: 45–59.

Pinedo, M., 1982. Minimizing the expected makespan in stochastic flow shops. *Oper. Res.,* 30: 148–162.

Pinedo, M., 1983. Stochastic scheduling with release dates and due dates. *Oper. Res.,* 31: 559–572.

Pinedo, M. and G. Weiss, 1979. Scheduling of stochastic tasks on two parallel processors. *Naval Res. Logist. Quart.,* 26: 527–535.

Pinedo, M. and G. Weiss, 1987. The largest variance first policy in some stochastic scheduling problems. *Oper. Res.,* 35: 884–891.

Ramasesh, R. V., 1990. Dynamic job-shop scheduling: A survey of simulation research, *Omega,* 18: 43–57.

Sculli, D., 1987. Priority dispatching rules in an assembly shop. *Omega,* 15 : 49–57.

Sen, T. and S. Gupta, 1984. A state-of-the-art survey of static scheduling research involving due dates. *Omega,* 12: 63–76.

Snyder, C. A., J. F. Cox, and S. J. Clark, 1985. Computerized production scheduling and control systems: major problem areas. *Methods Oper. Res.,* 55: 319–328.

Spachis, A. and J. King, 1979. Job-shop scheduling heuristics with local neighborhood search. *Int. J. Prod. Res.,* 17: 507–526.

Walpole, R. and R. Myers, 1978. *Probability and Statistics of Engineers and Scientists,* 2nd Edition, Macmillan, New York.

Weber, R. R., 1982. Scheduling jobs with stochastic processing requirements on parallel machines to minimize makespan or flow time. *J. Appl. Probab.,* 19: 167–182.

Weber, R. R., P. Varaiya, and J. Walrand, 1986. Scheduling jobs with stochastically ordered processing times on parallel machines to minimize expected flowtime. *J. Appl. Probab.,* 23: 841–847.

Wein, L. M. and P. B. Chevalier, 1992. A broader view of the job-shop scheduling problems. *Mgmt. Sci.,* 38: 1018–1033.

White, C., 1986. Production scheduling and materials requirements planning for flexible manufacturing. In A. Kusiak (Ed.), *Modeling and Design of Flexible Manufacturing Systems.* Elsevier, Amsterdam, pp. 229–248.

White, C. and N. A. J. Hastings, 1986. Scheduling the factory of the future. Paper presented to the 27th Int. Meeting of the Institute of Management Sciences (TIMS), July 1986.

White, C., 1990. Factory control systems. Internal Report, Graduate School of Management, Faculty of Economics and Politics, Monash University, Clayton, Victoria, Australia.

# Appendix

## Determining Conditional Mean Values

As outlined above, the problem is to determine the mean value

$$E_\alpha(x/x \geq \Delta),$$ 
(B1)

where $\alpha$ is a random variable with a cumulative probability function

$$F_\alpha(x) = \int_{-\infty}^{x} f_\alpha(y)\,dy,$$ 
(B2)

and $\Delta$ is a pregiven value within the distribution bounds. It can be well recognized (Walpole and Myers 1978) that

$$E_\alpha(x/x \geq \Delta) = \frac{1}{1 - F_\alpha(\Delta)} \cdot \int_{\Delta}^{\infty} x f_\alpha(x)\,dx$$ 
(B3)

holds.

We will determine value (B3) for three widely used probability distributions:

**I. Exponential distribution** with p.d.f. $F_\alpha(x) = \lambda e^{-\lambda x}$.

Since $F_\alpha(x) = 1 - e^{-\lambda x}$, value (B3) is as follows:

$$E_\alpha(x/x \geq \Delta) = \frac{\lambda \int_{\Delta}^{\infty} x e^{-\lambda x}\,dx}{e^{-\lambda \Delta}} = e^{\lambda \Delta} \cdot \lambda \int_{\Delta}^{\infty} x e^{-\lambda x}\,dx.$$

Using transformation $z = \lambda x$, we obtain

$$E_\alpha(x/x \geq \Delta) = \frac{1}{\lambda}e^{\lambda\Delta} \cdot \int_{\lambda\Delta}^{\infty} ze^{-z}dz$$

$$= \frac{1}{\lambda}e^{\lambda\Delta}\left(-\int_{\lambda\Delta}^{\infty} zde^{-z}\right) = \frac{1}{\lambda}e^{\lambda\Delta}\left(-ze^{-z}\Big|_{\lambda\Delta}^{\infty} + \int_{\lambda\Delta}^{\infty} e^{-z}dz\right)$$

$$= \frac{1}{\lambda}e^{\lambda\Delta}\left(-\lambda\Delta e^{-\lambda\Delta} - e^{-z}\Big|_{\lambda\Delta}^{\infty}\right) = \Delta + \frac{1}{\lambda}.$$

Thus, $E_\alpha(x/x \geq \Delta) = \Delta + \frac{1}{\lambda}$ in case of $F_\alpha(x) = \lambda e^{-\lambda x}$. $\qquad$ (B4)

**II. Uniform distribution** in [a, b], i.e., $F_\alpha(x) = \frac{1}{b-a}$ and $F_\alpha(x) = \frac{x-a}{b-a}$.
Value

$$E_\alpha(x/x \geq \Delta) = \frac{b-a}{b-\Delta}\int_\Delta^b x \cdot \frac{1}{b-a}dx = \frac{1}{b-\Delta}\frac{1}{2}x^2\Big|_\Delta^b = \frac{b+\Delta}{2}. \qquad (B5)$$

**III. Normal distribution** with parameters $(a, \sigma^2)$. The p.d.f. $f_\alpha(x) = \frac{1}{\sqrt{2\pi}}e^{\frac{-(x-a)^2}{2\sigma^2}}$, while the cumulative probability function $f_\alpha(x) = \frac{1}{\sqrt{2\pi}\cdot\sigma}\int_{-\infty}^x e^{\frac{-(x-a)^2}{2\sigma^2}}dy$. Using transformation $x = a + \sigma z$, $dx = \sigma dz$, and taking into account $1 - F_\alpha(\Delta) = 1 - \phi(\frac{\Delta-a}{\sigma})$, where $\phi(x) = \frac{1}{\sqrt{2\pi}}\int_{-\infty}^x e^{\frac{-y^2}{2}}dy$.
we obtain

$$E_\alpha(x/x \geq \Delta) = \frac{1}{1-\phi\left(\frac{\Delta-a}{\sigma}\right)}\frac{1}{\sqrt{2\pi}\sigma}\int_\Delta^{\infty} xe^{-\frac{(x-a)^2}{2\sigma^2}}dx = \frac{1}{1-\phi\left(\frac{\Delta-a}{\sigma}\right)}\int_{\frac{\Delta-a}{\sigma}}^{\infty}(a+\sigma z)\varphi(z)dz$$

$$= \frac{1}{1-\phi\left(\frac{\Delta-a}{\sigma}\right)}\left\{a\left[1-\phi\left(\frac{\Delta-a}{\sigma}\right)\right] + \sigma\int_{z_0}^{\infty} z\varphi(z)dz\right\},$$

where

$$\varphi(z) = \frac{1}{\sqrt{2\pi}}e^{-\frac{z^2}{2}}, \qquad z_0 = \frac{\Delta-a}{\sigma}.$$

Since

$$\frac{1}{\sqrt{2\pi}}\int_{z_0}^{\infty} ze^{-\frac{z^2}{2}}dz = \frac{1}{\sqrt{2\pi}}\int_{z_0}^{\infty} e^{-\frac{z^2}{2}}d\left(\frac{z^2}{2}\right) = \frac{1}{\sqrt{2\pi}}\left\{-e^{-\frac{z^2}{2}}\Big|_{z_0}^{\infty}\right\} = \frac{1}{\sqrt{2\pi}}e^{-\frac{z^2}{2}},$$

we finally obtain

$$E_\alpha(x/x \geq \Delta) = a + \frac{\frac{\sigma}{\sqrt{2\pi}}e^{-\frac{(\Delta-a)^2}{2\sigma^2}}}{1-\phi\left(\frac{\Delta-a}{\sigma}\right)}. \qquad (B6)$$

Values (B4), (B5) and (B6) have to be taken into account by determining forecasted conditional average values $\bar{F}_{\eta_v k_v}$ in Sections 5.5 and 5.6.

# 6

# Failure Diagnosis in Computer-Controlled Systems

Udo Konradt
*University of Kiel*

## 6.1   Introduction

Companies are making greater efforts to increase competitiveness through enhanced labor productivity, flexibility, product quality, and automation. The need to adapt products to customer requirements and yet guarantee short delivery times has emerged a predictor equally as important as price and quality. At a time when competitive advantage is determined by the company's production, flexibility, and quality, the area of maintenance management has grown in importance for all kinds of industries (Paz, Leigh, and Rogers 1994). Maintenance management must look at new ways of ensuring high standards and reliability of machinery and equipment. A prime factor hindering a rapid improvement of equipment effectiveness is a centralized maintenance organization reflecting a *specialist control model*. According to this model, mechanical and electrical specialists are responsible for maintaining and repairing the machinery and equipment (Wall, Corbett, Martin, Clegg, and Jackson 1990). The shop floor personnel is involved only to a minimal extent, with little or no responsibility for diagnosis, service, or repair tasks. However, the specialist control model suffers from various drawbacks:

- The maintenance personnel is often not available.
- The maintenance personnel has to cover a long distance between maintenance workshop and machine location.
- The tools for diagnosis and repair are stored in the maintenance workshop.
- The required spare parts are not available and must be ordered.
- Work on one machine must be interrupted because of the breakdown of another machine with a more urgent priority.

The reallocation of maintenance and repair devices to the factory floor personnel offers a first step toward overcoming these problems. Studies have shown that operators—though it is not their responsibility—usually provide the maintenance specialist with important information about symptoms and failure characteristics (Bereiter and Miller 1989, Konradt 1995). In contrast to the specialist model, the *operator control model* stresses multiskilled operators who are responsible for maintenance tasks. The operator control model does not imply that maintenance specialists are no longer required but that they are responsible for those diagnosis and repair tasks that are too complex for operators.

The idea of an operator control model of maintenance is part of the concept of Total Productive Maintenance (TPM). TPM is a comprehensive management system for maintenance, dealing with technical, organizational, and personal factors. TPM covers aspects of designing equipment as well as the daily autonomous maintenance done by operators (Nakajima 1988, Nakajima and Yamashina, 1994). It features improvement of equipment effectiveness through (semi-) autonomous maintenance by operators, scheduled maintenance by maintenance personnel, education and training, and the development of an early equipment management program. Further, TPM is an all-embracing philosophy integrated in a company-wide manufacturing excellence program. The importance of personal and organizational factors in maintenance efficiency was stressed by Wall et al. (1990). They showed that aspects of downtime (amount, incidence, and downtime per incident) for operators of cutting-machine tools in basic maintenance and fault correction were predicted by knowledge and skill. Results of a representative survey in metallurgical companies in Germany show that in 1991, 8.2% of the companies reported that operators on the shop floor carried out all or most maintenance and service tasks. In 1992 this increased to 11.9 and in 1993 to 15.9%. The results reflect that companies are discovering the increasing significance of integrating maintenance and repair tasks into the work of operators for productivity improvements (Davis 1995, Harmon 1992). However, compared with quality assurance tasks—in 1993, 76% of quality assurance tasks were fulfilled to a greater part by operational personnel —the responsibility for troubleshooting and repair tasks has been shifted to a much lesser extent to operational personnel.

Following the idea of TPM, a new allocation of tasks must be found. Operators on the shop floor have to maintain basic equipment conditions (e.g., cleaning, lubrication, bolting), maintain operating conditions (proper operation and visual inspection) and discover deterioration (visual inspection and early identification of signs of abnormalities during operation). This task enrichment should enhance skills such as equipment operation, setup, and adjustment. Maintenance personnel is responsible for providing technical support for the production department's autonomous maintenance activities, for repairing thoroughly and accurately, using inspections, condition monitoring, and overhaul for clarifying operating standards by tracing design weaknesses and making appropriate improvements. In order to foster operators' competencies, maintenance personnel should enhance their maintenance skills for checkups, condition monitoring, inspections, and overhaul. It should be noted that not all maintenance functions can be assigned them to the operators. Operations involving a high safety risk or requiring special technical knowledge must remain the responsibility of maintenance technicians. This allocation of functions leads to an organization in which the operational personnel has the major responsibility for the planning, execution, and controlling of basic failure diagnosis, troubleshooting of simple failures, and change and control of wearing parts (Davis 1995). The underlying organizational concept is the implementation of semi-autonomous work groups. In semi-autonomous work groups, tasks usually separated from the manufacturing processes are partially or wholly integrated in a working group. The work group is responsible for production planning, time-scheduling of material, manufacturing, quality control, maintenance, and repair.

Operational and maintenance personnel have several techniques at their disposal for analyzing and eliminating the causes of breakdowns, stoppages, and defects. The implementation of effective maintenance systems requires these techniques to be identified in order to evaluate efficiency, develop training programs, or provide adequate information support to the operators.

Everyday experience and studies of real-life situations suggest that assessment and reasoning from underlying principles often leads to appropriate solutions to problems in this domain. However, there is a lack of empirical studies on the structures and kinds of application of this type of knowledge in real-life settings. Moreover, it remains unclear how this knowledge can be systematically used to increase the effectiveness and the efficiency of technical problem solving. This chapter deals with technical problem solving and consists of four main sections. In the first section the task of maintenance is defined. Some critical cognitive structures—i.e., strategies—and their acquisition are discussed from a cognitive engineering point of view. The second section explores the techniques used by mechanical and electrical maintenance technicians in real-life troubleshooting. Section 6.3 analyses the conditions for using these techniques, depending on problem complexity, familiarity with the problem, and job experience. Section 6.4 describes and evaluates how the techniques in failure diagnosis can be employed to develop a knowledge-based information system that assists operators during maintenance tasks. A software design approach that uses cognitive structures will be introduced and an example of a computer-based information system for the support of TPM structures presented.

## 6.2 Maintenance: Task Characteristics, Cognitive Structures, and Measurement

From a cognitive engineering point of view, the analysis of maintenance tasks has to start with an understanding of the decision situation. The following example of the troubleshooting of a rotary-table drilling center may serve to uncover some important aspects of maintenance tasks:

*At a rotary-table drilling center that had recently undergone maintenance, the three hydraulically operated devices became loose for a split second when the stations were in the loading position. The work piece was in danger of dropping out and therefore a locking mechanism stopped the machine. The hydraulic technician attached different manometers to the branches of the hydraulic system following the hydraulic plan. He recognized that the pressure in the system dropped when the device swung into the loading position. He assumed that the fluid distributor had been displaced slightly during the last maintenance which would account for the momentary relaxation in pressure. The necessary repair work would take one or two shifts, so the hypothesis is discussed with the supervisor and the chief supervisor and finally rejected because in the case of a displaced fluid distributor the defect should appear at all devices instead of only at one. In the third step an electrician is involved, who finds out by inspecting the wiring scheme that there is a fault in the synchronization between the limit switches that are operated by contractors and the switching rates of the hydraulic steering. As a result the hydraulic fluid is carried into different parts of the machine for a short time. The actual reason proves to be a cam-wheel that was attached slightly out of place at the last maintenance. The cam-wheel is responsible for the synchronized operating of the limit switches. The failure can be remedied within fifteen minutes now by readjusting the cam-wheel, which is easily accessible.*

This example reveals several features of maintenance tasks typical for problems in this domain. Typically maintenance tasks exhibit one or more of the following features:

- the problem is multidimensional;
- there is vast amount of information available, which is partly contradictory;
- the expected consequences are only partly known;
- the sub-goals and preferences are not fully specified;
- the cognitive resources are exceeded/exhausted;
- the task must be completed under time pressure;
- the problem is part of a dynamic process with continuously changing parameters.

Tasks with these features are denoted as complex, semi-structured, or ill-defined problems. The process of problem solving involves the acquisition and handling of appropriate information and additional goal specification. In short, the troubleshooter has to develop a cognitive representation of the problem situation, representing properties of the goal and procedures for accomplishing it. However, the lack of clarity makes it difficult for the problem solver to generate the appropriate structure of goals and procedures. Moreover the solving process not only depends on the decision situation, but also on individual abilities, knowledge, experiences, and habits leading to different subjective representations of the external reality. A commonly used denotation for the user's internal representations that are based on previous experience as well as current observations and which dictate the level of task performance is *mental model*. Rouse and Morris (1986) defined mental models as "the mechanisms whereby humans are able to generate descriptions of system purpose and form, explanations of the system functioning and observed system states, and predictions of the future system states" (p. 351). Although there is confusion about the definition and utilization of this concept (for a critical review see Wilson and Rutherford 1989), this definition should be taken as a basis, because of its wide acceptance.

Another model for analyzing the cognitive demands posed by a diagnosis task is the theory of action-regulation (Hacker 1973, 1994, Volpert 1982). The theory of action-regulation takes mental representations as well as overt behavior into consideration and is thus able to describe concrete working actions and deduce the underlying cognitive regulation processes. It makes two central assumptions: actions are goal-related and the goals are hierarchically structured. Actions are directed towards a goal to fulfill the task. The psychological structure of actions is derived from subgoals that give the behavior a functionally related hierarchical order.

Figure 6.1 shows the hierarchical structure of failure diagnosis goals for the domain of cutting-machine tools. If the operator notices for example a drop in pressure, he may check the hydraulic system. Appropriate and allowable subgoals may be to check the tank, investigate the distributors, or measure the voltage at the magnet of a hydraulic switch immediately. The task requires him to make a choice from a sequence of particular operations, means and methods to deduce subgoals, as well as means and methods for general goal attainment. Action regulation theory assumes that the choice of subgoals, methods, and operations is controlled by metacognitive structures like plans and strategies.

## Strategies

According to Bruner et al. (1956), the term *cognitive strategy* denotes the capabilities of humans to control their attention, perceiving, encoding, and retrieval processes. Besides declarative and procedural knowledge, strategic knowledge is meta-cognitive and encompasses plans, goals, and rules. General search strategies such as the means-end analysis, reasoning by analogy, and backward chaining played an important role in the "General Problem Solver" of Newell and Simon (1972).

The strategies typically used by decision makers depend upon the properties of the task, i.e., task complexity and familiarity with the task/task-class (e.g., Payne, 1982). For example, it has often been found that low task complexity leads decision makers to use full processing strategies whereas an increase in task complexity results in the use of reduced processing strategies. The following strategies were identified in a number of investigations:

- *Additive Compensatory Strategy.* The decision maker processes all alternatives using cues by attaching weights to all cues and then summing the weights to arrive at an overall value for each alternative.
- *Additive Difference Strategy.* The decision maker processes information cues of two alternatives, evaluates the differences, and selects a binary choice for each alternative pair.
- *Elimination by Aspects.* The decision maker first compares all information using a selection of cues or the most important cue, and then eliminates those alternatives having unsatisfactory values.
- *Mixed Strategies.* This type of strategy typically involves different compensatory and noncompensatory strategies.

Paquette and Kida (1988) investigated whether reduced processing strategies result in less efficient decisions. Decision accuracy and the time taken to make a decision were assessed when subjects used different strategies. Efficiency was achieved by saving time without losing decision accuracy. Results show that decision makers employing reduced processing strategies—such as elimination by aspects—were

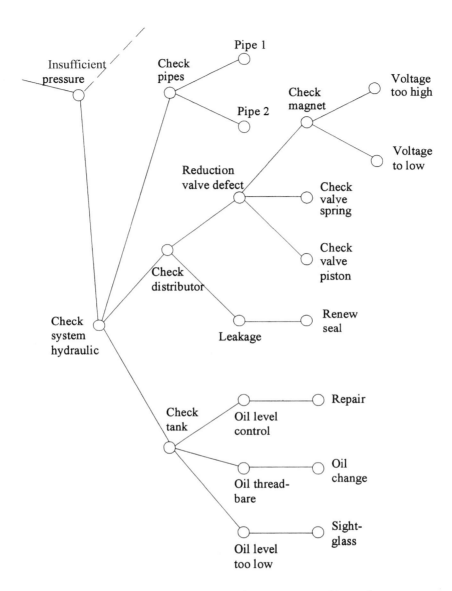

**FIGURE 6. 1**  Hierarchical structure of failure diagnosis goals at a cutting-machine-tool.

more efficient when faced with a complex decision task. This suggests that decision makers can benefit from using simplifying strategies when confronted with cognitively complex tasks. Reduced strategies require less cognitive effort.

Several authors have investigated the meaning and function of strategic knowledge in planning. In complex situations, such as diagnostic problem solving, planning leads to the selection of a suitable sequence of test procedures and to the evaluation of situation-specific attributes. Strategic knowledge is intrinsic to those tasks in which (1) alternative decisions are possible; (2) criteria for choices are incompatible or contradictory, as for example with the criteria rapidity, economy, ease, and accuracy; or (3) the expected outcomes are not clearly appreciable (Gruber 1988). In contrast to an algorithmic procedure, the choice among possible actions is not determinate and the task cannot be accomplished automatically. In such tasks, the use of strategic knowledge leads to an initially reduced choice of steps and less analysis in advance, thereby shortening the diagnosis process (Rouse 1983). Strategies were investigated in the domains of geometry (Greeno 1980), architectural design (Simon 1973), troubleshooting (Rasmussen and Jensen 1974, Bereiter and Miller 1989, Reed and Johnson 1993), programming (Jeffries et al. 1981, Soloway and Ehrlich 1984, Vessey 1985), and process control tasks (Hoc 1989). Jeffries et al. (1981) showed that expert software designers decomposed the problem into finer subparts than novices and considered several different methods of solving the task. Soloway and Ehrlich (1984) found that one typical method of program planning is stepwise refinement, in which the problem is broken into several subproblems that are subsequently solved recursively. Vessey (1985) compared novices with experts in program debugging and found that experts used a breadth-first search. Strategic planning is moreover a high-level activity on which experts and novices differ greatly. Experts are more likely than novices to examine the general goal of program and high-level program modules in debugging, decompose the program into finer parts and explore more alternative solutions, and show more differences in comprehending plan-like vs. unplanned-like programs. Hoc's analysis (1989) of operators' strategies in a simulated continuous process task revealed that operators usually used "historical information" instead of examining separate components in process diagnosis. In a case study of electronic troubleshooting, Rasmussen (Rasmussen and Jensen 1974, Rasmussen 1981) identified two types of generalized strategies: the symptomatic search and the topographical search. The *symptomatic search* comprises a set of symptoms collected by the operator that is matched to a set of symptoms under normal system state. If the search is done in the actual system or physical domain, it is called a *topographic search*. Whereas a topographical search is based on a binary test of parameters of system components, a symptomatic search is based on the mental representation of the system. This representation, however, depicts only the topology of the physical system to a certain extent. Rasmussen showed that a symptomatic search is used by troubleshooters in well-known and routine tasks.

In expert systems, complex problem-solving structures are used for implementation in the inference machine (Milne 1987). "Deep knowledge" is used in expert systems for electronic troubleshooting (Davis 1983), mechanical fault diagnosis (Fink and Lusth 1987), and medical diagnosis (Chandrasekaran and Mittal 1983). In expert systems of the first generation, the knowledge base does not have any representative structures according to the model of human problem solvers. In expert systems of the second generation, a structuring of the represented model is added which is compatible with the models of human problem-solvers. The knowledge base takes the form of an explicit psychological model of the problem-solver, his task relevant knowledge, and effective strategies to solve problems in the domain. The structure of the knowledge base allows hierarchical analysis of isolated parts of the problem space. The selection problem is reduced to the composition of rules, which allow the user to draw from a hierarchy of choices among different complexes of functions.

"Knowledge Acquisition and Documentation Structuring" (KADS) (Breuker et al. 1988) is a model-based acquisition technique that models the human problem-solver. KADS describes expertise on four layers: the domain layer, the inference layer, the task layer, and the strategy layer. The *domain layer* contains the static domain knowledge (concepts, relations, structures). The *inference layer* consists of structures (meta-classes) and types of inference (knowledge sources). The third layer is the *task layer*, containing goals and tasks as basic objects. Tasks are possible combinations of knowledge sources to achieve a particular goal. Whereas a task structure is a fixed strategy, the flexibility of human behavior is represented on the *strategy layer*. The characteristics of experts are represented at this level, which encompasses the choice of a problem-solving strategy and the flexible change between different strategies. This leads to the early recognition of dead ends and errors. The strategy controls goal and task structures that apply meta-classes and knowledge sources described by concepts and their relation to the domain. KADS is very similar to the process-oriented action regulation model, but emphasizes the structural aspects.

Whereas the vast majority of investigations into problem solving are undertaken in experimental settings and simulation studies, strategies in real-life are relatively seldom investigated. Although real-life studies in complex work settings suffer from a lack of precise control and systematic variation of variables, they are nevertheless necessary for generating hypotheses and understanding what information is actually used during task fulfillment. To support operators' diagnosis, detection, and task-planning, a thorough analysis is needed to address how operators make decisions, what information they use, and how it is organized mentally. In particular, research on models of human problem solving for the user-centered design of human-computer systems is required, with attention given to strategies and their limitations (Rasmussen 1985, 1994).

## Acquisition of Strategies

Plans and strategies are mentally represented and must be inferred from observable behavior. According to the literature, three standard methods can be used to analyze plans and strategies: asking users to think aloud, asking users to answer questions, and observation techniques. In the following, the interview technique is stressed because it is an economical way for strategic knowledge acquisition.

According to Rasmussen (1983), goal-related activities are regulated on three different levels. On the *knowledge-based* level the goal is the result of the explicit analysis of the situation parameters. *Rule-based* activities consist of a composed sequence of subordinate operations that are generated in and applied to well-known working situations. *Skill-based* behavior consists of a highly integrated behavior pattern that is observed in familiar or routine situations, e.g., manual control of fuel rod insertion and withdrawal in a nuclear power plant. Skill-based behavior is overlearned, meaning that there is no explicit or conscious analysis of the significance of the situation. It is a schema-driven process which leads to the recognition of cues consistent with that schema, even if they were not explicitly presented (Arkes and Harkness 1980). Schema-driven problem-solving strategies are automatically invoked by a problem schema containing information about the typical goals, constraints, and solution processes that have been useful in the past for that type of problem (Gick 1986). In contrast, knowledge-based behavior is controlled by conscious processes. Because of the underlying learning processes, knowledge structures and processing characteristics of experts and novices are different. Task characteristics and the user's experience with the task determine the level of control (Morris and Rouse 1985, Hoc 1989, Rasmussen 1983). If the task is complex and poses high cognitive demands on the problem-solver, or if the task is unknown, i.e., a sequence of algorithms is not known, the expert has to draw on a knowledge-based control level.

The availability and accessibility of information are basic assumptions for the application of a verbal technique of knowledge acquisition. Verbal data reflect only those cognitive processes that are conscious (Ericsson and Simon 1985). Interviews suffer from the problem that experts may not always be aware of what they know or in what specific way they solve a problem (Nisbett and Wilson 1977). They use procedures in problem solving that are often highly automated and not directly accessible for inspection. Complex tasks, such as failure diagnosis in advanced manufacturing systems, usually require a conscious preparation for the processing sequence. Therefore, the assumption is drawn that a person is able to verbalize their strategies explicitly.

Empirical results stress that strategies can be recorded reliably through verbal reports. In an evaluation study, Hoc and Leplat (1983) investigated different modalities of verbalization. They show that precise verbal data were elicited by an aided subsequent verbalization method. Rasmussen and Jensen (1974) investigated electronic troubleshooting and found that electronic technicians were able to provide accurate verbal information about their troubleshooting strategies. In an extensive literature review about verbal reports on operators' knowledge, Bainbridge (1979) noted that general information on control strategies could be elicited from interviews. Olson and Rueter (1987) argued that inference rules, relations, and objects can be revealed by interviews. There is even empirical evidence that strategy-related data obtained by concurrent verbalization proves to be more valid than data obtained by retrospective verbalization; an advantage of the latter method is that it has no reactivity on the reports on cognitive processes (Brinkman 1993).

## 6.3   Strategies of Maintenance Technicians

This section serves to present an investigation on techniques in failure diagnosis. Because a cognitive engineering point of view is held, the techniques are denoted as strategies. Failure diagnosis strategies of maintenance and repair technicians who diagnose and repair cutting-machine tools were recorded with a retrospective verbal knowledge acquisition technique. Interviews were conducted in quiet rooms near the shop-floor. The investigation was divided into three parts. First, the technician was asked to freely recall failures that appeared at cnc-machines in his work domain. Failures that were assessed as occurring frequently related to deviations in the surface of work pieces, dimensional imperfection, problems with the reference position, or total malfunction of the machine. After a brainstorming session, failures were selected and a half-standardized interview was conducted. The interview was performance-oriented, and the questions reflected the sequence of steps in addressing the failure, from its recognition to its remedy. A case-oriented interview structure was intended to facilitate the technician's recall of his approach in diagnosis. After the interview was completed, demographic data, work qualifications, and operational fields of the technician were collected. Their job experience, the degree of complexity (simple, complex), and the degree of familiarity (novel, familiar, routine) with the failures was rated by the technicians.

Based upon level of experience, the technicians were separated into three groups: novices, advanced, and experts. Technicians with over 20 years job experience were classified as expert, while the novice had up to 6 years job experience. To test this classification, the relation between these groups and the operational field was analyzed on a ten-point rating scale reflecting task complexity. Novices with job experience of up to 6 years were assigned to easy tasks, e.g., easy maintenance tasks and localized breakdowns. Advanced maintenance technicians were assigned mostly to tasks with high demands, whereas experts were exclusively assigned to the most complicated tasks, such as difficult maintenance tasks, repair of replacement parts, and modification of machinery and equipment.

## Analysis of the Data

The interviews were recorded and transcribed, and a content-analysis was performed (for more details see Konradt 1995). In constructing the category system, the diagnosis strategies reported in the literature were compiled. According to Christensen and Howard (1981), these are: 1) an examination of direct sense-impressions; 2) historical information; 3) an inquiry of the most common errors; 4) an examination of the errors with the highest probability; 5) a pattern of symptoms (syndromes); 6) execution of indicative alterations and observation of the progression of signals; 7) systematic division; 8) encircle-ment; 9) execution of the simplest examination; and 10) examination that eliminates the largest number of possible sources of error. Each strategy was defined and elaborated upon. The category system was determined by definitions and benchmark examples. In those cases in which a clear classification of a category was not possible, a coding rule for the classification decision was developed. After a first analysis-trial, the categories were revised and sixteen types of strategies were identified:

(1) **Conditional probability.** Check of the failure with the highest probability under a given condition. In contrast to "Systematic narrowing," short if-then sequences are applied. Example: "If the operator says that the machine doesn't take his commands, then we know that it could only be the keyboard." "[...] if the operator tells us "Crash," we immediately check the [...]."

(2) **Diagnosis software.** Consultation of program or system messages to identify the failure. Example: "Then I have a look at the failure message; you can identify the failure through the control message and the number of the control lamp."

(3) **Exclusion.** Elimination of failures at machine elements and groups that are at the same abstraction level. In contrast to category 8, expressions of this category are "exclude," "eliminate," or "it isn't possible that this is the cause." Example: "First I make this check, then I can exclude the slide carriage."

(4) **Frequency.** Check of the most frequent failures. Words of this category are "often," "frequent," "most important cause," or "principle." Example: "A frequent failure is lack of oil because of a broken pipe." "The ratio is at best one to ten that another element is defect, so I first check the control element."

(5) **Historical information.** Use of maintenance records, failure statistics, or contact with colleagues. The same or similar failures happened before. Experience with failures, particular machine tools and types of machine tools. Example: "[ ... ] this happened before, that means we have much experience with this." "With these types of motors this happened before, especially with the older ones."

(6) **Information uncertainty.** Check that eliminates the greatest number of failure causes. Application in complex situations. Probability information is accessible. Example: "We have more electrical failures than mechanical ones. If the machine doesn't work at all, then the failure is electrical."

(7) **Least effort.** Use easy checks first. Words of this category are "easy," "simple," "quick." Example: "First I have a look at what one can see at the first glance, without disassembling any of the assembly groups."

(8) **Manuals.** The procedure is oriented towards wiring diagrams and machine plans. The manuals are used to picture the construction of the machine tool and advise where to address the cause of failure. Example: "The plan advises me how to go on."

(9) **Pattern of symptoms.** Direct confirmation of a hypothesis. Diagnosis through pattern recognition. Example: "An indication for a broken cable is that the $x$-axis doesn't work. Then I measure whether the transistor has voltage and observe whether the motor doesn't work simultaneously. Then I check the motor for a broken cable."

(10) **Reconstruction.** Asking the operator about the progression of failure. Example: "I ask the operator to inform me; the more information I get from him, the more directed I can go on. Information from the operator is the most important."

(11) **Sensory checks.** Perception of symptoms, for example loose connections, odors, sound, play. It is not necessary to use measuring instruments. Example: "The play in the tenth part is noticeable. Then you can feel disturbances in the true running."

(12) **Signal tracing.** Enter changing conditions to show the course, e.g., of signals. Direct manipulation of the system. Example: "The operator drives the sliding carriage and I check if the pulse is transmitted."

(13) **Split half.** Application, if no probability information is available. Procedure, if two or more alternatives are equally probable. Example: "You have to separate the mechanical from the electrical part. Then you try to test the isolated parts. Afterwards you can say precisely whether the failure is mechanical or electrical."

(14) **Systematic narrowing.** From general to specific answers. Words of this category are "systematic," "first [...], then [...], after [...]," "approach," or "in sequence." In contrast to Conditional probability several sequential "if-then-clauses" appear. Example: "Systematic test of all failure causes until the cause is identified."

(15) **Topographic search.** Imagination of design and function of the machine tool. Disturbed functions are related mentally to disturbed machine elements, in contrast to 'Manuals.' Example: "Depending on the locus or the machine stops, I can see which control element is effected. Therefore, it is necessary to know the internal working processes in the machine."

(16) **Miscellaneous.** The relation to another category is not possible.

## Strategies of Maintenance Technicians

In sum, 182 strategies were analyzed in 69 interviews. Table 6.1 represents the frequencies of failure diagnosis strategies. It shows that the most frequent failure diagnosis strategies were "Historical information" (29%), "Least effort" (11.5%), "Reconstruction" (9.8%), and "Sensory check" (8.7%). Strategies such as "Historical information," which involves checking available information about the failure history, and "Least effort" are low-cost checks which shorten the time needed for diagnosis activities. In contrast, strategies such as "Split half," leading to a binary reduction of the problem space, and "Information uncertainty" play only a minor role in real-life failure diagnosis of machine tools (1.1%).

In fault diagnosis in advanced manufacturing systems, four typical strategies are found:

1. Restriction of diagnosis to components which are known to be susceptible to failures (Historical information);
2. Performing tests that result in least efforts (Least effort);
3. Reconstruction of the conditions that led to the failure (Reconstruction);
4. Perception of symptoms, i.e., loose connections, odors, sound, play (Sensory checks).

**TABLE 6.1**    Observed Frequencies of
Diagnosis Strategies

| Strategies | Percent |
|---|---|
| *Topographic strategies* | |
|     Conditional probability | 5.5 |
|     Exclusion | 3.3 |
|     Manuals | 3.3 |
|     Patterns of symptoms | 2.7 |
|     Signal tracing | 6.0 |
|     Split half | 1.1 |
|     Topographic search | 2.2 |
| *Symptomatic strategies* | |
|     Reconstruction | 9.8 |
|     Sensory checks | 8.7 |
|     Systematic narrowing | 6.6 |
| *Case-based strategies* | |
|     Frequency | 4.9 |
|     Historical information | 29.0 |
|     Information uncertainty | 1.1 |
| *Other* | |
|     Diagnosis software | 1.6 |
|     Least effort | 11.5 |
|     Miscellaneous | 2.7 |

These strategies appeared in about 60% of the total observed strategies. The primary strategy was Historical information. This result corresponds to the results of an empirical analysis of blast furnace conductors strategies in a simulation of the process (Hoc 1989). Experimental problem solving research has repeatedly shown that the technique of tracing analogous problems is an appropriate way of solving ill-defined problems (Klein and Weitzenfeld 1978).

In information theory, strategies such as Information uncertainty, which eliminates the greatest number of failure causes, or Split half, which results in a binary splitting of the problem space, are economical ways of shortening the problem space. In contrast, the current results show that maintenance technicians did not use these elaborate and efficient strategies. Overall, these results correspond with a replication study that analyzed heuristics of maintenance technicians (Wiedemann 1995).

## 6.4    Conditions of Application

Experts verbalized more strategies than novices and advanced workers, although this effect was not statistically significant. When comparing extreme groups, technicians working in complex domains verbalized more strategies than those fulfilling easy tasks. In Figure 6.2, the length of the strategy-sequence for the 69 interviews analyzed is presented. In 73.8% of the cases, up to three strategies were applied and not more than two changes in strategies occurred. Eighty-four percent of the cases consisted of as many as four strategies. The length of the sequence did not depend on work-experience, degree of familiarity, task-complexity, and operational domain.

Domain-specific differences between mechanical and electrical maintenance technicians in the frequency of application of particular strategies were not found except in the case of Sensory checks, which were used more often by mechanics. The familiarity with the problem influenced the choice of strategy. In novel and familiar failure tasks, systematic search from general to specific aspects (Systematic narrowing) and elimination of failures at machine elements that are on the same level of abstraction (Exclusion) were used by technicians. Only in one case were these strategies used in routine tasks. On the other hand, the strategy Conditional probability was used predominantly in routine failures.

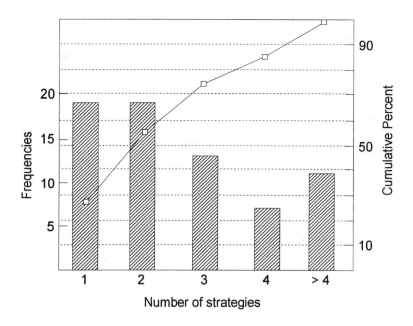

**FIGURE 6.2** Number of strategies applied from maintenance technicians during a troubleshooting session.

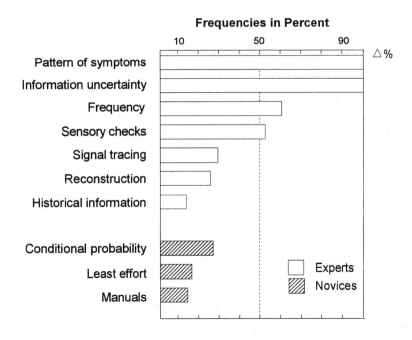

**FIGURE 6.3** Primary differences between experts and novices in using failure diagnosis strategies.

In Figure 6.3, the primary differences strategies between experts and novices in using failure diagnosis are presented. Delta percent represents how much of the observed frequency of occurrence of a single strategy was attributed to experts or novices. The dominant strategies of novices were "Conditional probability," "Least effort," and "Manuals." The dominant strategies of experts were "Patterns of symptoms,"

"Information uncertainty," and "Frequencies." Strategies such as "Pattern of symptoms" and "Information uncertainty" were exclusively used by experts.

As noted earlier, Rasmussen (1981) proposes two different ways of performing a diagnostic search: a Symptomatic search and a Topographic search. Results of his study suggest a further class of generalized strategies that are called "case-based." In case-based strategies, symptoms observed in the diagnostic situation are collected and compared to those in similar cases. The observed information is matched either to external information (maintenance reports, weak point statistics) or to the individual's knowledge. Strategies are aggregated to form three upper-level groups of strategies: symptomatic strategy (Conditional probability, Pattern of symptoms, Sensory checks), topographic strategy (Diagnosis software, Exclusion, Manuals, Reconstruction, Signal tracing, Split half, Systematic narrowing, Topographic search), and case-based strategy (Frequency, Historical information, Information uncertainty) (see Table 6.1). This grouping is chosen because in a Topographic search the diagnostic context is represented as a physical or functional anatomy of the system.

Using a hierarchical log-linear model, the number of cases in each cell of the contingency table was interpreted as a function of generalized strategy (symptomatic, topographic, case-based), level of familiarity (novel, familiar, routine), job experience (novice, advanced, expert), and task complexity (easy, complex). Log-linear models are a special class of statistical techniques for uncovering complex relationships among categorical variables in a multiway cross-tabulation. The goal of a log-linear analysis is to identify a model that explains the variation of the observed cell frequency. After a model has been specified that can explain the pattern of observed frequency distributions, the expected frequencies are calculated under the assumption that the model yields accurate predictions. Through the comparison of the expected and the observed frequencies, the model will either be discarded and a new one specified, or it will be accepted and the model parameters will be estimated.

The solution model shows no significant differences between the expected frequencies predicted by the model and the observed ones. The partial chi-square values and significance levels for each effect of the contingency table were calculated. The interaction between generalized strategy and level of familiarity with the problem should be depicted. Results show that in novel failures, symptomatic and case-based strategies seldom occurred; Topographic search dominated instead. In contrast, in routine failures, case-based strategies were most frequently used by maintenance technicians, and the importance of topographical strategies diminished. Topographical and case-based strategies were dominant in familiar failures. We found that in real-life failure diagnosis, even maintenance experts with more than twenty years experience seldom used these strategies. One reason may be that the use of these strategies requires information about conditional probabilities and a fully described problem space that cannot be supposed for troubleshooting in complex manufacturing systems.

In novel failures, the diagnosis process is cognitively regulated on a knowledge-based level (Rasmussen 1983). If the task poses high demands, or the decision situation contains new features, even experts have to fall back on a knowledge-based level (Rouse and Morris 1986, Yoon and Hammer 1988). In novel failures, technicians seldom used symptomatic and case-based strategies, instead opting for topographical search strategies. In routine failures, however, case-based strategies dominated whereas in known failures, topographical and case-based strategies were most frequently observed.

## 6.5 Using Cognitive Strategies for the Development of an Information System

Strategic knowledge allows the user-centered development of information and planning tools which are adaptive to human performance. Adaptive aiding stresses that a human operator should only be supported by technical tools to the extent of need (Rouse 1988). Adaptive systems should provide ways for adaptive user interface design, especially for decision support systems in diagnosis and supervisory control (Morrison and Duncan 1988). While application needs have been clearly identified, concrete industrial applications remain largely undeveloped. From a human-factors point of view, possible causes may be that appropriate reasoning techniques, as well as modes of explaining and representing have not yet been developed. In summary, from the results of this study, several general conclusions can be drawn for an adaptive design:

- Maintenance technicians in electrical and mechanical failure diagnosis use topographical, symptomatic, and case-based strategies. The system should have options for the support of short "if-then-sequences," patterns of causes and symptoms and database functions.
- Novices in failure diagnosis prefer short if-then-sequences (Conditional probability, Least effort, and Manuals). A more extended dialogue should be presented to the less-practiced user which encompasses an explanation of the construction and functioning of assembly elements and assembly groups.
- More experienced users usually have hypotheses about possible reasons for the failures. Thus, they have a demand for specific information about nominal values of system elements and the sequence of failures. The system should allow the user to shorten the number of interactive steps.
- The expert's knowledge is mostly derived from cases and concrete episodes. The more experienced troubleshooter has a demand for information about the failure history of the machine.
- There are a variety of quick opening strategies, such as checking the failure history (Historical information) or Least effort, before entering more systematic checks and tests. The system should support economical ways of shortening the time needed for diagnosis activities on request.

The analysis of strategies in failure diagnosis leads to the general design conclusions that it is essential to store failure cases in a data bank and provide flexible data access. A support system for maintenance and repair should serve as an information source and database assisting and extending the task-relevant skills of users. Additionally, the system should support the planning and decision processes by providing short if-then sequences for beginners, as well as case-based and topographical information for experts. Systems of this type are denoted as decision support systems (DSS). DSS include a body of represented knowledge that can be acquired and maintained, selected for either presentation or problem solving and which interact directly with a decision maker in order to describe some aspects of the decision-maker's world, specifying how to accomplish various tasks, indicating what conclusions are valid in various circumstances, and so forth (Holsapple and Whinston 1992).

## A Theoretical Software Design Method

The development of software systems that create the software life cycle is composed of different phases. Although the models differ in types of classification, procedures, and details, they usually contain requirement analysis, design, implementation, and evaluation. Requirement analysis is a central point in software design, because it determines the decomposition of problems, the hierarchical structure of the system, the specification of main features, and the design of user interfaces. A core problem in software design is that data from psychological analyses are often too complex, although the quality and the ecological validity of the software product could be improved by taking it into account. Another problem is that the results of (psychological) requirement analysis cannot be transformed into design. Object-oriented analysis and design (OOAD) is a methodology to support the translation of user demands more smoothly into prototypes. In OOAD, data structure is defined in user language, and actions are matched. Domains of users are represented through specification of a set of semantic objects that are revealed in cognitive task analysis. In this study objects were derived from the cognitive analysis of strategies (for more details see Konradt 1996). The methodology is called strategy-based software design (SSD). The aim of SSD is to design flexible software using a model of human strategies. This software should be compatible with the operator and an appropriate tool in working tasks. Recent developments in software design support the conversion of cognitive structures into programming-code, as with the Object Behaviour Analysis of Rubin and Goldberg (1992). Following this approach, a continuous transition between analysis and design is guaranteed.

Figure 6.4 presents how a strategy leads to interface and software design issues. The informational support of the strategy Sensory checks first requires the implementation of objects that are generally

| Objects | Operation | Representation |
|---|---|---|
| Components of the machine hierarchical, linear | Search/Selection of components<br>Relate symptoms to components | graphical, textual |
| Symptoms | Input of symptoms, not localized (internal)<br>Input of symptoms (external)<br>Search/Selection of symptoms<br>Relate symptoms to components, hierarchical | graphical, textual, acoustically |
| List of identified components and symptoms | Search/Selection of components/symptoms, additive<br>Put in components/symptoms (not localized, internal, external) | graphical, textual |
| Failures | Search/Selection of components/symptoms, substractive<br>Search/Selection of failures<br>Search/Selection of (further) symptoms | graphical, textual |
| Cause of failure | Search/Selection of failure causes<br>Search/Selection of components<br>Relate symptoms to components<br>Input of symptoms, not localized (internal)<br>Input of symptoms (external) | graphical, textual |

**FIGURE 6.4**   Principle of the strategy-based software design model (SSD).

**TABLE 6.2**   List of the Classes of Objects

| Object Class | Description |
|---|---|
| Components | A description of the system and its structure (for example machine tools) that correspond to the units of examination in the error model. |
| Symptoms | Certain conditions which indicate an error and which can, in turn, themselves be errors. |
| Error | Group of all deviations from the desired value which appear in components. |
| Facts | List of identified symptoms and components that are given as input through the diagnosis. |
| Causes of errors | Connections between the symptoms and the error (for example, in the form of rules or matrices). |
| Choice strategies | Different criteria for sorting the error distribution (for example according to the criteria of the least effort). |
| Test procedure | Description of test possibilities for the identification of symptoms or the testing of an error. |
| Dates and conditions | Events arising in the past (for example, change in a machine or system breakdown, stoppages, and defects). |
| Cases | Description of errors arising in the past, the procedures and processors of diagnosis and repair. |
| Statistics | Reading and notification of particular developments in the features of the components. |

used by the operator pursuing this strategy. A list of object classes is presented in Table 6.2. Following the strategy sensory checks relevant objects are classes of components of the machine, symptoms, lists of identified components and symptoms, failures and causes of failure. There is evidence from the interviews that these objects are used by technicians pursuing the strategy of sensory checks. Second, allowable operations are defined to process the objects. Such operations are selection, search, input, and relation to. In a third step the kind of representation has to determined, that means textual, graphical or acoustical information forms.

When considering the elements established with the aid of strategies, it becomes clear that several strategies require the same elements. It is therefore logical to take advantage of common elements from different strategies rather than handling each strategy separately when transposing and constructing a flexible and applicable system for the user. The resulting network allows the user to switch between different strategies and combine them with one other. The elements of the strategies provide the bonds in a hyperlink structure whose connections arise out of the individual strategies. Thus, it is ensured that the system supports all strategies applied by a user group.

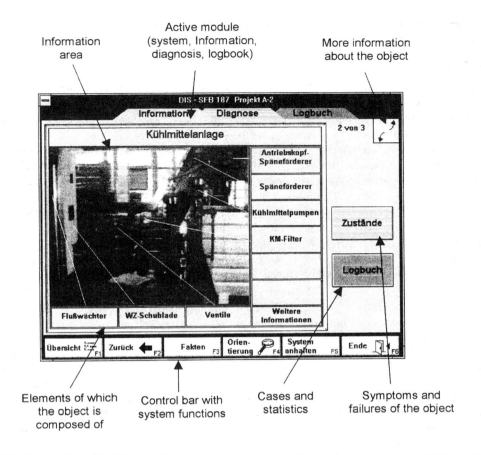

**FIGURE 6.5**   Interface of the Diagnosis Information System showing hydraulic components at a CNC-machine.

The SSD methodology was applied for the development of the Diagnosis Information System (DIS). DIS is mouse-driven and implemented under Microsoft-Windows™. The application is implemented for a machining center of a German producer and consists of about 400 pages of information and about 10,000 lines of source code. DIS is object-oriented, contains multimedia components and is easily expandable by the user. In case of a breakdown or an unallowed deviation within the working system, DIS informs the operator upon request about possible types and patterns of symptoms and failures, examines and explains their causes and reveals the ensuing failures. It also serves as a data base and offers the failure history of machines on the shop-floor. It provides the user with information about the most common causes of failure as well as with machine documentation and functional drawings. Components of the machine are represented in pictures which are identical to the real machine and are not displayed in abstract forms (Figure 6.5). Thereby context-relevant marks and information should be used. During the development process DIS was continuously refined by user participation.

## 6.6   Conclusion

Strategies in troubleshooting are important cognitive structures to model how people actually search for information when solving a problem. An understanding of these general skills and techniques can provide support for the adaptable design of an information system that does not restrict but rather fosters flexibility in problem solving. Often cognitive structures and behavioral aspects are not incorporated in system development due to the gap between requirements analysis and design. To bridge this gap, a software design method is presented.

# References

Arkes, H. R. & A. R. Harkness (1980). Effect of making a diagnosis on subsequent recognition of symptoms. *Journal of Experimental Psychology: Human Learning and Memory,* 6, 568–575.

Bainbridge, L. (1979). Verbal reports as evidence of the process operator's knowledge. *International Journal of Man-Machine Studies,* 11, 411–436.

Bereiter, S. R. & S. M. Miller (1989). A field based study of troubleshooting in computer-controlled manufacturing systems. *IEEE Transactions on Systems, Man, and Cybernetics,* 19, 205–219.

Breuker, J. A., B. W. Wielinga, M. van Someren, R. de Hoog, G. Schreiber, P. De Greef, B. Bredeweg, J. Wielemaker, J. Billaut, M. Davoodo, & S. Hayward (1988). *Modeldriven Knowledge Acquisition: Interpretation Models.* Deliverable task A1, Esprit Project 1098. University of Amsterdam.

Brinkman, J. A. (1993). Verbal protocol accuracy in fault diagnosis. *Ergonomics,* 36, 1381–1397.

Bruner, J. S., J. J. Goodnow, & G. A. Austin (1956). *A Study of Thinking.* New York: Wiley.

Chandrasekaran, B. & S. Mittal (1983). Deep versus compiled knowledge approaches to diagnostic problem-solving. *International Journal of Man-Machine Studies,* 19, 425–436.

Christensen, J. M. & J. M. Howard (1981). Field experience in maintenance. In J. Rasmussen & W. B. Rouse (Eds.), *Human Detection and Diagnosis of System Failures,* pp. 111–133. New York: Plenum Press.

Davis, R. (1983). Reasoning from first principles in electronic troubleshooting. *International Journal of Man-Machine Studies,* 19, 403–423.

Davis, R. K. (1995). *Productivity Improvements through TPM.* Englewood Cliffs, NJ: Prentice-Hall.

Ericsson, K. A. & H. A. Simon (1985). *Protocol Analysis: Verbal Report as Data.* London: MIT Press.

Fink, P. K. & J. C. Lusth (1987). Expert systems and diagnostic expertise in the mechanical and electrical domains. *IEEE Transactions on Systems, Man, and Cybernetics,* 17, 340–349.

Gick, M. L. (1986). Problem-solving strategies. *Educational Psychologist,* 21, 99–120.

Greeno, J. (1980). Some examples of cognitive task analysis with instructional implications. In R. Snow, P. A. Frederico & W. Montague (Eds.), *Aptitude, Learning, and Instruction: Cognitive Process Analyses of Learning and Problem Solving* (Vol. 2). Hillsdale, NJ: Lawrence Erlbaum.

Gruber, T. (1988). Acquiring strategic knowledge from experts. *International Journal of Man-Machine Studies,* 29, 579–598.

Hacker, W. (1973). *General Working and Engineering Psychology* (in German). Bern: Huber.

Hacker, W. (1994). Action regulation theory and occupational psychology. Review of German empirical research since 1987. *The German Journal of Psychology,* 18, 91–120.

Harmon, R. L. (1992). *Reinventing the Factory II. Managing the World Class Factory.* New York: The Free Press.

Hoc, J.-M. (1989). Strategies in controlling a continuous process with long response latencies: needs for computer support to diagnosis. *International Journal of Man-Machine Studies,* 30, 47–67.

Hoc, J.-M. & J. Leplat (1983). Evaluation of different modalities of verbalisation in a sorting task. *International Journal of Man-Machine Studies,* 18, 283–306.

Holsapple, C.W. & A. B. Whinston (1992). Decision support systems. In G. Salvendy (Ed.), *Handbook of Industrial Engineering* (pp. 109–141). New York: Wiley.

Jeffries, R., A. Turner, P. Polson, & M. Atwood (1981). The processes involved in designing software. In J. Anderson (Ed.), *Cognitive Skills and Their Acquisition* (pp. 255–283). Hillsdale, NJ: Erlbaum.

Klein, G. A. & J. Weitzenfeld (1978). Improvement of skills for solving ill-defined problems. *Educational Psychologist,* 13, 31–40.

Konradt, U. (1996). A task oriented framework for the design of information support systems and Organizational Change. In: R. J. Koubek & W. Karwowski (Eds.), *Manufacturing Agility and Hybrid Automation* (Vol. 1) (pp. 159–162). Louisville: IEA Press.

Milne, R. (1987). Strategies for diagnosis. *IEEE Transactions on Systems, Man, and Cybernetics,* 17, 333–339.

Morris, N. M. & W. B. Rouse (1985). The effects of type of knowledge upon problem solving in a process control task. *IEEE Transactions on Systems, Man, and Cybernetics,* 16, 503–530.

Morrison, D. L. & K. D. Duncan (1988). Strategies and tactics in fault diagnosis. *Ergonomics*, 31, 761–784.

Nakajima, S. (1988). *Introduction to TPM: Total Productive Maintenance.* Cambridge, MA: Productivity Press.

Nakajima, S. & H. Yamashina (1994). Human factors in AMT maintenance. In G. Salvendy and W. Karwowski (Eds.). *Design of Work and Development of Personnel in Advanced Manufacturing* (pp. 403–430). New York: Wiley.

Newell, A. & H. A. Simon (1972). *Human Problem Solving.* Englewood Cliffs, NJ: Prentice-Hall.

Nisbett, R. E. & T. D. Wilson (1977). Telling more than we can know: Verbal reports on verbal processes. *Psychological Review,* 84, 231–259.

Olson, J. R. & H. H. Rueter (1987). Extracting expertise from experts: Methods for knowledge acquisition. *Expert Systems,* 4, 152–168.

Paquette, L. & T. Kida (1988). The effect of decision strategy and task complexity on decision performance. *Organizational Behavior and Human Decision Processes,* 41, 128–142.

Payne, J. (1982). Contingent decision behavior. *Psychological Bulletin,* 92, 382–402.

Paz, N. M., W. R. Leigh, & R. Rogers (1994). The development of knowledge for maintenance management using simulation. *IEEE Transactions on Systems, Man, and Cybernetics,* 24, 574–593.

Rasmussen, J. (1981). Models of mental strategies in process plant diagnosis. In J. Rasmussen & W. B. Rouse (Eds.), *Human Detection and Diagnosis of System Failures* (pp. 241–258). New York: Plenum Press.

Rasmussen, J. (1983). Skills, rules, knowledge; signals, signs, and symbols, and other distinctions in human performance models. *IEEE Transactions on Systems, Man, and Cybernetics,* 13, 257–267.

Rasmussen, J. (1985). The role of hierarchical knowledge representation in decision making and system management. *IEEE Transactions on Systems, Man, and Cybernetics,* 15, 234–243.

Rasmussen, J. (1994). Taxonomy for work analysis. In G. Salvendy & W. Karwowski (Eds.), *Design of Work and Development of Personnel in Advanced Manufacturing* (pp. 41–77). New York: Wiley.

Rasmussen, J. & A. Jensen (1974). Mental procedures in real-life tasks: A case study of electronic troubleshooting. *Ergonomics,* 17, 293–307.

Reed, N. E. & P. E. Johnson (1993). Analysis of expert reasoning in hardware diagnosis. *International Journal of Man-Machine Studies,* 38, 251–280.

Rouse, W. B. (1983). Models of human problem solving: Detection, diagnosis, and compensation for system failures. *Automatica,* 19, 613–625.

Rouse, W. B. (1988). Adaptive aiding for human-computer control. *Human Factors,* 30, 431–443.

Rouse, W. B. & N. M. Morris (1986). On looking into the black box: Prospects and limits in the search for mental models. *Psychological Bulletin,* 100, 349–363.

Rubin, K. S. & A. Goldberg (1992). Object behavior analysis. *Communications of the ACM,* 35, 48–62.

Simon, H. A. (1973). The structure of ill-structured problems. *Artificial Intelligence,* 4, 181–201.

Soloway, E. & K. Ehrlich (1984). Empirical studies of programming knowledge. *IEEE Transactions on Software Engineering,* 10, 595–609.

Vessey, I. (1985). Expertise in debugging computer programs: A process analysis. *International Journal of Man-Machine Studies,* 23, 459–494.

Volpert, W. (1982). The model of the hierarchical-sequential organisation of action. In W. Hacker, W. Volpert, & W. M. V. Cranach (Eds.), *Cognitive and Motivational Aspects of Action* (pp. 35–51). Amsterdam: North-Holland.

Wall, T. D., J. M. Corbett, R. Martin, C. W. Clegg, & P. R. Jackson (1990). Advanced manufacturing technology, work redesign, and performance: A change study. *Journal of Applied Psychology,* 75, 691–697.

Wiedemann, J. (1995). *Investigation of Demands for Qualification.* Münster: Waxmann (in German).

Wilson, J. R. & A. Rutherford (1989). Mental models: Theory and application in human factors. *Human Factors,* 31, 617–634.

Yoon, W. C. & J. M. Hammer (1988). Aiding the operator during novel fault diagnosis. *IEEE Transactions on Systems, Man, and Cybernetics,* 18, 142–148.

# 7

# Analysis of Stochastic Models in Manufacturing Systems Pertaining to Repair Machine Failure

Rakesh Gupta
*Ch. Charan Singh University*

Alka Chaudhary
*Meerut College*

This chapter deals with three stochastic models *A*, *B*, and *C*, each consisting of two nonidentical units in standby network. One unit is named as the priority unit (*p*-unit) and the other as the nonpriority or ordinary unit (*o*-unit). In each model, the *p*-unit gets priority in operation over the *o*-unit. A single server is available to repair a failed unit and a failed repair machine (R.M.). The R.M. is required to do the repair of a failed unit. In models *A* and *C*, the *o*-unit gets priority in repair over the *p*-unit, whereas in model-*B* the priority in repair is also given to the *p*-unit over the *o*-unit. In each model it is assumed that the R.M. may also fail during its working and then the preference in repair is given to R.M. over any of the units. In models *A* and *B*, the failure and repair times of each unit are assumed to be uncorrelated independent random variables (r.vs.), whereas in model *C* these two r.vs. are assumed to be correlated having bivariate exponential distribution. In each model we have obtained various economic measures of system effectiveness by using the regenerative point technique.

## 7.1   Introduction

Two-unit standby systems have been widely studied in the literature of reliability due to their frequent and significant use in modern business and industry. Various authors including [1, 3, 8–10, 17–23, 25] have studied two-unit standby systems with different sets of assumptions and obtained various characteristics of interest by using the theories of semi-Markov process, regenerative process, Markov-renewal process and supplementary variable technique. They have given equal priority to both the units in respect of operation and repair. But realistic situations may arise when it is necessary to give priority to the main unit in respect of operation and repair as compared to the ordinary (standby) unit. A very good example of this situation is that of a system consisting of two units, one power supply and the other generator. The priority is obviously given to the power through power station rather than generator. The generator will be used only when the power supply through power station is discontinued. Further, due to costly operation of the generator, the priority in repair may be given to power station rather than the generator.

Keeping the above concept in view, Nakagawa and Osaki [24] have studied the behavior of a two-unit (priority and ordinary) standby system with two modes of each unit — normal and total failure. Goel et al. [2] have obtained the cost function in respect of a two-unit priority standby system with imperfect switching device. They have assumed general distributions of failure and repair times of each unit. Recently, Gupta and Goel [11] investigated a two-unit priority standby system model under the assumption that whenever an operative unit fails, a delay occurs in locating the repairman and having him available to repair a failed unit/system. Some other authors including [12–15] have also investigated two-unit priority standby system models under different sets of assumptions. The common assumption in all the above models is that a single repairman is considered and the preference with respect to operation and repair is given to priority (*p*) unit over the ordinary (*o*) unit. However, situations may also arise when one is to provide preference to priority (*p*) unit only in operation and not in repair. Regarding the repair, either the preference may be given to *o*-unit over the *p*-unit or the repair discipline may be first come first serve (FCFS). So, more recently Gupta et al. [16] investigated a two nonidentical unit cold standby system model assuming that the preference in operation is given to the first unit (*p*-unit) while in repair the preference is given to the second unit (*o*-unit). The system model under this study can be visualised by a very simple example: Suppose in a two-unit cold standby system model two nonidentical units are an air conditioner (A.C.) and an air cooler. Obviously the preference in operation will be given to the A.C. and air cooler will get the preference in repair as the repair of A.C. is costly and time-consuming. The case of standby redundant system is not seen in the literature of reliability when the preference in operation is given to *p*-unit but in repair the policy is FCFS.

All the above discussed authors have analysed the system models under the assumptions that the machine/device used for repairing a failed unit remains good forever. In real situations this assumption is not always practicable as the repair machine (R.M.) may also have a specified reliability and can fail during the repair process of a failed unit. For example, in the case of nuclear reactors, marine equipments, etc., the robots are used for the repair of such type of systems. It is evident that a robot, a machine, may fail while performing its intended task. In this case obviously the repairman first repairs the repair machine and then takes up the failed unit for repair.

In this chapter we discuss three system models, *A*, *B*, and *C*, each consisting of two nonidentical units named as *p*-unit and *o*-unit. It is assumed that in each model the *p*-unit gets priority in operation as only one unit is sufficient to do the required job. A repair machine (R.M.) is required to do the repair of a failed unit which can also fail during its operation. Further, a single repairman is available to repair a failed unit as well as a failed R.M. and in each model the priority is given to R.M. over any of the failed units. Regarding the repair of failed units, it is assumed in model *B* that the *p*-unit gets preference in repair over the *o*-unit, whereas in models *A* and *C* the priority in repair is given to the *o*-unit rather than to the *p*-unit. In models *A* and *B*, the basic assumption is that the failure and repair times are taken uncorrelated independent r.vs. However, a common experience of system engineers and managers reveals that in many system models there exists some sort of correlation between failure and repair times. It is observed that in most of the system models an early (late) failure leads to early (delayed) repair. The concept of linear relationship is the main point of consideration. Therefore, taking this concept in view, in model *C*, the joint distribution of failure and repair times is assumed to be bivariate exponential (B.V.E) of the form suggested by "Paulson" ($0 \leq r \leq 1$). The p.d.f. of the B.V.E. is

$$f(x, y) = \alpha \beta (1 - r) e^{-\alpha x - \beta y} I_0(2\sqrt{\alpha \beta r \times y})$$
$$x, y, \alpha, \beta > 0, \quad 0 \leq r < 1 \tag{1}$$

where $I_0(z) = \sum_{k=0}^{\infty} \frac{(Z/2)^{2K}}{(K!)^2}$ is the modified Bessel function of type *I* and order Zero. Some authors including [4–7,16] have already analysed system models by using the above mentioned concept.

Using regenerative point technique in the Markov renewal process, the following reliability characteristics of interest to system designers and operation managers have been obtained for models *A*, *B*, and *C*.

   (i) reliability of the system and mean time to system failure (MTSF);
  (ii) pointwise and steady state availabilities of the system;
 (iii) the probability that the repairman is busy at an epoch and in steady state;
 (iv) expected number of repairs by the repairman in $(0, t)$ and in steady state; and
  (v) expected profit incurred by the system in $(0, t)$ and in steady state.

Some of the above characteristics have also been studied and compared through graphs and important conclusions have been drawn in order to select the most suitable model under the given conditions.

## 7.2  System Description and Assumptions

   (i) The system is comprised of two nonidentical units and a repair machine (R.M.). The units are named as priority (*p*) unit and ordinary (*o*) unit. The operation of only one unit is sufficient to do the job.
  (ii) In each model the *p*-unit gets priority in operation over the *o*-unit. The *o*-unit operates only when *p*-unit has failed. So, initially the p-unit is operative and o-unit is kept as cold standby which cannot fail during its standby state.
 (iii) Each unit of the system has two modes normal (*N*) and total failure (*F*). A switching device is used to put the standby unit into operation and its functioning is always perfect and instantaneous.
 (iv) A single repairman is available with the system to repair a failed unit and failed R.M. In models *A* and *C*, the *o*-unit gets priority in repair over the *p*-unit, whereas in model *B*, the priority in repair is given to the *p*-unit over the *o*-unit. Further, the R.M. gets the preference in repair over both the units.
  (v) The R.M. repairs a failed unit and it can also fail during the repair of a unit. In such a situation the repair of the failed unit is discontinued and the repairman starts the repair of the R.M. as a single repairman is available. Each repaired unit and R.M. work as good as new.
 (vi) The R.M. is good initially and it cannot fail until it begins functioning.
(vii) In models *A* and *B*, the failure times and repair times of a unit and R.M. are assumed to be independent and uncorrelated r.vs., whereas in model *C* the failure and repair times of the units are correlated r.vs.

(viii) In models $A$ and $B$, the failure time distributions of each unit and the R.M. are taken to be negative exponential with different parameters while all the repair time distributions are general having different probability density function (p.d.f). Further, in these models it is assumed that the restarted repair (after interruption) of a unit is preemptive repeat type having the same p.d.f as that of the fresh repair.

(ix) In model $C$, the failure and repair times of each unit are jointly distributed having the bivariate exponential density function of the form (1.1) with different parameters. Due to the priority in repair, the repairs of the $p$-unit and the $o$-unit are interrupted many times. When such a unit is again taken up for repair, then the time required to complete the repair now is known as residual repair time of the unit concerned. The residual repair time of each unit need not depend on its failure time and the random variable denoting it also has a negative exponential distribution.

(x) The failure and repair time distributions of R.M. in model $C$ are taken to be negative exponential with different parameters.

## 7.3   Notation and States of the System

$E_0$ : initial state of the system i.e., the state at time
   $t = 0$

$E$ : set of regenerative states

$\bar{E}$ : complementary set of $E$

$q_{ij}(.), Q_{ij}(.)$ : p.d.f. and c.d.f. of one step or direct transition time from state $S_i \in E$ to $S_j \in E$.

$p_{ij}$ : [1]steady-state transition probability from state $S_i$ to $S_j$ such that
   $p_{ij} = Q_{ij}(\infty) = \int q_{ij}(u)\, du$

$q_{ij}^{(k,l)}(.), Q_{ij}^{(k,l)}(.)$ : p.d.f. and c.d.f. of transition time from state $S_i \in E$ to $S_j \in E$ via states $S_k \in \bar{E}$ and $S_l \in \bar{E}$

$p_{ij}^{(k,l)}$ : steady-state transition probability from state $S_i \in E$ to $S_j \in E$ via states $S_k \in \bar{E}$ and $S_l \in \bar{E}$ such that

$$p_{ij}^{(k,l)} = Q_{ij}^{(k,l)}(\infty) = \int q_{ij}^{(k,l)}(u)\, du$$

$Q_{ij|x}(.)$ : c.d.f. of transition time from state $S_i \in \bar{E}$ to $S_j \in E$ given that the unit under repair in state $S_i$ entered into F-mode after an operation of time $x$ (for model-$C$)

$p_{ij|x}$ : steady-state probability of transition from state $S_i \in \bar{E}$ to $S_j \in E$ given that the unit under repair in state $S_i$ entered into F-mode after an operation of time $x$ (for model-$C$)
   $= \lim\limits_{t \to \infty} Q_{ij|x}(t) = \lim\limits_{s \to 0} \tilde{Q}_{ij|x}(s)$

$Z_i(t)$ : probability that the system sojourns in state $S_i$ up to time $t$.

$\psi_i$ : mean sojourn time in state $S_i$
   $= \int Z_i(t)\, dt = \lim\limits_{s \to 0} Z_i^*(s)$

$\psi_{i|x}$ : mean sojourn time in state $S_i \in \bar{E}$ given that the unit under repair in this state entered into F-mode after an operation of time $x$ (for model-$C$)

$*, \sim$ : symbols for Laplace and Laplace Stieltjes Transforms (LT and LST).

---

[1]The limits of the integration are not mentioned throughout the chapter whenever they are 0 to $\infty$.

©,Ⓢ : symbols for ordinary and Stieltjes convolutions

$$i.e., \quad A(t)©B(t) = \int_0^t A(u)B(t - u)du$$

$$A(t)Ⓢ B(t) = \int_0^t dA(u)B(t - u)$$

$\alpha_1, \alpha_2$ : constant failure rates of $p$ and $o$-unit, respectively (for models $A \& B$)

$G_1(.), G_2(.)$ : c.d.f. of the repair time of $p$ and $o$-unit, respectively (for models $A \& B$)

$\lambda$ : constant failure rate of R.M.

$H(.)$ : c.d.f. of time to repair of R.M. (for models $A \& B$)

$\mu$ : constant repair rate of R.M. (for model $C$)

$X_1, Y_1$ : r.vs. denoting the life time and repair time, respectively, of $p$-unit (for model $C$)

$X_2, Y_2$ : r.vs. denoting the life time and repair time, respectively, of $o$-unit (for model $C$)

$Y_1', Y_2'$ : r.vs. denoting the residual repair times of $p$ and $0$-unit, respectively each having the negative exponential distribution with parameters $\beta_1$ and $\beta_2$ (for model $C$)

$f_i(x, y)$ : joint p.d.f. of $(X_i, Y_i)$

$i = 1, 2 \quad = \alpha_i \beta_i (1 - r_i) e^{-(\alpha_i x + \beta_i y)} I_0(2\sqrt{\alpha_i \beta_i r_i xy})$

so that the conditional p.d.f. of $Y_i$ given $X_i = x$ is

$$\beta_i e^{-(\beta_i y + \alpha_i r_i x)} I_0(2\sqrt{\alpha_i \beta_i r_i xy})$$

where

$$I_0(2\sqrt{\alpha_i \beta_i r_i xy}) = \sum_{j=0}^{\infty} \frac{(\alpha_i \beta_i r_i xy)^j}{(j!)^2}$$

$$x \geq 0, y \geq 0, \quad 0 \leq r_i \leq 1$$

$q_i(x)$ : marginal p.d.f. of $X_i$

$$= \alpha_i(1 - r_i) e^{-\alpha_i(1 - r_i)x}$$

**Symbols used to represent the states of the system:**

$N_{10}, N_{20}$ : $p$ and $o$-unit in $N$-mode and operative

$N_{2s}$ : $o$-unit in $N$-mode and standby

$F_{1r}, F_{2r}$ : $p$ and $o$-unit in $F$-mode and under repair

$F_{1w}, F_{2w}$ : $p$ and $o$-unit in $F$-mode and waiting for repair

$RM_g, PM_o, RM_r$ : R.M. in good condition, operative, and under repair

$F_{1r'}, F_{2r'}$ : $p$ and $o$-unit in $F$-mode and again taken up for repair after interruption

$F_{1w'}, F_{2w'}$ : $p$ and $o$-unit in $F$-mode and waiting for repair after interruption

Considering the above symbols for the two units and the assumptions stated earlier, we have the following states of system models $A$, $B$, and $C$.

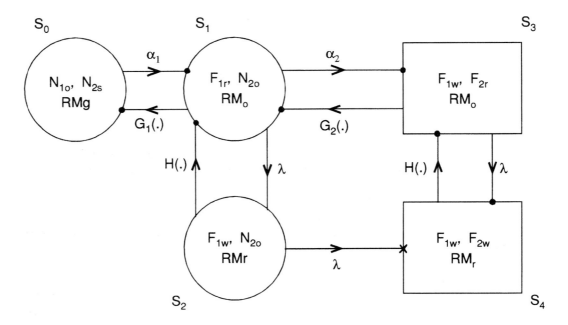

**FIGURE 7.1**   Transition diagram of system model with failure/repair rates or repair time cdf.

### (i) Model A

$$\text{Up states}: \quad S_0 \equiv \begin{pmatrix} N_{10}, N_{2S} \\ RM_g \end{pmatrix}, \qquad S_1 \equiv \begin{pmatrix} F_{1r}, N_{20} \\ RM_o \end{pmatrix}, \qquad S_2 \equiv \begin{pmatrix} F_{1w}, N_{20} \\ RM_r \end{pmatrix}$$

$$\text{Failed states}: \quad S_3 \equiv \begin{pmatrix} F_{1w}, F_{2r} \\ RM_o \end{pmatrix}, \qquad S_4 \equiv \begin{pmatrix} F_{1w}, F_{2w} \\ RM_r \end{pmatrix}$$

The epoch of transition from $S_2$ to $S_4$ is nonregenerative while all the other entrance epochs into the states are regenerative. The transition diagram of the system model along with failure/repair rates or repair time cdf is shown in Figure 7.1.

### (ii) Model B

$$\text{Up states}: \quad S_0 \equiv \begin{pmatrix} N_{10}, N_{2S} \\ RM_g \end{pmatrix}, \quad S_1 \equiv \begin{pmatrix} F_{1r}, N_{20} \\ RM_o \end{pmatrix}, \quad S_2 \equiv \begin{pmatrix} F_{1w}, N_{20} \\ RM_r \end{pmatrix}$$

$$S_5 \equiv \begin{pmatrix} N_{10}, F_{2r} \\ RM_o \end{pmatrix}, \quad S_6 \equiv \begin{pmatrix} N_{10}, F_{2w} \\ RM_r \end{pmatrix}$$

$$\text{Failed states}: \quad S_3 \equiv \begin{pmatrix} F_{1r}, F_{2w} \\ RM_o \end{pmatrix}, \quad S_4 \equiv \begin{pmatrix} F_{1w}, F_{2w} \\ RM_r \end{pmatrix}$$

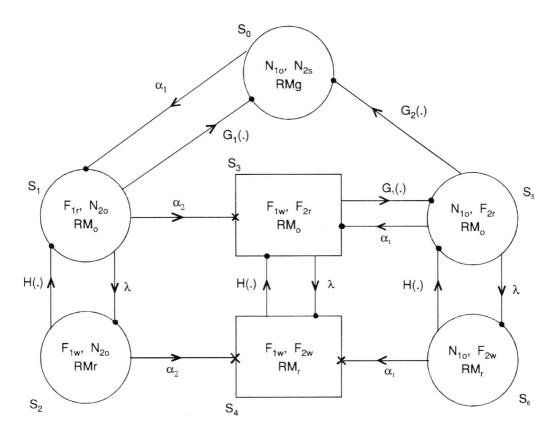

**FIGURE 7.2** Transition diagram of the system model with failure/repair rates or repair time cdf.

The epochs of transition entrance into the states $S_3$ from $S_1$ and $S_4$ from $S_2$, $S_6$ are nonregenerative. The transition diagram of the system model along with failure/repair rates or repair time cdf is shown in Figure 7.2.

**(iii) Model C**

$$\textbf{Up states :} \ S_o \equiv \begin{pmatrix} N_{10}, N_{2S} \\ RM_g \end{pmatrix}, \quad S_1 \equiv \begin{pmatrix} F_{1r}, N_{20} \\ RM_o \end{pmatrix}$$

$$S_2 \equiv \begin{pmatrix} F_{1w'}, N_{20} \\ RM_r \end{pmatrix}, \quad S_5 \equiv \begin{pmatrix} F_{1r'}, N_{20} \\ RM_o \end{pmatrix}$$

$$\textbf{Failed States:} \ S_3 \equiv \begin{pmatrix} F_{1w'}, N_{2r} \\ RM_o \end{pmatrix}, \quad S_4 \equiv \begin{pmatrix} F_{1w'}, F_{2w'} \\ RM_r \end{pmatrix}$$

$$S_6 \equiv \begin{pmatrix} F_{1w'}, F_{2w} \\ RM_r \end{pmatrix}, \quad S_7 \equiv \begin{pmatrix} F_{1w'}, F_{2r'} \\ RM_o \end{pmatrix}$$

The epochs of entrance from $S_0$ to $S_1$, $S_1$ to $S_3$, $S_5$ to $S_3$, $S_2$ to $S_6$, and $S_6$ to $S_3$ are nonregenerative while all the other entrance epochs into the states are regenerative. The transition diagram of the system model along with failure/repair times or failure/repair rates is shown in Figure 7.3.

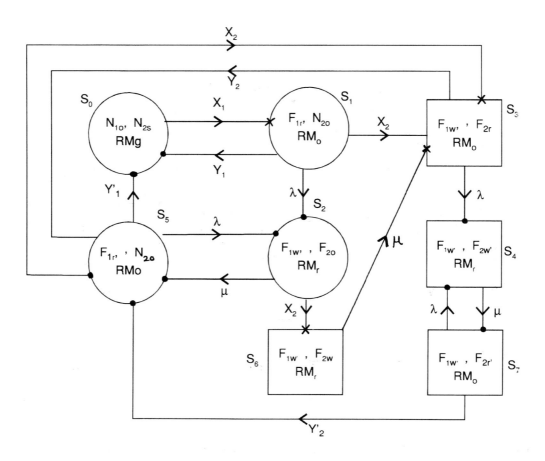

**FIGURE 7.3**  Transition diagram of the system model with failure/repair times or failure/repair rates.

## 7.4  Model A

### Transition Probabilities and Sojourn Times

Let $T_0 (\equiv 0)$, $T_1$, $T_2$, ... denote the epochs at which the system enters any state $S_i \in E$, and let $X_n$ be the state visited at epoch $T_n +$, i.e., just after the transition at $T_n$. Then $\{X_n, T_n\}$ is a Markov renewal process with state space $E$. If

$$Q_{ij}(t) = P[X_{n+1} = S_j, T_{n+1} - T_n \leq t \mid X_n = S_i]$$

Then the transition probability matrix (t.p.m.) is given by

$$P = (\mu_{ij}) = [Q_{ij}(\infty)] = Q(\infty)$$

(i)  By simple probabilistic reasoning the nonzero elements of $Q = (Q_{ij}(t))$ may be obtained as follows:

For the system to reach state $S_1$ from $S_0$ on or before time $t$, we suppose that the system transits from $S_0$ to $S_1$ during $(u, u + du)$, $u \leq t$. The probability of this event is $\alpha_1 e^{-\alpha u} du$. Since $u$ varies from $o$ to $t$, therefore,

$$Q_{01}(t) = \int_0^t \alpha_1 e^{-\alpha_1 u} du = 1 - e^{-\alpha_1 t}$$

Similarly,

$$Q_{10}(t) = \int_0^t e^{-(\alpha_2 + \lambda)u} dG_1(u)$$

$$Q_{12}(t) = \lambda \int_0^t e^{-(\alpha_2 + \lambda)u} \overline{G}_1(u) du$$

$$Q_{13}(t) = \alpha_2 \int_0^t e^{-(\alpha_2 + \lambda)u} \overline{G}_1(u) du$$

$$Q_{21}(t) = \int_0^t e^{-\alpha_2 u} dH(u)$$

$$Q_{24}(t) = \alpha_2 \int_0^t e^{-\alpha_2 u} \overline{H}(u) du$$

$$Q_{31}(t) = \int_0^t e^{-\lambda u} dG_2(u)$$

$$Q_{34}(t) = \lambda \int_0^t e^{-\lambda u} \overline{G}_2(u) du$$

$$Q_{43}(t) = \int_0^t dH(u) = H(t) \tag{1-8}$$

To derive an expression for $Q_{23}^{(4)}(t)$, we suppose that the system transits from state $S_2$ to $S_4$ during the time interval $(u, u + du)$, $u \leq t$; the probability of this event is $\alpha_2 e^{-\alpha_2 u} \overline{H}(u) du$. Further, suppose that the system passes from state $S_4$ to $S_3$ during the interval $(v, v + dv)$ in $(u, t)$; the probability of this event is $dH(v)/\overline{H}(u)$. Thus,

$$Q_{23}^{(4)}(t) = \int_0^t \alpha_2 e^{-\alpha_2 u} \overline{H}(u) du \int_u^t dH((v)/\overline{H})(u)$$

$$= \alpha_2 \int_0^t dH(v) \int_0^v e^{-\alpha_2 u} du \quad \text{(by change of order of integration)}$$

$$= \int_0^t (1 - e^{-\alpha_2 v}) dH(v) \tag{9}$$

(ii) The steady-state transition probabilities are given by

$$p_{ij} = \lim_{t \to \infty} Q_{ij}(t)$$

Therefore,

$$p_{01} = p_{43} = 1$$

$$p_{10} = \tilde{G}_1(\lambda + \alpha_2), \quad p_{12} = \lambda[1 - \tilde{G}_1(\lambda + \alpha_2)]/(\lambda + \alpha_2)$$

$$p_{13} = \alpha_2[1 - \tilde{G}_1(\lambda + \alpha_2)]/(\lambda + \alpha_2)$$

$$p_{21} = \tilde{H}(\alpha_2), \quad p_{23}^{(4)} = p_{24} = 1 - \tilde{H}(\alpha_2)$$

$$p_{31} = \tilde{G}(\lambda), \quad p_{34} = 1 - \tilde{G}_2(\lambda) \tag{10–17}$$

It is easily verified that

$$p_{10} + p_{12} + p_{13} = 1$$

$$p_{21} + p_{23}^{(4)}(p_{24}) = 1$$

$$p_{31} + p_{34} = 1 \tag{18–20}$$

(iii) Mean sojourn time $\psi_i$ in state $S_i$ is defined as the expected time for which the system stays in state $S_i$ before transiting to any other state. To calculate the mean sojourn time $\psi_0$ in state $S_0$. We observe that so long as the system is in state $S_0$, there is no transition to $S_1$. Hence if $T_0$ denotes the sojourn time in $S_0$, then

$$\psi_0 = \int p(T_0 > t)dt = \int e^{-\alpha_1 t}dt = 1/\alpha_1 \tag{21}$$

Similarly,

$$\psi_1 = [1 - \tilde{G}_1(\lambda + \alpha_2)]/(\lambda + \alpha_2), \quad \psi_2 = [1 - \tilde{H}(\alpha_2)]/\alpha_2$$

$$\psi_3 = [1 - \tilde{G}_2(\lambda)]/\lambda, \quad \psi_4 = \int \overline{H}(t)dt \tag{22–25}$$

(iv) We define $m_{ij}$ as the mean sojourn time by the system in state $S_i$ when the system is to transit to regenerative state $S_j$, i.e.,

$$m_{ij} = \int t dQ_{ij}(t) = \int t q_{ij}(t) dt$$

Therefore,

$$m_{01} = \alpha_1 \int te^{-\alpha_1 t}dt = 1/\alpha_1.$$

$$m_{10} = \int te^{-(\alpha_2 + \lambda)t}dG_1(t).$$

$$m_{12} = \lambda \int te^{-(\alpha_2 + \lambda)t}\overline{G}_1(t)dt.$$

$$m_{13} = \alpha_2 \int te^{-(\alpha_2 + \lambda)t}\overline{G}_1(t)dt.$$

$$m_{21} = \int te^{-\alpha_2 t}dH(t).$$

$$m_{23}^{(4)} = \int t(1 - e^{-\alpha_2 t})dH(t).$$

$$m_{31} = \int te^{-\lambda t}dG_2(t).$$

$$m_{34} = \lambda \int te^{-\lambda t}\overline{G}_2(t)dt.$$

$$m_{43} = \int t \, dH(t). \tag{26–34}$$

The following relations among $m_{ij}$'s are observed

$$m_{01} = \psi_0$$

$$m_{10} + m_{12} + m_{13} = \int te^{-(\alpha_2+\lambda)t} dG_1(t) + \lambda \int te^{-(\alpha_2+\lambda)t} \overline{G}_1(t) + \alpha_2 \int te^{-(\alpha_2+\lambda)t} \overline{G}_1(t)$$

$$= [1 - \tilde{G}_1(\lambda + \alpha_2)]/(\lambda + \alpha_2) = \psi_1$$

$$m_{21} + m_{23}^{(4)} = \int te^{-\alpha_2 t} dH(t) + \int t(1 - e^{-\alpha_2 t}) dH(t)$$

$$= \int t\, dH(t) = \int \overline{H}(t)\, dt = \psi_4$$

$$m_{31} + m_{34} = \psi_3, \quad m_{43} = \int t\, dH(t) = \psi_4 \tag{35–39}$$

## Analysis of Reliability and Mean Time to System Failure

Let the random variable $T_i$ be the time to system failure (TSF) when the system starts its operation from state $S_i \in E$, then the reliability of the system is given by

$$R_i(t) = P(T_i > t)$$

To determine $R_i(t)$, we assume that the failed states ($S_3$ and $S_4$) of the system as absorbing. By probabilistic arguments, we have

$$R_0(t) = Z_0(t) + q_{01}(t) \copyright R_1(t)$$
$$R_1(t) = Z_1(t) + q_{10}(t) \copyright R_0(t) + q_{12}(t) \copyright R_2(t) \tag{1–3}$$
$$R_2(t) = Z_2(t) + q_{21}(t) \copyright R_1(t)$$

where

$$Z_0(t) = e^{-\alpha_1 t}, \quad Z_1(t) = e^{-(\alpha_2+\lambda)t}\overline{G}_1(t), \quad Z_2(t) = e^{-\alpha_2 t}\overline{H}(t)$$

As an illustration, $R_0(t)$ is the sum of the following contingencies:

(i) System remains up in state $S_0$ without making any transition to any other state up to time $t$. The probability of this contigency is

$$e^{-\alpha_1 t} = Z_0(t)$$

(ii) System first enters the state $S_1$ from $S_0$ during $(u, u + du)$, $u \leq t$ and then starting from $S_1$, it remains up continuously during the time $(t - u)$, the probability of this contingency is

$$\int_0^t q_{01}(u)\, du R_1(t - u) = q_{01}(t) \copyright R_1(t)$$

Taking Laplace Transform (LT) of the relations (1–3), we can write the solution of resulting set of algebraic equations in the matrix form as follows:

$$\begin{bmatrix} R_0^* \\ R_1^* \\ R_2^* \end{bmatrix} = \begin{bmatrix} 1 & -q_{01}^* & 0 \\ -q_{01}^* & 1 & -q_{12}^* \\ 0 & -q_{21}^* & 1 \end{bmatrix}^{-1} \begin{bmatrix} Z_0^* \\ Z_1^* \\ Z_2^* \end{bmatrix} \tag{4}$$

For brevity, we have omitted the argument "s" from $q_{ij}^*(s)$, $Z_i^*(s)$ and $R_i^*(s)$. Computing the above matrix equation for $R_0^*(s)$, we get

$$R_0^*(s) = N_1(s)/D_1(s) \tag{5}$$

where

$$N_1(s) = (1 - q_{12}^* q_{21}^*)Z_0^* + q_{01}^*(Z_1^* + Z_2^* q_{12}^*)$$

and

$$D_1(s) = 1 - q_{12}^* q_{21}^* - q_{01}^* q_{10}^*$$

Taking the Inverse Laplace Transform (ILT) of (7.5), we can get the reliability of the system when initially it starts from $S_0$.

The mean time to system failure (MTSF) can be obtained using the well-known formula

$$E(T_0) = \int R_0(t)dt = \lim_{s \to 0} R_0^*(s) = N_1(0)/D_1(0) \tag{6}$$

To determine $N_1(0)$ and $D_1(0)$ we first obtain $Z_1^*(0)$, using the result

$$\lim_{s \to 0} Z_i^*(s) = \int Z_i(t)dt$$

Therefore

$$Z_0^*(0) = \psi_0, \quad Z_1^*(0) = \psi_1, \quad Z_2^*(0) = \psi_2$$

Thus, using $q_{ij}^*(0) = p_{ij}$ and the above results, we get

$$N_1(0) = \psi_0(1 - p_{12}^* p_{21}^*)\psi_0^* + \psi_1 + p_{12}\psi_2$$

and

$$D_1(0) = 1 - p_{10} - p_{12}p_{21}$$

## Availability Analysis

Let $A_i^P(t)$ and $A_i^0(t)$ be the probabilities that the system is up (operative) at epoch $t$ due to $p$-unit and $o$-unit, respectively, when initially it starts from state $S_i \in E$.

By simple probabilistic laws, $A_0^P(t)$ is the sum of the following mutually exclusive contigencies:

(i) The system continues to be up in state $S_0$ until epoch $t$. The probability of this event is $e^{-\alpha_1 t}$.

(ii) The system transits to $S_1$ from $S_0$ in $(u, u + du)$, $u \leq t$ and then starting from $S_1$ it is observed to be up at epoch $t$, with probability $A_i^P(t - u)$. Therefore,

$$A_0^P(t) = e^{-\alpha_1 t} + \int_0^t q_{01}(u)du\, A_1^P(t - u)$$

$$= Z_0(t) + q_{01}(t)©A_1^P(t)$$

Similarly,

$$
\begin{aligned}
A_1^P(t) &= q_{10}(t) \copyright A_0^P(t) + q_{12}(t) \copyright A_2^P(t) + q_{13}(t) \copyright A_3^P(t) \\
A_2^P(t) &= q_{21}(t) \copyright A_1^P(t) + q_{(23)}^{(4)}(t) \copyright A_3^P(t) \\
A_3^P(t) &= q_{31}(t) \copyright A_1^P(t) + q_{34}(t) \copyright A_4^P(t) \\
A_4^P(t) &= q_{43}(t) \copyright A_3^P(t)
\end{aligned}
\tag{1-5}
$$

Taking LT of relations (7.1–7.5) and solving the resulting set of algebraic equations for $A_0^{P*}(s)$, we get

$$
A_0^{P*}(s) = N_2(s)/D_2(s) \tag{7}
$$

where

$$
N_2(s) = (1 - q_{12}^*(q_{21}^* + q_{23}^{(4)*} q_{31}^*) - q_{13}^* q_{31}^* - q_{34}^* q_{43}^*(1 - q_{12}^* q_{21}^*)] Z_0^* \tag{8}
$$

and

$$
\begin{aligned}
D_2(s) &= 1 - q_{12}^*(q_{21}^* + q_{23}^{(4)*} q_{31}^*) - q_{13}^* q_{31}^* - q_{34}^* q_{43}^*(1 - q_{12}^* q_{21}^*) \\
&\quad - q_{01}^* q_{10}^*(1 - q_{34}^* q_{43}^*)
\end{aligned}
\tag{9}
$$

Similarly, the recurrence relations among pointwise availabilities $A_i^0(t)$, i.e., when system is up due to o-unit, can also be obtained and are as follows:

$$
\begin{aligned}
A_0^0(t) &= q_{01}(t) \copyright A_1^0(t) \\
A_1^0(t) &= Z_1(t) + q_{10}(t) \copyright A_0^0(t) + q_{12}(t) \copyright A_1^0(t) + q_{13}(t) \copyright A_3^0(t) \\
A_2^{(t)}(t) &= Z_2(t) + q_{21}(t) \copyright A_1^0(t) + q_{23}^{(4)}(t) \copyright A_3^0(t) \\
A_3^0(t) &= q_{31}(t) \copyright A_1^0(t) + q_{34}(t) \copyright A_4^0(t) \\
A_4^0(t) &= q_{43}(t) \copyright A_3^0(t)
\end{aligned}
\tag{10-14}
$$

where

$$
Z_1(t) = e^{-(\alpha_2 + \lambda)t} \overline{G_1}(t) \qquad \text{and} \qquad Z_2(t) = e^{-\alpha_2 t} \overline{H}(t)
$$

Taking LT of the relations (7.10–7.14) and solving for $A_0^{0*}(s)$, we get

$$
A_0^{0*} = N_3(s)/D_2(s) \tag{15}
$$

where

$$
N_3(s) = q_{01}^*(1 - q_{34}^* q_{43}^*)(Z_1^* + q_{12}^* Z_2^*) \tag{16}
$$

Now to obtain the steady-state probabilities that the system will be operative due to p-unit and o-unit, we proceed as follows:

$$
Z_i^*(0) = \int Z_i(t)dt = \psi_i \qquad (i = 0, 1, 3)
$$

Using the result $q_{ij}^*(0) = p_{ij}$, we have

$$D_2(0) = 0$$

Therefore, the steady-state probability that the system will be operative due to p-unit is given by

$$
\begin{aligned}
A_0^p &= \lim_{t \to \infty} N_2(t)/D_2(t) \\
&= \lim_{s \to 0} s\, N_2(s)/D_2(s) \qquad\qquad (17) \\
&= N_2(0)/D_2'(0)
\end{aligned}
$$

where

$$N_2(0) = p_{31} p_{10} \psi_0$$

To obtain $D_2'(0)$, we collect the coefficients of $-q_{ij}^{*'}(0)(=m_{ij})$ in $D_2'(0)$ for various values of $i$ and $j$ as follows:

(i)   Coefficient of $m_{01}$ $= p_{10}(1 - p_{34})$
          $= p_{31} p_{10}$
(ii)  Coefficient of $m_{10}$ $= 1 - p_{34} = p_{31}$
(iii) Coefficient of $m_{12}$ $= p_{21} + p_{23}p_{31} - p_{34}p_{21}$
          $= p_{121}(1 - p_{34}) + p_{23}p_{31}$
          $= p_{31}$
(iv)  Coefficient of $m_{13}$ $= p_{31}$
(v)   Coefficient of $m_{21}$ $= p_{12} - p_{12}p_{34}$
          $= p_{31}p_{12}$
(vi)  Coefficient of $m_{23}^{(4)}$ $= p_{31}p_{12}$
(vii) Coefficient of $m_{31}$ $= p_{12}p_{23}^{(4)} + p_{13}$
(viii)Coefficient of $m_{34}$ $= 1 - p_{12}p_{21} - p_{10}$
          $= p_{12}p_{23}^{(4)} + p_{13}$
(ix)  Coefficient of $m_{43}$ $= p_{34}(1 - p_{12}p_{21}) - p_{10}p_{34}$
          $= p_{34}(1 - p_{10} - p_{12}p_{21})$
          $= p_{34}(p_{12}p_{23}^{(4)} + p_{13})$

Thus,

$$D_2'(0) = P_{31}(p_{10}\psi_0 + \psi_1 + p_{12}\psi_4) + (p_{13} + p_{12}p_{23}^{(4)})(\psi_3 + p_{34}\psi_4) \qquad (18)$$

Similarly, the steady-state probability that the system will be operative due to o-unit is given by

$$A_0^o = \lim_{t \to \infty} N_3(t)/D_2(t) = N_3(0)/D_2'(0) \qquad (19)$$

where

$$N_3(0) = p_{31}(\psi_1 + p_{12}\psi_2)$$

and $D_2'(0)$ has already been defined by equation (18).

## Busy-Period Analysis

Let $B_i^p(t)$, $B_i^o(t)$, and $B_i^m(t)$ be the respective probabilities that the repairman is busy in the repair of $p$-unit, $o$-unit, and RM when system initially starts from state $S_i \in E$. Using elementary probabilistic arguments in respect to the above definition of $B_i^p(t)$, we have the following recursive relations:

$$
\begin{aligned}
B_0^p(t) &= q_{01}(t) © B_1^p(t) \\
B_1^p(t) &= Z_1(t) + q_{10}(t) © B_0^p(t) + q_{12}(t) © B_2^p(t) + q_{13}(t) © B_3^p(t) \\
B_2^p(t) &= q_{21}(t) © B_1^p(t) + q_{23}^{(4)}(t) © B_3^p(t) \\
B_3^p(t) &= q_{31}(t) © B_1^p(t) + q_{34}(t) © B_4^p(t) \\
B_4^p(t) &= q_{43}(t) © B_3^p(t)
\end{aligned}
\tag{1-5}
$$

where

$$
Z_1(t) = e^{-(\alpha_2 + \lambda)t} \overline{G}_1(t)
$$

For an illustration, $B_1^p(t)$ is the sum of the following mutually exclusive contingencies:

(i) The repairman remains busy in state $S_1$ continuously up to time $t$. The probability of this event is $e^{-(\alpha_2 + \lambda)t} \overline{G}_1(t)$

(ii) The system transits from state $S_i$ to $S_j$ ($j = 0, 2, 3$) during time $(u, u + du)$, $u \le t$ and then starting from state $S_j$ the repairman may be observed to be busy with $p$-unit at epoch $t$. The probability of this contingency is

$$
\int_0^t q_{ij}(u)\,du\, B_j^p(t - u) = q_{ij}(t) © B_j^p(t)
$$

Taking LT of relations (1–5) and solving the resulting set of algebraic equations for $B_0^{p*}(s)$ we get

$$
B_0^{p*}(s) = N_4(s)/D_2(s)
\tag{6}
$$

where

$$
N_4(s) = q_{01}^*(1 - q_{34}^* q_{43}^*)Z_1^*
$$

and $D_2(s)$ is the same as given by equation (3.9).

Similarly, the recursive relations in $B_i^o(t)$ and $B_i^m(t)$ may be developed as follows:

$$
\begin{aligned}
B_0^0(t) &= q_{01}(t) © B_1^0(t) \\
B_1^0(t) &= q_{10}(t) © B_0^0(t) + q_{12}(t) © B_2^0(t) + q_{13}(t) © B_3^0(t) \\
B_2^0(t) &= q_{21}(t) © B_1^0(t) + q_{23}^{(4)}(t) © B_3^0(t) \\
B_3^0(t) &= Z_3(t) + q_{31}(t) © B_1^0(t) + q_{34}(t) © B_4^0(t) \\
B_4^0(t) &= q_{43}(t) © B_3^0(t)
\end{aligned}
\tag{7-11}
$$

and

$$B_0^m(t) = q_{01}(t) \copyright B_1^m(t)$$

$$B_1^m(t) = q_{10}(t) \copyright B_0^m(t) + q_{12}(t) \copyright B_2^m(t) + q_{13}(t) \copyright B_3^m(t)$$

$$B_2^m(t) = Z_2(t) + q_{24}(t) \copyright Z_4(t) + q_{21}(t) \copyright B_1^m(t) + q_{23}^{(4)}(t) \copyright B_3^m(t)$$

$$B_3^m(t) = q_{31}(t) \copyright B_1^m(t) + q_{34}(t) \copyright B_4^m(t)$$

$$B_4^m(t) = Z_4(t) + q_{43}(t) \copyright B_3^m(t)$$

where

$$Z_3(t) = e^{-\lambda t} \overline{G}_2(t), \quad Z_4(t) = \overline{H}(t)$$

and $Z_2(t)$ has already been defined. Taking LT of (7.71) and (12–16) and solving for $B_0^{o*}(s)$ and $B_0^{m*}(s)$, one gets

$$B_0^{o*}(s) = N_5(s)/D_2(s)$$

and

$$B_0^{m*}(s) = N_6(s)/D_2(s) \qquad\qquad (17\text{–}18)$$

where

$$N_5(s) = q_{01}^*(q_{12}^* q_{23}^* + q_{13}^*) Z_3^*$$

and

$$N_6(s) = q_{01}^* q_{12}^* (1 - q_{34}^* q_{43}^*)(Z_2^* + q_{24}^* Z_4^*) + q_{01}^* q_{34}^* (q_{13}^* + q_{12}^* q_{23}^*) Z_4^*$$

Now to obtain the steady-state probabilities $B_0^p$, $B_0^o$, and $B_0^m$ that the repairman is busy with the *p*-unit, *o*-unit, and RM, respectively, we use the results

$$Z_i^*(0) = \psi_i, \quad i = 0 \quad \text{to} \quad 4 \quad \text{and} \quad q_{ij}^*(0) = p_{ij}$$

and observe that

$$B_0^p = p_{31} \psi_1 / D_2'(0)$$

$$B_0^o = (p_{13} + p_{12} p_{23}^{(4)}) \psi_3 / D_2'(0)$$

and

$$B_0^m = [p_{31} p_{12}(\psi_2 + p_{24} \psi_4) + (p_{13} + p_{12} p_{23}^{(4)}) p_{34} \psi_4]/D_2'(0)$$

where $D_2'(0)$ is same as defined by equation (3.78).

## Expected Number of Repairs During (0, *t*)

Let $N_i^p(t)$, $N_i^o(t)$, and $N_i^m(t)$ be the expected number of repairs of the $p$-unit, $o$-unit, and RM, respectively, during $(0, t) \mid E_0 = S_i$. Here, using the above definition, the recursive relations in $N_i^p(t)$ are as follows:

$$N_0^p(t) = Q_{01}(t) \,\textcircled{S}\, N_1^p(t)$$

$$N_1^p(t) = Q_{10}(t) \,\textcircled{S}\, [1 + N_0^p(t)] + Q_{12}(t) \,\textcircled{S}\, N_2^p(t) + Q_{13}(t) \,\textcircled{S}\, N_3^p(t)$$

$$N_2^p(t) = Q_{21}(t) \,\textcircled{S}\, N_1^p(t) + Q_{23}^{(4)}(t) \,\textcircled{S}\, N_3^p(t)$$

$$N_3^p(t) = Q_{31}(t) \,\textcircled{S}\, N_1^p(t) + Q_{34}(t) \,\textcircled{S}\, N_4^p(t)$$

$$N_4^p(t) = Q_{43}(t) \,\textcircled{S}\, N_3^p(t)$$

As an illustration, $N_1^p(t)$ is the sum of the following contingencies:

(i) The system transits from $S_1$ to $S_0$ in the interval $(u, u + du)$, $u \le t$ having completed one repair and then starting from $S_0$ at epoch $u$ we may count the expected number of repairs during time $(t - u)$. The probability of this event is

$$\int_0^t dQ_{10}(u)[1 + N_0^p(t - u)] = Q_{10}(t) \,\textcircled{S}\, [1 + N_0^p(t)]$$

(ii) The system transits from $S_1$ to $S_j$ ($j = 2, 3$) during time $(u, u + du)$, $u \le t$ and then starting from $S_j$ at epoch $u$ we may count the expected number of repairs during time $(t - u)$. The probability of this contingency is

$$\int_0^t dQ_{ij}(u)N_j^p(t - u) = Q_{ij}(t) \,\textcircled{S}\, N_j^p(t)$$

Taking Laplace-Stieltjes transform (LST) of the above equations (1–5) and solving for $\tilde{N}_0^p(s)$ we have

$$\tilde{N}_0^p(s) = N_7(s)/D_3(s) \tag{6}$$

where

$$N_7(s) = \tilde{Q}_{01}\tilde{Q}_{10}(1 - \tilde{Q}_{34}\tilde{Q}_{43})$$

and $D_3(s)$ can be written simply on replacing $q_{ij}^*$ and $q_{ij}^{(k)*}$, respectively, by $\tilde{Q}_{ij}$ and $\tilde{Q}_{ij}^{(k)}$ in $D_2(s)$ given by equations (3.87).

Similarly, the recurrence relations in $N_i^o(t)$ and $N_i^m(t)$ can be obtained to get the expected number of repairs of $o$-unit and RM, respectively, and are as follows:

$$N_0^o(t) = Q_{01}(t) \,\textcircled{S}\, N_1^o(t)$$

$$N_1^o(t) = Q_{10}(t) \,\textcircled{S}\, N_0^o(t) + Q_{12}(t) \,\textcircled{S}\, N_2^o(t) + Q_{13}(t) \,\textcircled{S}\, N_3^o(t)$$

$$N_2^o(t) = Q_{21}(t) \,\textcircled{S}\, N_1^o(t) + Q_{23}^{(4)}(t) \,\textcircled{S}\, N_3^o(t) \tag{7-11}$$

$$N_3^o(t) = Q_{31}(t) \,\textcircled{S}\, [1 + N_1^o(t)] + Q_{34}(t) \,\textcircled{S}\, N_4^o(t)$$

$$N_4^o(t) = Q_{43}(t) \,\textcircled{S}\, N_3^o(t)$$

and

$$N_0^m(t) = Q_{01}(t) \circledS N_1^m(t)$$

$$N_1^m(t) = Q_{10}(t) \circledS N_0^m(t) + Q_{12}(t) \circledS N_2^m(t) + Q_{13}(t) \circledS N_3^m(t)$$

$$N_2^m(t) = Q_{21}(t) \circledS [1 + N_1^m(t)] + Q_{23}^{(4)}(t) \circledS N_3^m(t) \qquad (12\text{--}16)$$

$$N_3^m(t) = Q_{31}(t) \circledS N_1^m(t) + Q_{34}(t) \circledS N_4^m(t)$$

$$N_4^m(t) = Q_{43}(t) \circledS [1 + N_3^m(t)]$$

Taking LST of the above relations (7–11) and (12–16) and solving for $\tilde{N}_0^o(s)$ and $\tilde{N}_0^m(s)$, we get

$$\tilde{N}_0^o(s) = N_8(s)/D_3(s)$$

and

$$\tilde{N}_0^m(s) = N_9(s)/D_3(s)$$

where,

$$N_8(s) = \tilde{Q}_{01}\tilde{Q}_{31}(\tilde{Q}_{13} + \tilde{Q}_{12}\tilde{Q}_{23}^{(4)})$$

and

$$N_9(s) = \tilde{Q}_{01}[\tilde{Q}_{12}\tilde{Q}_{21}(1 - \tilde{Q}_{34}\tilde{Q}_{43}) + \tilde{Q}_{34}\tilde{Q}_{43}(\tilde{Q}_{13} + \tilde{Q}_{12}\tilde{Q}_{23}^{(4)})]$$

Now let $N_0^p$, $N_0^o$, and $N_0^m$, be the expected number of repairs of $p$-unit, $o$-unit, and RM in steady state, respectively. Then their expressions can be obtained as follows:

$$N_0^p = \lim_{t \to \infty} N_0^p(t) = \lim_{s \to 0} \tilde{N}_0^p(s)$$

$$= P_{31}P_{10}/D_3'(0)$$

Similarly,

$$N_0^o = p_{31}(p_{13} + p_{12}p_{23}^{(4)})/D_3'(0)$$

$$N_0^m = [p_{12}p_{21}p_{31} + p_{34}(p_{13} + p_{12}p_{23}^{(4)})]/D_3'(0)$$

Since

$$q_{ij}^*(0) = \tilde{Q}_{ij}(0) = p_{ij}, \quad q_{ij}^{(k)*}(0) = \tilde{Q}_{ij}^{(k)}(0) = p_{ij}^{(k)}$$

and

$$q_{ij}^{*'}(0) = \tilde{Q}_{ij}'(0) = -m_{ij}, \quad q_{ij}^{(k)*'}(0) = \tilde{Q}_{ij}^{(k)'}(0) = -m_{ij}^{(k)}$$

therefore the value $D_3'(0)$ comes out to be same as that of $D_2'(0)$ given by equation (3.18).

## Particular Case

In this section, we consider a case when the repair time distributions are also negative exponential, i.e.,

$$g_1(t) = \beta_1 e^{-\beta_1 t}, \quad g_2(t) = \beta_2 e^{-\beta_2 t}$$

$$h(t) = \mu e^{-\mu t}$$

Then in results (1.11–1.17) and (1.22–1.25) we have the following changes:

$$p_{10} = \beta_1/(\beta_1 + \alpha_2 + \lambda), \qquad p_{12} = \lambda/(\beta_1 + \alpha_2 + \lambda)$$

$$p_{13} = \alpha_2/(\beta_1 + \alpha_2 + \lambda), \qquad p_{21} = \mu/(\mu + \alpha_2), \qquad p_{23}^{(4)} = \alpha_2/(\mu + \alpha_2)$$

$$p_{31} = \beta_2/(\lambda + \beta_2), \qquad p_{34} = \lambda/(\lambda + \beta_2), \qquad \psi_1 = 1/(\beta_1 + \alpha_2 + \lambda)$$

$$\psi_2 = 1/(\mu + \alpha_2), \qquad \psi_3 = 1/(\lambda + \beta_2), \qquad \psi_4 = 1/\mu$$

## 7.5  Model B

### Transition Probabilities and Sojourn Times

(i)  By simple probabilistic arguments as discussed in Section 7.4 the nonzero elements of $Q = (Q_{ij}(t))$ are as follows:

$$Q_{01}(t) = \int_0^t \alpha_1 e^{-\alpha_1 us}\, du$$

$$Q_{10}(t) = \int_0^t e^{-(\alpha_2 + \lambda)}\, dG_1(u)$$

$$Q_{12}(t) = \lambda \int_0^t e^{-(\alpha_2 + \lambda)u} \overline{G}_1(u)\, du$$

$$Q_{13}(t) = \alpha_2 \int_0^t e^{-(\alpha_2 + \lambda)u} \overline{G}_1(u)\, du$$

$$Q_{21}(t) = \int_0^t e^{-\alpha_2 u}\, dH(u)$$

$$Q_{24}(t) = \alpha_2 \int_0^t e^{-\alpha_2 u} \overline{H}(u)\, du$$

$$Q_{34}(t) = \lambda \int_0^t e^{-\lambda u} \overline{G}_1(u)\, du$$

$$Q_{35}(t) = \int_0^t e^{-\lambda u}\, dG_1(u)$$

$$Q_{43}(t) = \int_0^t dH(u) = H(t)$$

$$Q_{50}(t) = \int_0^t e^{-(\alpha_1 + \lambda)u}\, dG_2(u)$$

$$Q_{53}(t) = \alpha_1 \int_0^t e^{-(\alpha_1 + \lambda)u} \overline{G}_2(u)\, du$$

$$Q_{56}(t) = \lambda \int_0^t e^{-(\alpha_1 + \lambda)u} \overline{G}_2(u)\, du$$

$$Q_{64}(t) = \alpha_1 \int_0^t e^{-\alpha_1 u} \overline{H}(u)\, du$$

$$Q_{65}(t) = \int_0^t e^{-\alpha_1 u}\, dH(u) \tag{1-14}$$

(ii) To derive an expression for $Q_{14}^3(t)$, we suppose that the system transits from state $S_1$ to $S_4$ during the time interval $(u, u + du)$, $u \le t$, the probability of this event is $\alpha_2 e^{-(\alpha_1 + \lambda)u} \overline{G}_1(u)du$. Further, suppose that the system passes from state $S_3$ to $S_4$ during the interval $(v, v + dv)$ in $(u, t)$; the probability of this event is

$$\lambda e^{-\lambda(v - u)} dv\, \overline{G} \cdot (v)/\overline{G}_1(u)$$

Thus,

$$Q_{14}^3(t) = \int_0^t \alpha_2 e^{-(\alpha_2 + \lambda)u} \overline{G}_1(u)du \int_0^t \lambda e^{-\lambda(v - u)} dv\, \overline{G}_1(v)/\overline{G}_1(u)$$

$$= \alpha_2 \lambda \int_0^t e^{-\lambda v} \overline{G}_1(v)dv \int_0^v e^{-\alpha_2 u} du$$

$$= \lambda \int_0^t e^{-\lambda v}[1 - e^{-\alpha_2 v}]\overline{G}_1(v)dv$$

Similarly,

$$Q_{15}^3(t) = \int_0^t e^{-\lambda v}[1 - e^{-\alpha_2 v}]dG_1(v)$$

$$Q_{23}^4(t) = \int_0^t [1 - e^{-\alpha_2 v}]dH(v) \qquad (15\text{–}18)$$

$$Q_{63}^4(t) = \int_0^t [1 - e^{-\alpha_1 v}]dH(v)$$

(iii) Using the result

$$p_{ij} = \lim_{t \to \infty} Q_{ij}(t)$$

the steady-state transition probabilities are as follows:

$$p_{01} = p_{43} = 1$$
$$p_{10} = \tilde{G}_1(\alpha_2 + \lambda)$$
$$p_{12} = \lambda[1 - \tilde{G}_1(\alpha_2 + \lambda)]/(\alpha_2 + \lambda)$$
$$p_{13} = \alpha_2[1 - \tilde{G}_1(\alpha_2 + \lambda)]/(\alpha_2 + \lambda)$$
$$p_{14}^{(3)} = 1 - \tilde{G}_1(\lambda) - \lambda[1 - \tilde{G}_1(\alpha_2 + \lambda)]/(\alpha_2 + \lambda)$$
$$p_{15}^{(3)} = \tilde{G}_1(\lambda) - \tilde{G}_1(\alpha_2 + \lambda)$$
$$p_{21} = \tilde{H}(\alpha_2), \quad p_{24} = p_{23}^{(4)} = 1 - \tilde{H}(\alpha_2) \qquad (19\text{–}33)$$
$$p_{34} = 1 - \tilde{G}_1(\lambda), \quad p_{35} = \tilde{G}_1(\lambda)$$
$$p_{50} = \tilde{G}_2(\alpha_1 + \lambda)$$
$$p_{53} = \alpha_1[1 - \tilde{G}_2(\alpha_1 + \lambda)]/(\alpha_1 + \lambda)$$
$$p_{56} = \lambda[1 - \tilde{G}_2(\alpha_1 + \lambda)]/(\alpha_1 + \lambda)$$
$$p_{64} = p_{63}^{(4)} = 1 - \tilde{H}(\alpha_1), \quad p_{65} = \tilde{H}(\alpha_1)$$

It is easily verified that

$$p_{10} + p_{12} + p_{13}(p_{14}^{(3)} + p_{15}^{(3)}) = 1, \quad p_{21} + p_{24}(=p_{23}^{(4)}) = 1$$

$$p_{34} + p_{35} = 1, \quad p_{50} + p_{53} + p_{56} = 1 \qquad (34\text{–}38)$$

$$p_{64}(=p_{63}^{(4)}) + p_{65} = 1$$

(iv) By similar probabilistic arguments as in Section 7.4, the mean sojourn times in various states are as follows:

$$\psi_0 = \int e^{-\alpha_1 t} dt = 1/\alpha_1$$

$$\psi_1 = \int e^{-(\alpha_2 + \lambda)t} \overline{G}_1(t) dt$$

$$\psi_2 = \int e^{-\alpha_2 t} \overline{H}(t) dt$$

$$\psi_3 = \int e^{-\lambda t} \overline{G}_1(t) dt \qquad (39\text{–}45)$$

$$\psi_4 = \int \overline{H}(t) dt$$

$$\psi_5 = \int e^{-(\alpha_2 + \lambda)t} \overline{G}_2(t) dt$$

$$\psi_6 = \int e^{-\alpha_1 t} \overline{H}(t) dt$$

(v) $m_{ij}$, the mean sojourn time by the system in state $S_i$ when the system is to transit to regenerative state $S_j$ is given by

$$m_{ij} = \int t \, dQ_{ij}(t) = \int t q_{ij}(t) dt$$

therefore,

$$m_{01} = \alpha_1 \int t e^{-\alpha_1 t} dt = 1/\alpha_1$$

$$m_{10} = \int t e^{-(\alpha_2 + \lambda)t} dG_1(t)$$

$$m_{12} = \lambda \int t e^{-(\alpha_2 + \lambda)t} \overline{G}_1(t) dt$$

$$m_{13} = \alpha_2 \int t e^{-(\alpha_2 + \lambda)t} \overline{G}_1(t) dt$$

$$m_{14}^{(3)} = \lambda \int t e^{-\lambda t}(1 - e^{-\alpha_2 t}) \overline{G}_1(t) dt$$

$$m_{15}^{(3)} = \int t e^{-\lambda t}(1 - e^{-\alpha_2 t}) dG_1(t)$$

$$m_{21} = \int t e^{-\alpha_2 t} dH(t)$$

$$m_{24} = \alpha_2 \int t e^{-\alpha_2 t} \overline{H}(t) dt$$

$$m_{23}^4 = \int t[1 - e^{-\alpha_2 t}] dH(t)$$

$$m_{34} = \lambda \int te^{-\lambda t}\overline{G}_1(t)\,dt$$

$$m_{35} = \int te^{-\lambda t}\,dG_1(t)$$

$$m_{43} = \int t\,dH(t)$$

$$m_{50} = \int te^{-(\alpha_1+\lambda)t}\,dG_2(t)$$

$$m_{53} = \alpha_1 \int te^{-(\alpha_1+\lambda)t}\overline{G}_2(t)\,dt \qquad\qquad (46\text{--}64)$$

$$m_{56} = \lambda \int te^{-(\alpha_1+\lambda)t}\overline{G}_2(t)\,dt$$

$$m_{64} = \alpha_1 \int te^{-\alpha_1 t}\overline{H}(t)\,dt$$

$$m_{63}^{4} = \int t[1 - e^{-\alpha_1 t}]\,dH(t)$$

$$m_{65} = \int te^{-\alpha_1 t}\,dH(t)$$

The following relations among $m_{ij}$'s and $\psi_i$'s are observed:

$$m_{01} = \psi_0, \quad m_{43} = \psi_4$$
$$m_{10} + m_{12} + m_{14}^{(3)} + m_{15}^{(3)} = \psi_3$$
$$m_{21} + m_{23}^{(4)} = \psi_4$$
$$m_{34} + m_{35} = \psi_3$$
$$m_{50} + m_{53} + m_{56} = \psi_5$$
$$m_{63}^{(4)} + m_{65} = \psi_4$$

## Analysis of Reliability and Mean Time to System Failure

To determine $R_i(t)$, we assume that the failed states $S_3$ and $S_4$ of the system as absorbing. By probabilistic arguments as used in Section 7.4, we observe that the following recursive relations hold good:

$$R_0(t) = Z_0(t) + q_{01}(t)\copyright R_1(t)$$
$$R_1(t) = Z_1(t) + q_{10}(t)\copyright R_0(t) + q_{12}(t)\copyright R_2(t) \qquad (1\text{--}3)$$
$$R_2(t) = Z_2(t) + q_{21}\copyright R_1(t)$$

where

$$Z_0(t) = e^{-\alpha_1 t}$$

$$Z_1(t) = e^{-(\alpha_2+\lambda)t}\overline{G}_1(t)$$

$$Z_2(t) = e^{-\alpha_2 t}\overline{H}(t)$$

Taking Laplace Transforms of relations (7.1–7.3) and solving the resulting set of algebraic equations for $R_0^*(s)$, we get

$$R_0^*(s) = N_1(s)/D_1(s) \qquad (4)$$

where

$$N_1(s) = (1 - q_{12}^* q_{21}^*)Z_0^* + q_{01}^*(Z_1^* + q_{12}^* Z_2^*)$$

and

$$D_1(s) = 1 - q_{12}^* q_{21}^* - q_{01}^* q_{10}^*$$

Taking the inverse LT of (4), we can get reliability of the system when initially it starts from state $S_0$. The mean time to system failure is given by

$$
\begin{aligned}
E(T_0) &= \int R_0(t)dt \\
&= \lim_{s \to 0} R_0^*(s) \qquad (5)\\
&= N_1(0)/D_1(0)
\end{aligned}
$$

To obtain $N_1(0)$ and $D_1(0)$, we use the results

$$\lim_{s \to 0} Z_0^*(s) = \int Z_1(t)dt = \psi_1 \quad \text{and} \quad q_{ij}^*(0) = p_{ij}$$

Therefore,

$$N_1(0) = (1 - p_{12}p_{21})\psi_0 + \psi_1 + p_{12}^*\psi_2$$

and

$$D_1(0) = 1 - p_{10} - p_{12}p_{21}$$

## Availability Analysis

In view of the definition of $A_i^P(t)$ and the probability arguments as used in Section 7.4, we have the following recursive relations:

$$
\begin{aligned}
A_0^P(t) &= Z_0(t) + q_{01}(t)©A_1^P(t) \\
A_1^P(t) &= q_{10}(t) © A_0^P(t) + q_{12}(t) © A_2^P(t) + q_{14}^{(3)}(t) © A_4^P(t) + q_{15}^{(3)}(t) © A_5^P(t) \\
A_2^P(t) &= q_{21}(t) © A_1^P(t) + q_{23}^{(4)}(t) © A_3^P(t) \\
A_3^P(t) &= q_{34}(t) © A_4^P(t) + q_{35}(t) © A_5^P(t) \qquad (1\text{--}7)\\
A_4^P(t) &= q_{43}(t) © A_3^P(t) \\
A_5^P(t) &= Z_5(t) + q_{50}(t) © A_0^P(t) + q_{53}(t) © A_3^P(t) + q_{56}(t) © A_6^P(t) \\
A_6^P(t) &= Z_6(t) + q_{63}^{(4)}(t) © A_3^P(t) + q_{65}(t) © A_5^P(t)
\end{aligned}
$$

where

$$Z_0(t) = e^{-\alpha_1 t}$$

$$Z_5(t) = e^{-(a_1 + \lambda)t} \overline{G_2}(t)$$

$$Z_6(t) = e^{-\alpha_1 t} \overline{H}(t)$$

Taking LT of relations (7.1–7.7) and solving the resulting set of algebraic equations for $A_0^{p*}(s)$, we get

$$A_0^{p*}(s) = N_2(s)/D_2(s) \tag{8}$$

where

$$N_2(s) = (1 - q_{12}^* q_{21}^*)[(1 - q_{34}^* q_{43}^*)(1 - q_{56}^* q_{65}^*) - q_{35}^*(q_{53}^* + q_{56}^* q_{63}^{(4)*})]Z_0^*$$
$$+ q_{01}^*[q_{12}^* q_{23}^{(4)*} q_{35}^* + q_{14}^{(3)*} q_{43}^* q_{35}^* + q_{15}^{(3)*}(1 - q_{34}^* q_{43}^*)](Z_5^* + q_{56}^* Z_6^*)$$

and

$$D_2(s) = (1 - q_{12}^* q_{21}^* - q_{10}^* q_{01}^*)[(1 - q_{34}^* q_{43}^*)(1 - q_{56}^* q_{65}^*) - q_{35}^*(q_{53}^* + q_{56}^* q_{63}^{(4)*})]$$
$$- q_{01}^* q_{50}^*[(q_{12}^* q_{23}^{(4)*} + q_{14}^{(3)*} q_{43}^*) q_{35}^* + q_{15}^{(3)*}(1 - q_{34}^* q_{43}^*)]$$

Similarly, the recursive relations among pointwise availability $A_i^o(t)$ (i.e., when system is up due to o-unit) can also be obtained and are as follows:

$$A_0^o(t) = q_{01}(t) \ \textcircled{c}\ A_1^o(t)$$
$$A_1^o(t) = Z_1(t) + q_{10}(t) \ \textcircled{c}\ A_o^o(t) + q_{12}(t) \ \textcircled{c}\ A_2^o(t) + q_{14}^{(3)}(t) \ \textcircled{c}\ A_4^o(t) + q_{15}^{(3)}(t) \ \textcircled{c}\ A_5^o(t)$$
$$A_2^o(t) = Z_2(t) + q_{21}(t) \ \textcircled{c}\ A_1^o(t) + q_{23}^{(4)}(t) \ \textcircled{c}\ A_3^o(t)$$
$$A_3^o(t) = q_{34}(t) \ \textcircled{c}\ A_4^o(t) + q_{35}(t) \ \textcircled{c}\ A_5^o(t)$$
$$A_4^o(t) = q_{43}(t) \ \textcircled{c}\ A_3^o(t)$$
$$A_5^o(t) = q_{50}(t) \ \textcircled{c}\ A_0^o(t) + q_{53}(t) \ \textcircled{c}\ A_3^o(t) + q_{56}(t) \ \textcircled{c} A_6^o(t)$$
$$A_6^o(t) = q_{63}^{(4)}(t) \ \textcircled{c}\ A_3^o(t) + q_{65}(t) \ \textcircled{c}\ A_5^o(t) \tag{9-15}$$

where

$$Z_1(t) = e^{-(\alpha_2 + \lambda)t} \overline{G_1}(t)$$

and

$$Z_2(t) = e^{-\alpha_2 t} \overline{H}(t)$$

Taking LT of the above relations and solving for $A_0^{0*}(s)$, we get

$$A_0^{0*}(s) = N_3(s)/D_2(s) \tag{16}$$

where

$$N_3(s) = (q_{01}^* Z_1^* + q_{01}^* q_{12}^* Z_2^*)[(1 - q_{34}^* q_{43}^*)(1 - q_{56}^* q_{65}^*) - q_{35}^*(q_{53}^* + q_{56}^* q_{63}^{(4)*})]$$

and $D_2(s)$ has already been defined.

To obtain the steady-state probabilities that the system will be operative due to $p$-unit and $o$-unit, we proceed as in Section 7.4 and find that

$$A_0^p = N_2(0)/D_2'(0) \tag{17}$$

$$A_0^0 = N_3(0)/D_2'(0) \tag{18}$$

where

$$N_2(0) = p_{35}[(1 - p_{12}p_{21})p_{50}\psi_0 + (p_{14}^{(3)} + p_{15}^{(3)} + p_{12}p_{23}^{(4)})(\psi_5 + p_{56}\psi_6)]$$
$$N_3(0) = p_{35}p_{50}(\psi_1 + p_{12}\psi_2)$$

and

$$\begin{aligned}
D_2'(0) = \ & p_{35}p_{50}[(1 - p_{12}p_{21})\psi_0 + \psi_3 + p_{12}\psi_4] \\
& + [(1 - p_{10} - p_{12}p_{21})(1 - p_{56}p_{65}) - p_{15}^{(3)}p_{50}](\psi_3 + p_{34}\psi_4) \\
& + p_{14}^{(3)}p_{35}p_{50}\psi_4 + p_{35}(1 - p_{10} - p_{12}p_{21})(\psi_5 + p_{56}\psi_6)
\end{aligned} \tag{19}$$

## Busy-Period Analysis

Using the same probabilistic arguments as in Section 7.4, we find that the probabilities $B_i^p(t)$ satisfy the following recursive relations:

$$\begin{aligned}
B_0^p(t) = \ & q_{01}(t) \, \copyright B_1^p(t) \\
B_1^p(t) = \ & Z_1(t) + q_{13}(t) \, \copyright \, Z_3(t) + q_{10}(t) \copyright B_0^p(t) + q_{12}(t) \copyright \, B_2^p(t) \\
& + q_{14}^{(3)}(t) \, \copyright \, B_4^p(t) + q_{15}^{(3)}(t) \, \copyright \, B_5^p(t) \\
B_2^p(t) = \ & q_{21}(t) \, \copyright \, B_1^p(t) + q_{23}^{(4)} \, \copyright \, B_3^p(t) \\
B_3^p(t) = \ & Z_3(t) \ + q_{34}(t) \, \copyright \, B_4^p(t) + q_{35}(t) \, \copyright \, B_5^p(t) \\
B_4^p(t) = \ & q_{43}(t) \, \copyright \, B_3^p(t) \\
B_5^p(t) = \ & q_{50}(t) \, \copyright \, B_0^p(t) + q_{53}(t) \, \copyright \, B_3^p(t) + q_{56}(t) \, \copyright \, B_6^p(t) \\
B_6^p(t) = \ & q_{63}^{(4)}(t) \, \copyright \, B_3^{(p)}(t) + q_{65}(t) \, \copyright \, B_5^p(t)
\end{aligned} \tag{1-7}$$

where $Z_3(t) = e^{-\lambda t}\overline{G}_1(t)$ and $Z_1(t)$ has already been defined.

Taking LT of relations (7.1–7.7) and solving the resulting set of equations for $B_0^{p*}(s)$, we get

$$B_0^{p*}(s) \;=\; N_4(s)/D_2(s) \tag{8}$$

where

$$
\begin{aligned}
N_4(s) \;=\; & [(1 - q_{34}^* q_{43}^*)(1 - q_{56}^* q_{65}^*) - q_{35}^*(q_{53}^* + q_{56}^* q_{63}^{(4)*})](Z_1^* + q_{13}^* Z_3^*) \\
& + q_{01}^*(1 - q_{56}^* q_{65}^*)(q_{12}^* q_{23}^{(4)*} + q_{14}^{(3)*} q_{43}^*)Z_3^* + q_{01}^* q_{15}^*(q_{53}^* + q_{56}^* q_{63}^{(4)*})Z_3^*
\end{aligned}
$$

Similarly, the recursive relations in $B_i^o(t)$ and $B_i^m(t)$ are as follows:

$$
\begin{aligned}
B_0^o(t) \;=\; & q_{01}(t) \; \copyright \; B_0^0(t) \\
B_1^o(t) \;=\; & q_{01}(t) \; \copyright \; B_0^0(t) + q_{12}(t) \; \copyright \; B_2^0(t) + q_{14}^{(3)}(t) \; \copyright \; B_4^0(t) + q_{15}^{(3)}(t) \; \copyright \; B_5^0(t) \\
B_2^o(t) \;=\; & q_{21}(t) \; \copyright \; B_1^0(t) + q_{23}^{(4)}(t) \; \copyright \; B_3^0(t) \\
B_3^o(t) \;=\; & q_{34}(t) \; \copyright \; B_4^0(t) + q_{35}(t) \; \copyright \; B_5^0(t) \\
B_4^o(t) \;=\; & q_{43}(t) \; \copyright \; B_3^0(t) \\
B_5^o(t) \;=\; & Z_5(t) + q_{50}(t) \; \copyright \; B_0^0(t) + q_{53}(t) \; \copyright \; B_3^0(t) + q_{56}^{(t)}(t) \; \copyright \; B_6^0(t) \\
B_6^o(t) \;=\; & q_{63}^{(4)}(t) \; \copyright \; B_3^0(t) + q_{65}(t) \; \copyright \; B_5^0(t)
\end{aligned}
\tag{9–15}
$$

and

$$
\begin{aligned}
B_0^m(t) \;=\; & q_{01}(t) \; \copyright \; B_0^m(t) \\
B_1^{m*}(t) \;=\; & q_{10}(t) \; \copyright \; B_0^m(t) + q_{12}(t) \; \copyright \; B_2^m(t) + q_{14}^{(3)}(t) \; \copyright \; B_4^m(t) + q_{15}^{(3)}(t) \; \copyright \; B_5^m(t) \\
B_2^m(t) \;=\; & Z_2(t) + q_{24}(t) \; \copyright \; Z_4(t) + q_{21}(t) \; \copyright \; B_1^m(t) + q_{23}^{(4)}(t) \; \copyright \; B_3^0(t) \\
B_3^m(t) \;=\; & q_{34}(t) \; \copyright \; B_4^m(t) + q_{35}(t) \; \copyright \; B_5^m(t) \\
B_4^m(t) \;=\; & Z_4(t) + q_{43}(t) \; \copyright \; B_3^m(t) \\
B_5^m(t) \;=\; & q_{50}(t) \; \copyright \; B_0^m(t) + q_{53}(t) \; \copyright \; B_3^m(t) + q_{56}(t) \; \copyright \; B_6^m(t) \\
B_6^m(t) \;=\; & Z_6(t) + q_{64}(t) \; \copyright \; Z_4(t) + q_{63}^{(4)}(t) \; \copyright \; B_3^m(t) + q_{65}(t) \; \copyright \; B_5^m(t)
\end{aligned}
\tag{16–22}
$$

where $Z_4(t) = \overline{H}(t)$ and $Z_i(t)$ ($i = 2, 5, 6$) have already been defined.

Taking LT of (7.9–7.15) and (7.16–7.22) and solving for $B_0^{0*}(s)$ and $B_0^{0*}(s)$ by the usual method, one gets

$$B_0^{0*}(s) \;=\; N_5(s)/D_2(s) \tag{23}$$

and

$$B_0^{m*}(s) \;=\; N_6(s)/D_2(s) \tag{24}$$

where

$$N_5(s) \;=\; q_{01}^* q_{12}^* q_{23}^{(4)*} q_{35}^* Z_5^* + q_{01}^* q_{14}^{(3)*} q_{43}^* q_{35}^* Z_5^* + q_{01}^* q_{15}^{(3)*}(1 - q_{34}^* q_{43}^*)Z_5^*$$

and

$$N_6(s) = q_{01}^* q_{12}^* (Z_2^* + q_{24}^* Z_4^*)[(1 - q_{34}^* q_{43}^*)(1 - q_{56}^* q_{65}^*) - q_{35}^*(q_{53}^* + q_{56}^* q_{63}^{(4)*})]$$
$$+ q_{01}^* q_{12}^* q_{23}^{(4)*} \{[q_{34}^*(1 - q_{56}^* q_{65}^*) + q_{35}^* q_{56}^* q_{64}^*]Z_4^* + q_{35}^* q_{56}^* Z_6^*\}$$
$$+ q_{01}^* q_{14}^{(3)*} \{[1 - q_{35}^* q_{53}^* - q_{56}^* q_{65}^* + (q_{64}^* q_{43}^* - q_{63}^{(4)*})q_{35}^* q_{56}^*]Z_4^* + q_{35}^* q_{56}^* q_{43}^* Z_6^*\}$$
$$+ q_{01}^* q_{15}^{(3)*} \{[q_{34}^* q_{53}^* + q_{56}^* q_{64}^* + (q_{64}^* q_{43}^* - q_{63}^{(3)*})q_{34}^* q_{56}^*]Z_4^* + q_{56}^*(1 - q_{34}^* q_{43}^*)Z_6^*\}$$

The steady-state probabilities $B_0^p$, $B_0^o$, and $B_0^m$ that the repairman is busy with $p$-unit, $o$-unit, and RM, respectively, can be easily obtained, and are as follows:

$$B_0^p = N_4(0)/D_2'(0)$$
$$B_0^o = N_5(0)/D_2'(0) \qquad\qquad (25\text{--}27)$$
$$B_0^m = N_6(0)/D_2'(0)$$

where

$$N_4(0) = p_{35} p_{50}(\psi_1 + p_{13}\psi_3) + (1 - p_{56} p_{65})(p_{14}^{(3)} + p_{12} p_{23}^{(14)})\psi_3$$
$$+ p_{15}^{(3)}(p_{53} + p_{56} p_{63}^{(4)})\psi_3$$
$$N_5(0) = p_{35}(p_{14}^{(3)} + p_{15}^{(3)} + p_{12} p_{23}^{(4)})\psi_5$$

and

$$N_6(0) = p_{12} p_{35} p_{50}(\psi_2 + p_{24}\psi_4) + [p_{12} p_{23}^{(4)}(p_{34} - p_{35} p_{56} - p_{56} p_{65})$$
$$+ p_{14}^{(3)}(1 - p_{35} p_{53} - p_{56} p_{65}) + p_{15}^{(3)}(p_{53} p_{34} + p_{56} p_{64})]\psi_4$$
$$+ p_{35} p_{56}(p_{12} p_{23} + p_{14}^{(3)} + p_{15}^{(3)})\psi_6$$

## Expected Number of Repairs during $(0, t)$

Similar probabilistic arguments as in Section 7.4 yield the recurrence relations for $N_i^p(t)$, $N_i^o(t)$, and $N_i^m(t)$ as follows:

$$N_0^p(t) = Q_{01}(t)\,ⓈN_1^p(t)$$
$$N_1^p(t) = Q_{01}(t)Ⓢ[1 + N_0^p(t)] + Q_{12}(t)ⓈN_2^p(t) + Q_{14}^{(3)}(t)ⓈN_4^p(t) + Q_{15}^{(3)}(t)Ⓢ[1 + N_5^p(t)]$$
$$N_2^p(t) = Q_{21}(t)ⓈN_1^p(t) + Q_{23}^{(4)}(t)ⓈN_3^p(t)$$
$$N_3^p(t) = Q_{34}(t)ⓈN_4^p(t) + Q_{35}(t)Ⓢ[1 + N_5^p(t)]$$
$$N_4^p(t) = Q_{43}(t)ⓈN_3^p(t)$$
$$N_5^p(t) = Q_{50}(t)ⓈN_0^p(t) + Q_{53}(t)ⓈN_3^p(t) + Q_{56}(t)ⓈN_6^p(t)$$
$$N_6^p(t) = Q_{63}^{(4)}(t)ⓈN_3^p(t) + Q_{65}(t)ⓈN_5^p(t) \qquad\qquad (1\text{--}7)$$

Similarly,

$$N_0^o(t) = Q_{01}(t) \circledS N_1^o(t)$$

$$N_1^o(t) = Q_{10}(t) \circledS N_0^o(t) + Q_{12}(t) \circledS N_2^o(t) + Q_{14}^{(3)}(t) \circledS N_4^o(t) + Q_{15}^{(3)}(t) \circledS N_5^o(t)$$

$$N_2^o(t) = Q_{21}(t) \circledS N_1^o(t) + Q_{23}^{(4)}(t) \circledS N_3^o(t)$$

$$N_3^0(t) = Q_{34}(t) \circledS N_4^p(t) + Q_{35}(t) \circledS N_5^o(t) \qquad (8\text{--}14)$$

$$N_4^o(t) = Q_{43}(t) \circledS N_3^o(t)$$

$$N_5^o(t) = Q_{50}(t) \circledS [1 + N_0^o(t)] + Q_{53}(t) \circledS N_3^o(t) + Q_{56}(t) \circledS N_6^o(t)$$

$$N_6^o(t) = Q_{63}^{(4)}(t) \circledS N_3^o(t) + Q_{65}(t) \circledS N_5^o(t)$$

and

$$N_0^m(t) = Q_{01}(t) \circledS N_1^m(t)$$

$$N_1^m(t) = Q_{10}(t) \circledS N_0^m(t) + Q_{12}(t) \circledS N_2^m(t) + Q_{14}^{(3)}(t) \circledS N_4^m(t) + Q_{15}^{(3)}(t) \circledS N_5^m(t)$$

$$N_2^m(t) = Q_{21}(t) \circledS [1 + N_1^m(t)] + Q_{23}^{(4)}(t) \circledS [1 + N_3^m(t)]$$

$$N_3^m(t) = Q_{34}(t) \circledS N_4^m(t) + Q_{35}(t) \circledS N_5^m(t)$$

$$N_4^m(t) = Q_{43}(t) \circledS [1 + N_3^m(t)]$$

$$N_5^m(t) = Q_{50}(t) \circledS N_0^m(t) + Q_{53}(t) \circledS N_3^m(t) + Q_{56}(t) \circledS N_6^m(t)$$

$$N_6^m(t) = Q_{63}^{(4)}(t) \circledS [1 + N_3^m(t)] + Q_{65}(t) \circledS [1 + N_5^m(t)] \qquad (15\text{--}20)$$

Taking LST of the above equations (1–7), (8–14), and (15–20) and solving them for $\tilde{N}_0^p(s)$, $\tilde{N}_0^o(s)$, and $\tilde{N}_0^m(s)$, we get

$$\tilde{N}_0^p(s) = N_7(s)/D_3(s)$$

$$\tilde{N}_0^0(s) = N_8(s)/D_3(s)$$

$$\tilde{N}_0^m(s) = N_9(s)/D_3(s) \qquad (21\text{--}23)$$

where

$$N_7(s) = [(1 - \tilde{Q}_{34}\tilde{Q}_{43})(1 - \tilde{Q}_{56}\tilde{Q}_{65}) - \tilde{Q}_{35}(\tilde{Q}_{53} + \tilde{Q}_{56}\tilde{Q}_{63}^{(4)})](\tilde{Q}_{10} + \tilde{Q}_{15}^{(3)})$$
$$\qquad + \tilde{Q}_{35}\tilde{Q}_{01}[(1 - \tilde{Q}_{56}\tilde{Q}_{65})(\tilde{Q}_{12}\tilde{Q}_{23}^{(4)} + \tilde{Q}_{14}^{(3)}\tilde{Q}_{43}) + \tilde{Q}_{15}^{(3)}(\tilde{Q}_{53} + \tilde{Q}_{56}\tilde{Q}_{63}^{(4)})]$$

$$N_8(s) = \tilde{Q}_{01}\tilde{Q}_{50}[\tilde{Q}_{12}\tilde{Q}_{23}^{(4)}\tilde{Q}_{35} + \tilde{Q}_{14}^{(3)}\tilde{Q}_{43}\tilde{Q}_{35} + \tilde{Q}_{15}^{(3)}(1 - \tilde{Q}_{34}\tilde{Q}_{43})]$$

$$N_9(s) = \tilde{Q}_{01}\tilde{Q}_{12}(\tilde{Q}_{21} + \tilde{Q}_{23}^{(4)})[(1 - \tilde{Q}_{34}\tilde{Q}_{43})(1 - \tilde{Q}_{65}\tilde{Q}_{65}) - \tilde{Q}_{35}(\tilde{Q}_{53} + \tilde{Q}_{56}\tilde{Q}_{63}^{(4)})]$$
$$\qquad + \tilde{Q}_{01}\tilde{Q}_{12}\tilde{Q}_{23}^{(4)}[\tilde{Q}_{34}\tilde{Q}_{43}(1 - \tilde{Q}_{56}\tilde{Q}_{65}) + \tilde{Q}_{35}\tilde{Q}_{56}(\tilde{Q}_{63}^{(4)} + \tilde{Q}_{65})]$$
$$\qquad + \tilde{Q}_{01}\tilde{Q}_{14}^{(3)}[\tilde{Q}_{43}(1 - \tilde{Q}_{35}\tilde{Q}_{53} - \tilde{Q}_{56}\tilde{Q}_{65}) + \tilde{Q}_{35}\tilde{Q}_{56}(\tilde{Q}_{63}^{(4)} + \tilde{Q}_{65})]$$
$$\qquad + \tilde{Q}_{01}\tilde{Q}_{15}^{(3)}[\tilde{Q}_{43}\tilde{Q}_{53}\tilde{Q}_{34} + \tilde{Q}_{56}(1 - \tilde{Q}_{34}\tilde{Q}_{43})(\tilde{Q}_{63}^{(4)} + \tilde{Q}_{65})]$$

and $D_3(s)$ can be written simply on replacing $q_{ij}^*$ in $D_2(s)$ by $\tilde{Q}_{ij}$.

In steady state the expected number of repairs of $p$-unit, $o$-unit, and RM per unit time are, respectively, as follows:

$$N_0^p = N_7(0)/D_2'(0)$$

$$N_0^o = N_8(0)/D_2'(0)$$

and

$$N_0^m = N_9(0)/D_2'(0)$$

where

$$N_7 = p_{35}[(p_{10} + p_{15}^{(3)})p_{50} + (1 - p_{56}p_{65})(p_{14}^{(3)} + p_{12}p_{23}^{(4)}) + p_{15}^{(3)}(p_{53} + p_{56}p_{63}^{(4)})]$$
$$N_8 = p_{35}p_{50}(p_{14}^{(3)} + p_{15}^{(3)} + p_{12}p_{23}^{(4)})$$

and

$$N_9 = p_{12}p_{35}p_{50} + p_{35}p_{56}(p_{14}^{(3)} + p_{15}^{(3)} + p_{12}p_{23}^{(4)}) + p_{12}p_{23}^{(4)}p_{34}(1 - p_{56}p_{65})$$
$$+ \; p_{14}^{(3)}(1 - p_{35}p_{53} - p_{56}p_{65}) + p_{15}^{(3)}p_{53}p_{34}$$

## Particular Case

As in Section 7.4, we consider a case when the repair time distributions are also negative exponential, i.e.,

$$g_1(t) = \beta_1 e^{(-\beta_1 t)}, \qquad g_2(t) = \beta_2 e^{(-\beta_2 t)}, \qquad h(t) = \mu e^{(-\mu t)}$$

then in the result (1.20–1.33), (1.40–1.45), we have the following changes:

$$p_{10} = \beta_1/(\beta_1 + \alpha_2 + \lambda), \qquad p_{12} = \lambda/(\beta_1 + \alpha_2 + \lambda), \qquad p_{13} = \alpha_2/(\beta_1 + \alpha_2 + \lambda)$$
$$p_{14}^{(3)} = 1 - \beta_1/(\beta_1 + \lambda) - \lambda/(\beta_1 + \alpha_2 + \lambda), \qquad p_{15}^{(3)} = \beta_1/(\beta_1 + \lambda) - \beta_1/(\beta_1 + \alpha_2 + \lambda)$$
$$p_{21} = \mu/(\mu + \alpha_2), \qquad p_{24}(p_{23}^{(4)}) = \alpha_2/(\mu + \alpha_2), \qquad p_{34} = \lambda/(\lambda + \beta_1)$$
$$p_{35} = \beta_1/(\lambda + \beta_1), \qquad p_{50} = \beta_2/(\beta_2 + \alpha_1 + \lambda), \qquad p_{53} = \alpha_1/(\beta_2 + \alpha_1 + \lambda)$$
$$p_{56} = \lambda/(\beta_2 + \alpha_1 + \lambda), \qquad p_{64}(p_{63}^{(4)}) = \alpha_1/(\mu + \alpha_1), \qquad p_{65} = \mu/(\mu + \alpha_1)$$
$$\psi_1 = 1/(\beta_1 + \alpha_2 + \lambda), \qquad \psi_2 = 1/(\mu + \alpha_2), \qquad \psi_3 = 1/(\lambda + \beta_1)$$
$$\psi_4 = 1/\beta_2, \; \psi_5 = 1/(\beta_2 + \alpha_1 + \lambda), \qquad \psi_6 = 1/(\mu + \alpha_1)$$

## 7.6   Model C

### Transition Probabilities and Sojourn Times

(i) By definition and simple probabilistic arguments as in earlier sections, the direct of one-step unconditional transition probabilities can be obtained as follows:

$$Q_{01}(t) = \int_0^t \alpha_1(1 - r_1)e^{-\alpha_1(1-r_1)\mu}\,du = 1 - e^{-\alpha_1(1-r_1)t}$$

$$Q_{25}(t) = \int_0^t \mu e^{-\mu u}\,du\, e^{-\alpha_2(1-r_2)u} = \mu[1 - e^{-\{\mu + \alpha_2(1-r_2)\}t}]/\{\mu + \alpha_2(1-r_2)\}$$

$$Q_{26}(t) = \alpha_2 1 - r_2[1 - e^{-\{\mu + \alpha_2(1-r_2)\}t}]/\{\mu + \alpha_2(1-r_2)\}$$

$$Q_{47}(t) = 1 - e^{-\mu t} = Q_{63}(t)$$

$$Q_{50}(t) = \beta_1[1 - e^{-\{\beta_1 + \lambda + \alpha_2(1-r_2)\}t}]/\{\beta_1 + \lambda + \alpha_2(1-r_2)\}$$

$$Q_{52}(t) = \lambda[1 - e^{-\{\beta_1 + \lambda + \alpha_2(1-r_2)\}t}]/\{\beta_1 + \lambda + \alpha_2(1-r_2)\}$$

$$Q_{53}(t) = \alpha_2(1 - r_2)[1 - e^{-\{\beta_1 + \lambda + \alpha_2(1-r_2)\}t}]/\{\beta_1 + \lambda + \alpha_2(1-r_2)\}$$

$$Q_{74}(t) = \mu[1 - e^{-(\mu + \beta_2)t}]/(\mu + \beta_2)$$

$$Q_{75}(t) = \beta_2[1 - e^{-(\mu + \beta_2)t}]/(\mu + \beta_2) \tag{1-9}$$

(ii) By the definition of $Q_{ij|x}$, the direct or one-step conditional transition probabilities from the states where the repair of a unit depends upon its failure time are as follows:

$Q_{10|x}(t) = $ Probability that the system transits from state $S_1$ to $S_0$ in time $\leq t$ given that the $p$-unit entered into $F$-mode after an operation of time $x$.

$= \int_0^t P$ [From state $S_1$ the system does not transit to state $S_3$ up to time $u$ and transit into $S_0$ during $(u, u + du)]$

$$= \int_0^t e^{-\{\lambda + \alpha_2(1-r_2)\}^\mu} \beta_1 e^{-(\beta_1 u + \alpha_1 r_1 x)}|_0(2\sqrt{\alpha_1\beta_1 r_1 xu})\,du$$

$$= \beta_1 e^{-\alpha_1 r_1 x}\sum_{j=0}^{\infty}\frac{\alpha_1\beta_1 r_1 x^j}{(j!)^2}\int_0^t e^{-\{\beta_1 + \lambda + \alpha_2(1-r_2)\}^u} u^j\,du$$

$$Q_{12|x}(t) = \int_0^t \lambda e^{-\{\lambda + \alpha_2(1-r_2)\}^u}\,du\left(\int_u^t \beta_1 e^{-(\beta_1 y + \alpha_1 r_1 x)}|_0(2\sqrt{\alpha_1\beta_1 r_1 xy})\,dy\right)$$

$$= \int_0^t \lambda e^{-\{\lambda + \alpha_2(1-r_2)\}^u}\,du\left(\beta_1 e^{-\alpha_1 r_1 x}\sum_{j=0}^{\infty}\frac{(\alpha_1\beta_1 r_1 x)^j}{(j!)^2}\int_u^t e^{-\beta_1 y}y^j\,dy\right)$$

$$= \lambda\beta_1 e^{-\alpha_1 r_1 x}\sum_{j=0}^{\infty}\frac{(\alpha_1\beta_1 r_1 x)^j}{(j!)^2}\int_0^t e^{-\beta_1 y}y^j\left(\int_0^y e^{-\{\lambda + \alpha_2(1-r_2)\}^u}\,du\right)dy$$

(On changing the limits of integration)

$$= \frac{\lambda\beta_1 e^{-\alpha_1 r_1 x}}{\lambda + \alpha_2(1-r_2)}\sum_{j=0}^{\infty}\frac{(\alpha_1\beta_1 r_1 x)^j}{(j!)^2}\int_0^t e^{-\beta_1 y}y^j[1 - e^{-\{\lambda + \alpha_2(1-r_2)\}y}]\,dy$$

Similarly,

$$Q_{13|x}(t) = \int_0^t \alpha_2(1 - r_2)e^{-\{\lambda + \alpha_2(1 - r_2)\}^u} du \left( \int_u^t \beta_1 e^{-(\beta_1 y + \alpha_1 r_1 x)} |_0(2\sqrt{\alpha_1 \beta_1 r_1 xy}) dy \right)$$

$$= \frac{\alpha_2(1 - r_2)\beta_1}{\lambda + \alpha_2(1 - r_2)} e^{-\alpha_1 r_1 x} \sum_{j=0}^{\infty} \frac{(\alpha_1 \beta_1 r_1 x)^j}{(j!)^2} \int_0^t e^{-\beta_1 y} y^j [1 - e^{-\{\lambda + \alpha_2(1 - r_2)\}^y}] dy$$

$$Q_{34|x}(t) = \int_0^t \lambda e^{-\lambda u} du \int_0^t \beta_2 e^{-(\beta_2 y + \alpha_2 r_2 x)} |_0(2\sqrt{\alpha_2 \beta_2 r_2 xy}) dy$$

$$= \beta_2 e^{-\alpha_2 r_2 x} \sum_{j=0}^{\infty} \frac{(\alpha_2 \beta_2 r_2 x)^j}{(j!)^2} \int_0^t e^{-\beta_2 y} y^j (1 - e^{-\lambda y}) dy$$

$$Q_{35|x}(t) = \int_0^t e^{-\lambda u} \beta_2 e^{-(\beta_2 u + \alpha_2 r_2 x)} |_0(2\sqrt{\alpha_2 \beta_2 r_2 xu}) du$$

$$= \beta_2 e^{-\alpha_2 r_2 x} \sum_{j=0}^{\infty} \frac{(\alpha_2 \beta_2 r_2 x)^j}{(j!)^2} \int_0^t e^{-(\beta_2 + \lambda)u} u^j du \tag{10–15}$$

(iii) Letting $t \to \infty$ in (1–9), the following unconditional steady-state transition probabilities are obtained:

$$
\begin{aligned}
p_{01} &= p_{47} = p_{63} = 1 \\
p_{25} &= \mu/\{\mu + \alpha_2(1 - r_2)\} \\
p_{26} &= \alpha_2(1 - r_2)/\{\mu + \alpha_2(1 - r_2)\} \\
p_{50} &= \beta_1/\{\beta_1 + \lambda + \alpha_2(1 - r_2)\} \\
p_{52} &= \lambda/\{\beta_1 + \lambda + \alpha_2(1 - r_2)\} \\
p_{53} &= \alpha_2(1 - r_2)/\{\beta_1 + \lambda + \alpha_2(1 - r_2)\} \\
p_{74} &= \lambda/(\beta_2 + \lambda), \qquad p_{75} = \beta_2/(\beta_2 + \lambda)
\end{aligned}
\tag{16–23}
$$

(iv) The steady-state conditional transition probabilities $p_{ij|x}$ can be obtained on taking $t \to \infty$ in (10–15) as follows:

$$p_{10|x} = \beta_1 e^{-\alpha_1 r_1 x} \sum_{j=0}^{\infty} \frac{(\alpha_1 \beta_1 r_1 x)^j}{(j!)^2} \frac{\Gamma(J + 1)}{\{\beta_1 + \lambda + \alpha_2(1 - r_2)\}^{j+1}}$$

$$= \beta_1' e^{-\alpha_1 r_1 x} \sum_{j=0}^{\infty} \left[ \frac{\alpha_1 \beta_1 r_1 x}{\beta_1 + \lambda + \alpha_2(1 - r_2)} \right]^j \Big/ j! = \beta_1' e^{-\alpha_1 r_1 x(1 - \beta_1')},$$

$$p_{12|x} = \frac{\lambda \beta_1}{\lambda + \alpha_2(1 - r_2)} e^{-\alpha_1 r_1 x} \sum_{j=0}^{\infty} \frac{(\alpha_1 \beta_1 r_1 x)^j}{(j!)^2} \left[ \frac{\Gamma(j + 1)}{\beta_1^{j+1}} - \frac{\Gamma(j + 1)}{\{\beta_1 + \lambda + \alpha_2(1 - r_2)\}^{j+1}} \right]$$

$$= \frac{\lambda e^{-\alpha_1 r_1 x}}{\lambda + \alpha_2(1 - r_2)} \left[ \sum_{j=0}^{\infty} \frac{(\alpha_1 r_1 x)^j}{j!} - \sum_{j=0}^{\infty} \beta_1' \frac{(\alpha \beta_1' r_1 x)^j}{j!} \right]$$

$$= \frac{\lambda}{\lambda + \alpha_2(1 - r_2)} [1 - \beta_1' e^{-\alpha_1 r_1 x(1 - \beta_1')}]$$

Similarly,

$$
p_{13|x} = \frac{\alpha_2(1 - r_2)}{\lambda + \alpha_2(1 - r_2)}[1 - \beta_1' \, e^{-\alpha_1 r_1 x(1 - \beta_1')}]
$$

$$
p_{34|x} = 1 - \beta_2' e^{-\alpha_2 r_2 x(1 - \beta')} \tag{24–28}
$$

$$
p_{35|x} = -\beta_2' e^{-\alpha_2 r_2 x(1 - \beta'_2)}
$$

where

$$
\beta_1' = \beta_1/\{\beta_1 + \lambda + \alpha_2(1 - r_2)\} \qquad \beta_2' = \beta_2/(\beta_2 + \lambda)
$$

We observe that

$$
p_{10|x} + p_{12|x} + p_{13|x} = 1 \tag{29}
$$

$$
p_{34|x} + p_{35|x} = 1 \tag{30}
$$

(v) It can be observed that the conditional transition probabilities (24–28) involve $r_i$ ($i = 1, 2$) and for $r_i = 0$ they give the corresponding unconditional transition probabilities with correlation coefficient zero. From (24–28), the unconditional transition probabilities with correlation coefficient can be obtained as follows:

$$
\begin{aligned}
p_{10} &= \int p_{10|x} q_1(x)\,dx \\
&= \beta_1' e^{-\alpha_1 r_1 x(1 - \beta'_1)} \alpha_1(1 - r_1) e^{-\alpha_1(1 - r_1)x}\,dx \\
&= \beta_1'(1 - r_1)/(1 - r_1 \beta'_1)
\end{aligned}
$$

Similarly,

$$
\begin{aligned}
p_{12} &= \int p_{12|x} q_1(x)\,dx \\
&= \frac{\lambda}{\lambda + \alpha_2(1 - r_2)}\left[1 - \frac{\beta_1'(1 - r_1)}{1 - r_1\beta_1'}\right] \\
p_{13} &= \frac{\alpha_2(1 - r_2)}{\lambda + \alpha_2(1 - r_2)}\left[1 - \frac{\beta_1'(1 - r_1)}{1 - r_1\beta_1'}\right] \\
p_{34} &= \int p_{34|x} q_2(x)\,dx \tag{31–35} \\
&= \int [1 - \beta_2' e^{-\alpha_2 r_2 x(1 - \beta'_2)}]\alpha_2(1 - r_2) e^{-\alpha_2(1 - r_2)x}\,dx \\
&= 1 - \beta_2'(1 - r_2)/(1 - r_2\beta_2') \\
p_{35} &= \int p_{35|x} q_2(x)\,dx \\
&= \beta_2'(1 - r_2)/(1 - r_2\beta_2')
\end{aligned}
$$

Obviously,

$$p_{10} + p_{12} + p_{13} = 1$$
$$p_{34} + p_{35} = 1$$

$$(36\text{--}37)$$

(vi) To obtain steady-state transition probabilities through nonregenerative state(s), we observe that

$$Q_{ij}^{(k)}(t) = Q_{ik}(t) \circledS Q_{kj}(t)$$

or

$$\lim_{s \to 0} \tilde{Q}_{ij}^{(k)}(s) = \lim_{s \to 0} \tilde{Q}_{ik}(s) \tilde{Q}_{kj}(s)$$

or

$$p_{ij}^{(k)} = p_{ik} \cdot p_{kj}$$

Similarly,

$$p_{ij}^{(k,l)} = p_{ik} p_{kl} p_{lj}$$

therefore,

$$p_{00}^{(1)} = p_{01} p_{10} = p_{10}$$
$$p_{02}^{(1)} = p_{01} p_{12} = p_{12}$$
$$p_{04}^{(1,3)} = p_{01} p_{13} p_{34} = p_{13} p_{34}$$
$$p_{05}^{(1,3)} = p_{01} p_{13} p_{35} = p_{13} p_{35}$$
$$p_{24}^{(6,3)} = p_{26} p_{63} p_{34} = p_{26} p_{34}$$
$$p_{25}^{(6,3)} = p_{26} p_{63} p_{35} = p_{26} p_{35}$$
$$p_{54}^{(3)} = p_{53} p_{34}$$
$$p_{55}^{(3)} = p_{53} p_{35}$$

The following relations are seen to be satisfied:

$$p_{00}^{(1)} + p_{02}^{(1)} + p_{04}^{(1,3)} + p_{05}^{(1,3)} = 1$$
$$p_{24}^{(6,3)} + p_{25} + p_{25}^{(6,3)} = 1$$
$$p_{50} + p_{52} + p_{54}^{(3)} + p_{55}^{(3)} = 1$$
$$p_{74} + p_{75} = 1$$

$$(38\text{--}41)$$

(vii) The unconditional mean sojourn times in states $S_i$ $(i = 0, 2, 4, 5, 6, 7)$ are given by

$$\psi_0 = \int e^{-\alpha_1(1 - r_1)t} dt = 1/\alpha_1(1 - r_1)$$

Similarly,

$$\psi_2 = 1/\{\mu + \alpha_2(1 - r_2)\}$$
$$\psi_4 = \psi_6 = 1/\mu$$
$$\psi_5 = 1/\{\beta_1 + \lambda + \alpha_2(1 - r_2)\}$$
$$\psi_7 = 1/(\beta_2 + \lambda)$$

$$(42\text{--}46)$$

The conditional mean sojourn times in states $S_1$ and $S_3$ are as follows:

$$\psi_{1|x} = \int e^{-\{\lambda + \alpha_2(1-r_2)\}^t}\left\{\int_t^{\infty} \beta_1 e^{-(\beta_1 u + \alpha_1 r_1 x)}\, I_0(2\sqrt{\alpha_1\beta_1 r_1 xu})\,du\right\}dt$$

$$= \beta_1 e^{-\alpha_1 r_1 x}\sum_{j=0}^{\infty}\frac{(\alpha_1\beta_1 r_1 x)^j}{(j!)^2}\int e^{-\beta_1 u} u^j\left[\int_0^u e^{-\{\lambda+\alpha_2(1-r_2)\}^t}dt\right]du \qquad (47)$$

$$= \frac{1}{\lambda+\alpha_2(1-r_2)}\left[1 - \beta_2' e^{-\alpha_1 r_1 x(1-\beta_1')}\right]$$

Similarly,

$$\psi_{3|x} = \frac{1}{\lambda}\left[1 - \beta_2' e^{-\alpha_2 r_2 x(1-\beta_2')}\right] \qquad (48)$$

From the conditional mean sojourn times in states $S_1$ and $S_3$, the unconditional mean sojourn times are given by

$$\psi_1 = \int \psi_{1!x}\, q_1(x)\,dx$$

$$= \{\lambda+\alpha_2(1-r_2)\}^{-1}\int\{1 - \beta_1' e^{-\alpha_1 r_1 x(1-\beta_1')}\}\alpha_1(1-r_1)e^{-\alpha_1(1-r_1)x}\,dx \qquad (49)$$

$$= \{\lambda+\alpha_2(1-r_2)\}^{-1}\left[1 - \frac{\beta'_1(1-r_1)}{1-r_1\beta'_1}\right]$$

$$\psi_3 = \int \psi_{3|x}\, q_2(x)\,dx$$

$$= \lambda^{-1}\{1 - \beta_1' e^{-\alpha_1 r_1 x(1-\beta_1')}\}\alpha_2(1-r_2)e^{-\alpha_2(1-r_2)x}\,dx \qquad (50)$$

$$= \lambda^{-1}\left[1 - \frac{\beta'_2(1-r_2)}{1-r_2\beta'_2}\right]$$

## Analysis of Reliability and Mean Time to System Failure

Assuring the failed states $S_3$, $S_4$, $S_6$, and $S_7$, as absorbing and enjoying the probabilistic arguments, the reliability of the system $R_i(t)$ satisfies the following relations:

$$R_0(t) = Z_0(t) + q_{01}(t) \copyright Z_1(t) + q_{00}^{(1)}(t) \copyright R_0(t) + q_{02}^{(1)}(t) \copyright R_2(t)$$
$$R_2(t) = Z_2(t) + q_{25}(t) \copyright R_5(t) \qquad (1-3)$$
$$R_5(t) = Z_5(t) + q_{50}(t) \copyright R_0(t) + q_{52}(t) \copyright R_2(t)$$

where

$$Z_0(t) = e^{-\alpha_1(1-r_1)t}$$

$$Z_1(t) = e^{-\{\lambda+\alpha_2(1-r_2)t\}}\int\left(\int_t^{\infty}\beta_1 e^{-(\beta_1 y+\alpha_1 r_1 x)}\, I_0(2\sqrt{\alpha_1\beta_1 r_1 xy})\,dy\right)q_1(x)\,dx$$

$$Z_2(t) = e^{-\{\mu+\alpha_2(1-r_2)\}t}$$

$$Z_5(t) = e^{-\{\beta_1+\lambda+\alpha_2(1-r_2)\}}$$

Taking Laplace Transform of relations (1–3) and solving the resulting set of algebraic equations for $R_0^*(s)$ we get

$$R_0^*(s) = N_1(s)/D_1(s) \tag{4}$$

where

$$N_1(s) = (Z_0^* + q_{01}^* Z_1^*)(1 - q_{25}^* q_{52}^*) + (Z_2^* + q_{25}^* Z_5^*)q_{02}^{(1)*}$$

and

$$D_1(s) = (1 - q_{00}^{(1)*})(1 - q_{25}^* q_{52}^*) - q_{02}^{(1)*} q_{25}^* q_{50}^*$$

Taking inverse LT of (4), we can get the reliability of the system when it starts initially from state $S_0$. The mean time to system failure can be obtained using the formula

$$E(T_0) = \int R_0(t)dt = \lim_{s \to 0} R_0^*(s)$$
$$= \frac{N_1(0)}{D_1(0)} \tag{5}$$

To obtain the RHS of (5), we note that

$$\lim_{s \to 0} Z_i^*(s) = \int Z_i(t)dt = \psi_i$$

$$\lim_{s \to 0} q_{ij}^*(s) = p_{ij}, \qquad \lim_{s \to 0} q_{ij}^{(k)*}(s) = p_{ij}^{(k)}$$

and

$$p_{00}^{(1)} = p_{10}, \qquad p_{02}^{(1)} = p_{12}, \qquad p_{01} = 1$$

Therefore,

$$E(T_0) = \frac{(\psi_0 + \psi_1)(1 - p_{25} p_{52}) + (\psi_2 + p_{25} \psi_5)p_{12}}{(1 - p_{10})(1 - p_{25} p_{52}) - p_{12} p_{25} p_{50}} \tag{6}$$

## Availability Analysis

According to the definition of $A_i^P(t)$, the elementary probabilistic arguments yield the following recursive relations:

$$A_0^P(t) = Z_0(t) + q_{00}^{(1)}(t) © A_0^P(t) + q_{02}^{(1)} © A_2^P(t) + q_{04}^{(1,3)}(t) © A_4^P(t) + q_{05}^{(1,3)}(t) © A_5^P(t)$$
$$A_2^P(t) = q_{24}^{(6,3)}(t) © A_4^P(t) + \{q_{25}(t) + q_{25}^{(6,3)}(t)\} © A_5^P(t)$$
$$A_4^P(t) = q_{47}(t) © A_7^P(t)$$
$$A_5^P(t) = q_{50}(t) © A_0^P(t) + q_{52}(t) © A_2^P(t) + q_{54}^{(3)}(t) © A_4^P(t) + q_{55}^{(3)}(t) © A_5^P(t)$$
$$A_7^P(t) = q_{74}(t) © A_4^P(t) + q_{75}(t) © A_5^P(t) \tag{1-5}$$

Similarly, the relations for $A_i^o(t)$ are:

$$
\begin{aligned}
A_0^o(t) &= q_{01}(t) \copyright Z_1(t) + q_{00}^{(1)}(t) \copyright A_0^o(t) + q_{02}^{(1)}(t) \copyright A_2^o(t) + q_{04}^{(1,3)}(t) \copyright A_4^o(t) \\
&\quad + q_{05}^{(1,3)}(t) \copyright A_5^o(t) \\
A_2^o(t) &= z_2(t) + q_{24}^{6,3}(t) \copyright A_4^o(t) + \{ q_{25}(t) + q_{25}^{(6,3)}(t) \} \copyright A_5^o(t) \\
A_4^o(t) &= q_{47}(t) \copyright A_7^o(t) \\
A_5^o(t) &= Z_5(t) + q_{50}(t) \copyright A_0^o(t) + q_{52}(t) \copyright A_2^o(t) + q_{54}^{(3)}(t) \copyright A_4^o(t) + q_{55}^{(3)} \copyright A_5^o(t) \\
A_7^o(t) &= q_{74}(t) \copyright A_4^o(t) + q_{75}(t) \copyright A_5^o(t)
\end{aligned}
\tag{6-10}
$$

Taking Laplace Transform of relations (1–5) and (6–10) and simplifying the resulting set of equations for $A_0^{p*}(s)$ and $A_0^{o*}(s)$, we get

$$
A_0^{p*}(s) = N_2(s)/D_2(s)
\tag{11}
$$

and

$$
A_0^{o*}(s) = N_3(s)/D_2(s)
\tag{12}
$$

where

$$
\begin{aligned}
N_2(s) &= J_1 Z_0, \qquad N_3(s) = q_{01}^* J_1 Z_1^* + J_2 Z_2^* + J_5 Z_5^* \\
D_2(s) &= (1 - q_{00}^{(1)*}) J_1 - q_{50}^* J_5 \\
J_1 &= (1 - q_{47}^* q_{74}^*)(1 - q_{55}^{(3)*} - q_{25}^* q_{52}^* - q_{25}^{(6,3)*} q_{52}^*) - q_{47}^* q_{75}^* (q_{54}^{(3)*} + q_{24}^{(6,3)*} q_{52}^*) \\
J_2 &= \{ (1 - q_{55}^{(3)*})(1 - q_{47}^* q_{74}^*) - q_{47}^* q_{75}^* q_{54}^{(3)*} \} q_{02}^{(1)*} + q_{05}^{(1,3)*} q_{52}^* (1 - q_{47}^* q_{74}^*) + q_{04}^{(1,3)*} q_{47}^* q_{75}^* q_{52}^*
\end{aligned}
$$

and

$$
J_5 = (q_{02}^{(1)*} q_{25}^* + q_{02}^{(1)*} q_{25}^{(6,3)*} + q_{05}^{(1,3)*})(1 - q_{47}^* q_{74}^*) + (q_{04}^{(1,3)*} + q_{02}^{(1)*} q_{25}^{(6,3)*}) q_{47}^* q_{75}^*
$$

In the long run, the probabilities that the system will be up (operative) due to $p$-unit and $o$-unit, respectively, are given by

$$
A_0^p = \lim_{s \to 0} s A_0^{p*}(s) = N_2(0)/D_2'(0)
\tag{13}
$$

$$
A_0^o = \lim_{s \to 0} s A_0^{o*}(s) = N_2(0)/D_2'(0)
\tag{14}
$$

Using the results

$$
Z_i^*(0) = \int Z_i^*(t)\,dt = \psi_i
$$

$$
q_{ij}^*(0) = p_{ij}, \qquad q_{ij}^{(k,l)*}(0) = p_{ij}^{(k,l)}
$$

$$
p_{ij}^{(k)} = p_{jk} p_{kj} \qquad \text{and} \qquad p_{ij}^{(k,l)} = p_{ik} p_{kl} p_{lj}
$$

we have

$$N_2(0) = p_{75}p_{50}\psi_0 \tag{15}$$

$$N_3(0) = p_{75}p_{50}\psi_1 + (p_{12} + p_{13})p_{75}\psi_5 + \{(p_{50} + p_{52})p_{12} + p_{13}p_{52}\}p_{75}\psi_2 \tag{16}$$

To obtain $D_2'(0)$, we first collect the coefficients of $m_{ij}$, $m_{ij}^{(k)}$, and $m_{ij}^{(k,l)}$ in $D_2'(0)$ and with calculations we find that

(i) Coefficient of $m_{00}^{(1)}$, $m_{02}^{(1)}$, $m_{04}^{(1,3)}$, and $m_{05}^{(1,3)}$ is $p_{50}p_{75} = \varphi_1$(say)

(ii) Coefficient of $m_{24}^{(6,3)}$ and $(m_{25} + m_{25}^{(6,3)})$ is

$$\{(p_{12} + p_{13})p_{52} + p_{12}p_{50}\}p_{75} = \varphi_2 \text{ (say)}$$

(iii) Coefficient of $m_{47}$ is

$$p_{34}\{(p_{12} + p_{13})(p_{53} + p_{52}p_{26}) + (p_{13} + p_{12}p_{26})p_{50}\} = \varphi_3 \text{ (say)}$$

(iv) Coefficient of $m_{50}$, $m_{52}$, $m_{54}^{(3)}$, and $m_{55}^{(3)}$ is

$$(p_{12} + p_{13})p_{75} = \varphi_4 \text{ (say)}$$

(v) Coefficient of $m_{74}$ and $m_{75}$ is

$$p_{34}\{(p_{12} + p_{13})(p_{53} + p_{52}p_{26}) + (p_{13} + p_{12}p_{26})p_{50}\} = \varphi_3$$

Therefore,

$$D_2'(0) = (m_{00}^{(1)} + m_{02}^{(1)} + m_{04}^{(1,3)} + m_{05}^{(1,3)})\varphi_1 + (m_{24}^{(6,3)} + m_{25} + m_{25}^{(6,3)})\varphi_2$$
$$+ m_{47}\varphi_3 + (m_{50} + m_{52} + m_{54}^{(3)} + m_{55}^{(3)})\varphi_4 + (m_{74} + m_{75})\varphi_3 \tag{17}$$

Further, we have

$$\sum_i m_{ij} = \psi_i \tag{18}$$

and

$$q_{ij}^{(k)}(t) = q_{ik}(t) \copyright q_{kj}(t)$$
$$\Rightarrow q_{ij}^{(k)*}(s) = q_{ik}^*(s) \cdot q_{kj}^*(s)$$
$$\Rightarrow -\frac{d}{ds}q_{ij}^{(k)*}(s)\big|_{s=0} = -\frac{d}{ds}\{q_{ik}^*(s) \cdot q_{kj}^*(s)\}\big|_{s=0} \tag{19}$$
$$\Rightarrow m_{ij}^{(k)} = p_{ik}m_{kj} + p_{kj}m_{ik}$$

Similarly,

$$m_{ij}^{(k,l)} = p_{ik}p_{kl}m_{lj} + p_{ik}p_{lj}m_{kl} + p_{kl}p_{lj}m_{ik} \tag{20}$$

Thus in view of the results (18–20), the expression (17) becomes

$$
\begin{aligned}
D_2'(0) &= (\psi_0 + \psi_1 + p_{13}\psi_3)\varphi_1 + (\psi_2 + p_{26}\psi_3 + p_{26}\psi_6)\varphi_2 \\
&\quad + (\psi_4 + \psi_7)\varphi_3 + (\psi_5 + p_{53}\psi_3)\varphi_4
\end{aligned}
\tag{21}
$$

## Busy-Period Analysis

We have already defined $B_i^p(t)$, $B_i^o(t)$, and $B_i^m(t)$ as probabilities that the repairman is busy with $p$-unit, $o$-unit, and RM at time $t$, respectively. When the system initially starts from regenerative state $S_i$ using the usual probabilistic reasoning for $B_i^p(t)$, we have the following relations:

$$
\begin{aligned}
B_0^p(t) &= q_{01}(t) \,\copyright\, Z_1(t) + q_{00}^{(1)}(t) \,\copyright\, B_0^p(t) + q_{02}^{(1)}(t) \,\copyright\, B_2^p(t) + q_{04}^{(1,3)}(t) \,\copyright\, B_4^p(t) + q_{05}^{(1,3)}(t) \,\copyright\, B_5^p(t) \\
B_2^p(t) &= q_{24}^{(6,3)}(t) \,\copyright\, B_4^p(t) + \{q_{25}(t) + q_{25}^{(6,3)}(t)\} \,\copyright\, B_5^p(t) \\
B_4^p(t) &= q_{47}(t) \,\copyright\, B_7^p(t) \\
B_5^p(t) &= Z_5(t) + q_{50}(t) \,\copyright\, B_0^p(t) + q_{52}(t) \,\copyright\, B_2^p(t) + q_{54}^{(3)}(t) \,\copyright\, B_4^p(t) + q_{55}^{(3)}(t) \,\copyright\, B_5^p(t) \\
B_7^p(t) &= q_{74}(t) \,\copyright\, B_4^p(t) + q_{75}(t) \,\copyright\, B_5^p(t)
\end{aligned}
\tag{1--5}
$$

The relations in $B_1^0(t)$ are as follows:

$$
\begin{aligned}
B_0^0(t) &= q_{01}(t) \,\copyright\, q_{13}(t) \,\copyright\, Z_3(t) + q_{00}^{(1)}(t) \,\copyright\, B_0^0(t) + q_{02}(t) \,\copyright\, B_2^0(t) \\
&\quad + q_{04}^{(1,3)}(t) \,\copyright\, B_4^0(t) + q_{05}^{(1,3)}(t) \,\copyright\, B_5^0(t) \\
B_2^0(t) &= q_{26}(t) \,\copyright\, q_{63}(t) \,\copyright\, Z_3(t) + q_{24}^{(6,3)}(t) \,\copyright\, B_4^0(t) + \{q_{25}(t) + q_{25}^{(6,3)}(t)\} \,\copyright\, B_5^0(t) \\
B_4^0(t) &= q_{47}(t) \,\copyright\, B_7^0(t) \\
B_5^0(t) &= q_{50}(t) \,\copyright\, B_0^0(t) + q_{52}(t) \,\copyright\, B_2^0(t) + q_{54}^{(3)}(t) \,\copyright\, B_4^0(t) + q_{55}^{(3)}(t) \,\copyright\, B_5^0(t) \\
B_7^0(t) &= Z_7(t) + q_{74}(t) \,\copyright\, B_4^0(t) + q_{75}(t) \,\copyright\, B_5^0(t)
\end{aligned}
\tag{6--10}
$$

For $B_i^m(t)$ we have

$$
\begin{aligned}
B_0^m(t) &= q_{00}^{(1)}(t) \,\copyright\, B_0^m(t) + q_{02}^{(1)}(t) \,\copyright\, B_2^m(t) + q_{04}^{(1,3)}(t) \,\copyright\, B_4^m(t) + q_{05}^{(1,3)}(t) \,\copyright\, B_5^m(t) \\
B_2^m(t) &= Z_2(t) + q_{26}(t) \,\copyright\, Z_6(t) + q_{24}^{(6,3)}(t) \,\copyright\, B_4^m(t) + \{q_{25}(t) + q_{25}^{(6,3)}(t)\} \,\copyright\, B_5^m(t) \\
B_4^m(t) &= Z_4(t) + q_{47}(t) \,\copyright\, B_7^m(t) \\
B_5^m(t) &= q_{50}(t) \,\copyright\, B_0^m(t) + q_{52}(t) \,\copyright\, B_2^m(t) + q_{54}^{(3)}(t) \,\copyright\, B_4^m(t) \\
B_7^m(t) &= q_{74}(t) \,\copyright\, B_4^m(t) + q_{75}(t) \,\copyright\, B_5^m(t)
\end{aligned}
\tag{11--15}
$$

Taking LT of relations (1–5), (6–10), and (11–15) and simplifying the resulting sets of equations for $B_0^{p*}(s)$, $B_0^{o*}(s)$, and $B_0^{m*}(s)$, we get

$$
B_0^{p*}(s) = N_4(s)/D_2(s)
\tag{16}
$$

$$
B_0^{o*}(s) = N_5(s)/D_2(s)
\tag{17}
$$

and

$$B_0^{m*}(s) \; = \; N_6(s)/D_2(s) \tag{18}$$

where

$$
\begin{aligned}
N_4(s) \; = \; & [(1 - q_{47}^* q_{74}^*)(1 - q_{55}^{(3)*} - q_{25}^* q_{52}^* - q_{25}^{(6,3)*} q_{52}^*) - q_{47}^* q_{75}^* (q_{54}^{(3)*} + q_{24}^{(6,3)*} q_{52}^*)] q_{01}^* Z_1^* \\
& + [(q_{02}^{(1)*} q_{25}^* + q_{02}^{(1)*} q_{25}^{(6,3)*} + q_{05}^{(1,3)*})(1 - q_{47}^* q_{74}^*) + (q_{04}^{(1,3)*} + q_{02}^{(1)*} q_{24}^{(6,3)*}) q_{47}^* q_{75}^*] Z_5^*
\end{aligned}
$$

$$
\begin{aligned}
N_5(s) \; = \; & [(1 - q_{47}^* q_{74}^*)(1 - q_{55}^{(3)*} - q_{25}^* q_{52}^* - q_{25}^{(6,3)*} q_{52}^*) - q_{47}^* q_{75}^* (q_{54}^{(3)*} + q_{24}^{(6,3)*} q_{52}^*)] q_{01}^* q_{13}^* Z_3^* \\
& + Z_3^* q_{26}^* q_{63}^* [q_2^{(1)*}(1 - q_{55}^{(3)*}) + q_{05}^{(1,3)*} q_{52}^*] + Z_3^* q_{26}^* q_{63}^* q_{47}^* [q_{52}^* q_{04}^{(1,3)*} q_{75}^* - q_{05}^{(1,3)*} q_{74}^*) \\
& - q_{02}^{(1)*}(q_{54}^{(3)*} q_{75}^* + q_{74}^*(1 - q_{55}^{(3)*}))] \\
& + q_{47}^* Z_7^* [q_{02}^{(1)*}(q_{24}^{(6,3)*}(1 - q_{55}\text{sup}(3)^*) + q_{54}^{(3)*} q_{25}^* + q_{54}^{(3)*} q_{25}^{(6,3)*}] \\
& + q_{04}^{(1,3)*}[(1 - q_{55}^{(3)*} - q_{52}^* q_{25}^* - q_{52}^* q_{25}^{(6,3)*}) + q_{05}^{(1,3)*}(q_{54}^{(3)*} + q_{52}^* q_{24}^{(6,3)*})]
\end{aligned}
$$

and

$$
\begin{aligned}
N_6(S) \; = \; & Z_4^* [q_2^{(1)*}(q_{24}^{(6,3)}(1 + - q_{55}^{(3)*}) - q_{54}^{(3)*} q_{25}^* + q_{54}^{(3)*} q_{25}^{(6,3)*}) \\
& + q_{04}^{(1,3)*}((1 - q_{55}^{(3)*} - q_{52}^* q_{25}^* - q_{52}^* q_{25}^{(6,3)}) + q_{05}^{(1,3)*}(q_{54}^{(1,3)*} + q_{52}^* q_{25}^{(6,3*)})] \\
& + (Z_2^* + q_{26}^* Z_6^*)[(q_2^{(1)*}(1 - q_{55m}^{(3)*} + q_{05}^{(1,3)*} q_{52}^*) \\
& + q_{47}^*(q_{52}^*(q_4^{(1,3)*} q_{75}^* - q_{05}^{(1,3)*} q_{74}^*) - q_{02}^{(1)*}(q_{54}^{(3)*} q_{75}^* + q_{74}^*(1 - q_{55}^{(3)*}))]
\end{aligned}
$$

Now using the result $q_{ij}^*(0) \; = \; p_{ij}$ and various relations among $p_{ij}$, the above three probabilities in steady state are given by

$$B_0^p \; = \; N_4(0)/D_2'(0)$$
$$B_0^o \; = \; N_5(0)/D_2'(0)$$

and

$$B_0^m \; = \; N_6(0)/D_2'(0) \tag{19–21}$$

where

$$
\begin{aligned}
N_4(s) \; = \; & p_{75} p_{50} \psi_1 + (p_{12} + p_{13}) p_{75} \psi_5 \\
N_5(0) \; = \; & (p_{13} + p_{26}) \psi_3 [p_{12} p_{75}(1 - p_{53}) + p_{13} p_{75} p_{52}] \\
& + p_{34} \psi_7 [p_{12} p_{26} + (1 - p_{12}) p_{26} p_{35} p_{53} + p_{13} + p_{25}(p_{53} - p_{13} p_{52})] \\
N_6(0) \; = \; & (\psi_2 + p_{26} \psi_6)[p_{12} p_{75}(1 - p_{53}) + p_{13} p_{75} p_{52}] \\
& + p_{35} \psi_4 [p_{12} p_{26} + (1 - p_{12}) p_{26} p_{35} p_{53} + p_{13} + p_{25}(p_{53} - p_{13} p_{52})]
\end{aligned}
$$

and $D_2'(0)$ has already been defined.

## Expected Number of Repairs during $(0, t)$

We have already defined $N_i^p(t)$ as the expected number of repairs of the $p$-unit during $(0, t)|E_0 = s_i$. For this model the recurrence relations among $N_i^p(t)$, $i = 0, 2, 4, 5, 7$ are as follows:

$$N_0^p(t) = Q_{00}^{(1)}(t)[1 + N_0^p(t)] + Q_{02}^{(1)}(t)\textcircled{S}N_2^p(t) + Q_{04}^{(1,3)}(t)\textcircled{S}N_4^p(t) + Q_{05}^{(1,3)}(t)\textcircled{S}N_5^p(t)$$

$$N_2^p(t) = Q_{24}^{(6,3)}(t)\textcircled{S}N_4^p(t) + [Q_{25}(t) + Q_{25}^{(6,3)}(t)]\textcircled{S}N_5^p(t)$$

$$N_4^p(t) = Q_{47}(t)\textcircled{S}N_7^p(t) \tag{1-5}$$

$$N_5^p(t) = Q_{50}(t)\textcircled{S}[1 + N_0^p(t)] + Q_{52}(t)\textcircled{S}N_2^p(t) + Q_{54}^{(3)}(t)\textcircled{S}N_4^p(t) + Q_{55}^{(3)}(t)\textcircled{S}N_5^p(t)$$

$$N_7^p(t) = Q_{74}(t)\textcircled{S}N_4^p(t) + Q_{75}(t)\textcircled{S}N_5^p(t)$$

Similarly, the recursive relations in $N_i^o(t)$ and $N_i^m(t)$ can be obtained to obtain the expected number of repairs of $o$-unit and RM, respectively. They are as follows:

$$N_0^o(t) = Q_{00}^{(1)}(t)\textcircled{S}N_0^o(t) + Q_{02}^{(1)}(t)\textcircled{S}N_2^o(t) + Q_{04}^{(1,3)}(t)\textcircled{S}N_4^o(t) + Q_{05}^{(1,3)}(t)\textcircled{S}[1 + N_5^o(t)]$$

$$N_2^o(t) = Q_{24}^{(6,3)}(t)\textcircled{S}N_4^o(t) + [Q_{25}(t) + Q_{25}^{(6,3)}(t)]\textcircled{S}[1 + N_5^o(t)]N_4^o(t)\textcircled{S}Q_{17}(t)\textcircled{S}N_6^o(t)$$

$$N_5^o(t) = Q_{50}(t)\textcircled{S}N_0^o(t) + Q_{52}(t)\textcircled{S}N_2^o(t) + Q_{54}^{(3)}(t)\textcircled{S}N_4^o(t) + Q_{55}^{(3)}(t)\textcircled{S}N_5^o(t)$$

$$N_7^o(t) = Q_{64}(t)\textcircled{S}N_4^o(t) + Q_{75}(t)\textcircled{S}1 + N_5^o(t) \tag{7-11}$$

and

$$N_0^m(t) = Q_{00}^{(1)}(t)\textcircled{S}N_0^m(t) + Q_{02}^{(1)}(t)\textcircled{S}N_0^m(t) + Q_{04}^{(1,3)}(t)\textcircled{S}N_4^m(t)$$
$$+ Q_{05}^{(1,3)}(t)\textcircled{S}N_5^m(t)$$

$$N_2^m(t) = Q_{24}^{(6,3)}(t)\textcircled{S}N_4^m(t) + [Q_{25}(t) + Q_{25}^{(6,3)}(t)]\textcircled{S}1 + N_5^m(t)]$$

$$N_4^m(t) = Q_{47}(t)\textcircled{S}[1 + N_7^m(t)]$$

$$N_5^m(t) = Q_{50}(t)\textcircled{S}N_0^m(t) + Q_{52}(t)\textcircled{S}N_2^m(t) + Q_{54}^{(3)}(t)\textcircled{S}N_4^m(t) + Q_{55}^{(3)}(t)\textcircled{S}N_5^m(t)$$

$$N_7^m(t) = Q_{74}(t)\textcircled{S}N_4^m(t) + Q_{75}(t)\textcircled{S}N_5^m(t) \tag{12-16}$$

Taking LST of the above equations (7-11) and (12-16) and solving, then $\tilde{N}_0^o(s)$ and $\tilde{N}_0^m(s)$ can be easily obtained.

In steady state the expected number of repairs of $p$-unit, $o$-unit, and RM per unit are, respectively,

$$N_0^p = N_7(0)/D_2'(0)$$

$$N_0^o = N_8(0)/D_2'(0)$$

and

$$N_0^m = N_9(0)/D'_2(0) \tag{17-19}$$

where

$$N_7(0) = p_{75}p_{50}/D_2'(0)$$

$$N_8(0) = (p_{13} + p_{36})p_{35}[p_{12}p_{75}(1 - p_{53}) + p_{13}p_{75}p_{52}]$$
$$+ p_{34}p_{75}[p_{12}p_{26} + (1 - p_{12})p_{26}p_{35}p_{53} + p_{13} + p_{25}(p_{53} - p_{13}p_{52})]$$

and

$$N_9(0) = (p_{25} + p_{26}p_{35})[p_{12}p_{75}(1 - p_{53}) + p_{13}p_{75}p_{52}]$$
$$+ p_{34}[p_{13} + p_{12}p_{26} + (1 - p_{12})p_{26}p_{35}p_{53} + p_{25}(p_{53} - p_{13}p_{52})]$$

## 7.7 Profit Analysis

The two profit functions $P_1(t)$ and $P_2(t)$ can be found easily for each of the three models *A*, *B*, and *C* with the help of the characteristics obtained in the earlier sections. The net expected total profit (gain) incurred during $(0, t)$ are

$$P_1(t) = \text{expected total revenue in } (0,t) - \text{expected total expenditure during } (0, t)$$
$$= K_0\mu_{\text{up}}^p + K_1\mu_{\text{up}}^0(t) - K_3\mu_b^0(t) - K_4\mu_b^m() \tag{1}$$

and

$$P_2(t) = K_0\mu_{\text{up}}^p(t) + K_1\mu_{\text{up}}^o(t) - K_5N_0^p(t) - K_6N_0^o(t) - K_7N_0^m(t) \tag{2}$$

where $K_0$ and $K_1$ are the revenues per-unit up-time due to *p*-unit and *o*-unit, respectively: $K_2$, $K_3$, and $K_4$ are the amounts paid to the repairman per-unit of time when he is busy in repairing the *p*-unit, *o*-unit, and RM, respectively: $K_5$, $K_6$, and $K_7$ are the per-unit repair costs of the *p*-unit, *o*-unit, and RM, respectively. Also the mean up-times of the system due to the operation of *p*-unit and *o*-unit during $(0, t)$ are given by

$$\mu_{\text{up}}^p(t) = \int_0^t A_0^p(u)\,du \tag{3}$$

and

$$\mu_{\text{up}}^o(t) = \int_0^t A_0^0(u)\,du \tag{4}$$

so that

$$\mu_{\text{up}}^{p*}(s) = A_0^{p*}(s)/s \tag{5}$$

and

$$\mu_{\text{up}}^{o*}(s) = A_0^{o*}(s)/s \tag{6}$$

Further, $\mu_b^p(t)$, $\mu_b^o(t)$, and $\mu_b^m(t)$ are the expected busy periods of the repairman with the *p*-unit, *o*-unit, and RM, respectively, in $(0, t)$ and are given by

$$\mu_b^p(t) = \int_0^t B_0^p(u)\,du \tag{7}$$

$$\mu_b^o(t) = \int_0^t B_0^p(u)\,du \tag{8}$$

and

$$\mu_b^m(t) = \int_0^t B_0^m(u)\,du \tag{9}$$

so that

$$\mu_b^{p*}(s) = B_0^{p*}(s)/s \tag{10}$$

$$\mu_b^{o*}(s) = B_0^{o*}(s)/s \tag{11}$$

and

$$\mu_b^{m*}(s) = B_0^{m*}(s)/s \tag{12}$$

Now the expected total profit per unit time in steady state is given by

$$P_1 = \lim_{t \to \infty} P_1(t)/t = \lim_{s \to \infty} s^2 P_1^*(s)$$

$$= K_0 \lim_{s \to 0} s^2 \mu_{up}^{p*}(s) + K_1 \lim_{s \to 0} s^2 \mu_{up}^{o*}(s) - K_2 \lim_{s \to 0} s^2 \mu_b^{p*}(s) - K_3 \lim_{s \to 0} s^2 \mu_b^{0*}(s) - K_4 \lim_{s \to 0} \mu_b^{m*}(s)$$

On using (5, 6, 10–12) we have

$$P_1 = K_0 \lim_{s \to 0} s A_0^{p*}(s) + K_1 \lim_{s \to 0} s A_0^{o*}(s) - K_2 \lim_{s \to 0} s B_0^{p*}(s)$$

$$- K_3 \lim_{s \to 0} s B_0^{o*}(s) - K_4 \lim_{s \to 0} s B_0^{m*}(s)$$

$$= K_0 A_0^p + K_1 A_0^o - K_2 B_0^p - K_3 B_0^o - K_4 B_0^m \tag{13}$$

Similarly,

$$P_2 = K_0 A_0^p + K_1 A_0^o - K_5 N_0^p - K_6 N_0^o - K_7 N_0^m \tag{14}$$

The values of $A_0^p$, $A_0^o$, $B_0^p$, $B_0^o$, $B_0^m$, $N_0^p$, $N_0^o$, and $N_0^m$ can be substituted in (13) and (14) from Sections 7.4, 7.5, and 7.6 for each of the three models $A$, $B$, and $C$, respectively.

## 7.8  Graphical Study of System Behaviour

For a more concrete study of system behaviour of models $A$ and $B$, we plot curves in Figures 7.4 to 7.7 for the profit functions $P_1$ and $P_2$ obtained in earlier section w.r.t. $\alpha_1$ for different values of $\beta_1$ while the other parameters are kept fixed as $\alpha_2 = 0.05$, $\beta_2 = 0.075$, $\lambda = 0.025$, $\mu = 0.075$, $K_0 = 250$, $K_1 = 100$, $K_2 = 75$, $K_3 = 25$, $K_4 = 50$, $K_5 = 100$, $K_6 = 50$, and $K_7 = 75$.

The comparison of profit functions $P_1$ and $P_2$ for model $A$ is shown in Figure 7.4. From the figure it is clear that both the profit functions decrease uniformly as the failure rate parameter of $p$-unit ($\alpha_1$) increases. Also with the increase in the value of repair rate parameter of $p$-unit ($\beta_1$), the profit functions $P_1$ and $P_2$ increase. Further, it is observed that the function $P_2$ provides the higher profit as compared to the function $P_1$ irrespective of the values of $\alpha_1$ and $\beta_1$.

Figure 7.5 provides the comparison of profit functions $P_1$ and $P_2$ for model $B$. Here also the same trends are observed for $P_1$ and $P_2$ as in Figure 7.4. One of the important features in this figure is that the function $P_1$ carries loss for $\alpha_1 > 0.08$ at $\beta_1 = 0.28$.

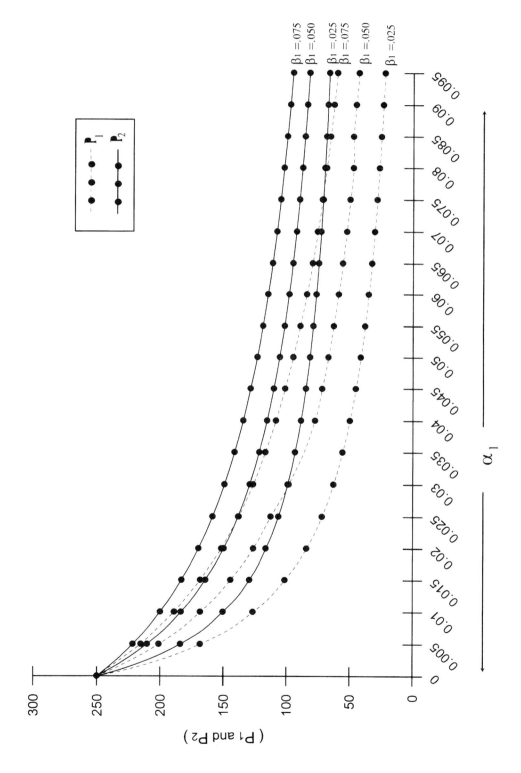

**FIGURE 7.4**   Comparison of profit functions $P_1$ and $P_2$ for model $A$.

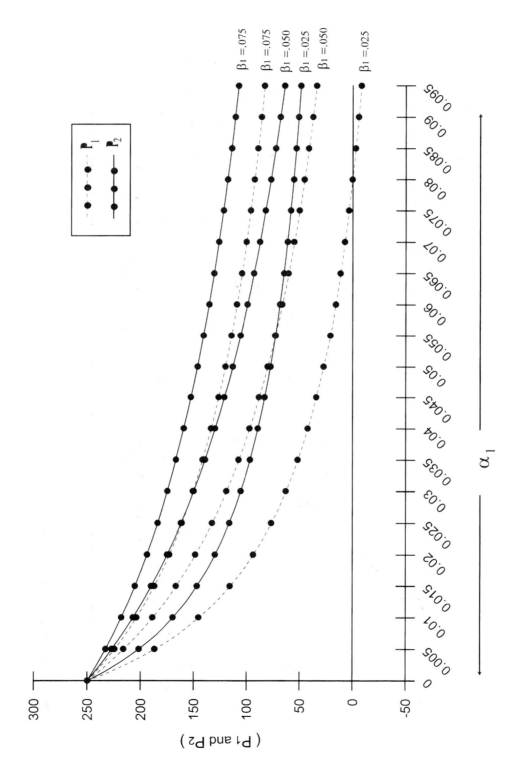

**FIGURE 7.5**   Comparison of profit functions $P_1$ and $P_2$ for model *B*.

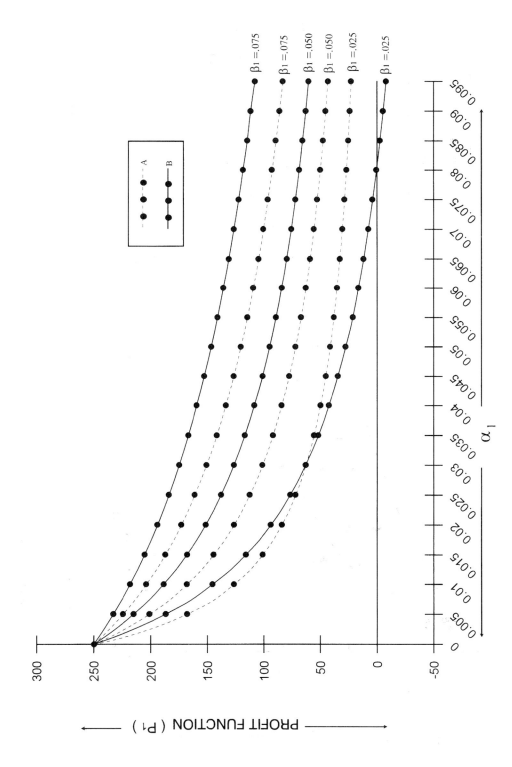

**FIGURE 7.6** Comparison of models $A$ and $B$ with respect to profit functions $P_1$.

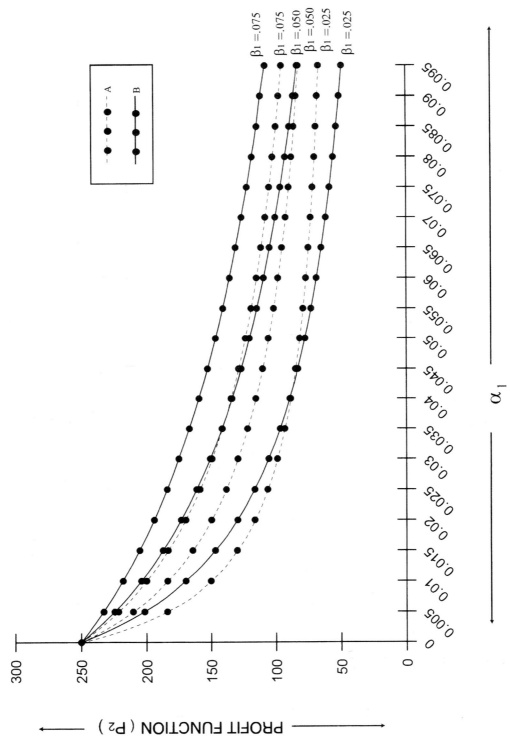

**FIGURE 7.7**    Comparative impression of function $P_2$ for models $A$ and $B$.

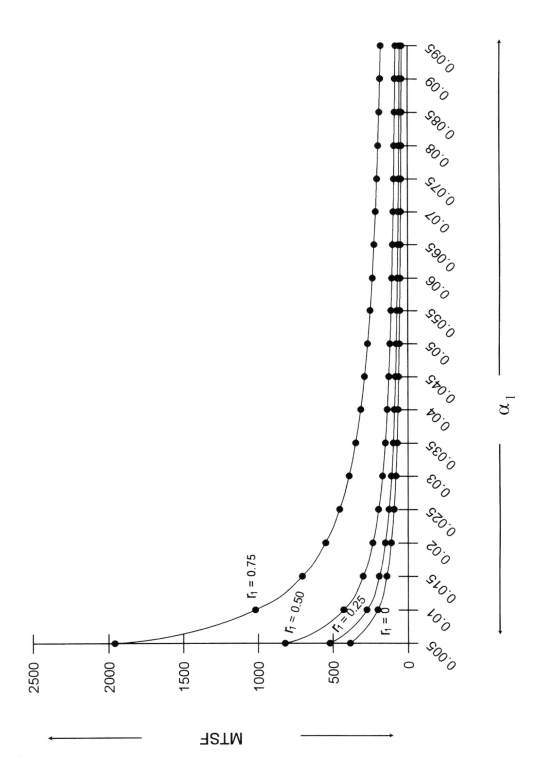

**FIGURE 7.8** Behaviours of MTSF for model C w.r.t. $\alpha_1$ for different values (0, 0.25, 0.50, 0.75) of correlation $r_1$.

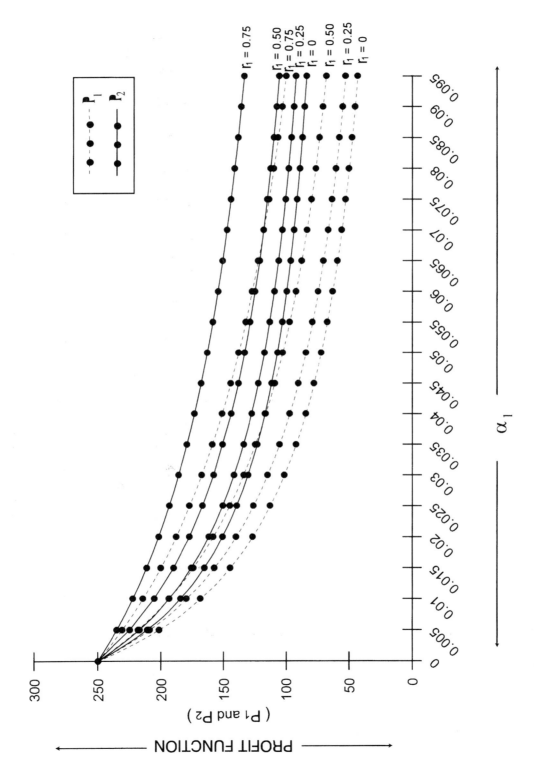

In Figure 7.6, the comparison of models $A$ and $B$ is shown with respect to profit function $P_1$. From the figure, it is observed that the profit in model $B$ is higher than model A for $\beta_1 = 0.050$ and $0.075$. But for $\beta_1 = 0.025$, as compared to model $A$, the profit in model $B$ is higher when $\alpha_1 \leqslant 0.03$ and this trend gets reversed when $\alpha_1 > 0.03$. Also, as $\beta_1$ decreases the rate of downfall in $P_1$ for model B is higher as compared to model A. Therefore, for $\beta_1 = 0.025$, the function $P_1$ provides loss when $\alpha_1 > 0.08$.

The comparative impression of function $P_2$ is shown for models $A$ and $B$ in Figure 7.7. It is implied that for $\beta_1 = 0.075$, $P_2$ provides higher profit for model B as compared to model A, and this variation reduces as $a_1$ increases. For $\beta_1 = 0.050$, $P_2$ is higher for model $B$ when $\alpha_1 < 0.095$ and the trend gets reversed when $\alpha_1 > 0.095$. But for $\beta_1 = 0.025$, $P_2$ is higher for model B when $\alpha_1 < 0.04$ and the trend becomes reversed when $\alpha_1 > 0.04$. Thus, $P_2$ in model B is higher then model A for high values of $\beta_1$ and low values of $\alpha_1$.

In Figures 7.8 and 7.9, the behaviours of MTSF and profit functions $P_1$ and $P_2$ for model $C$ are shown w.r.t. $\alpha_1$ for different values $(0, 0.25, 0.50, 0.75)$ of correlation $r_1$. Here, in addition of above-fixed values of the parameters, $\beta_1$ is also kept fixed on $\beta_1 = 0.05$. From Figure 7.8, it is observed that MTSF decreases uniformly as $\alpha_1$ increases and the rate of downfall is high for low values of $\alpha_1$. Further, the mean lifetime of the system in model $C$ increases as $r_1$ increases. The observations drawn from Figure 7.9 are that the functions $P_1$ and $P_2$ decrease uniformly as $\alpha_1$ increases and the function $P_2$ is profitable as compared to function $P_1$. Also the profit obtained by any of the functions increases as $r_1$ increases. Thus high correlation between the failure and repair times provides the better system performances.

# References

[1] Goel, L.R. and R. Gupta, A multi-component two-unit and standby system with three models. *Microelectron. Reliab.*, 23(5), 799–803 (1983).

[2] Goel, L.R., R. Gupta and S.K. Singh, Cost analysis of a two-unit priority standby system with imperfect switch and arbitrary distributions. *Microelectron. Reliab.*, 25(1), 65–69 (1985).

[3] Goel, L.R., R. Gupta and S.K. Singh, Cost analysis of a two-unit cold standby system with two types of operation and repair. *Microelectron. Reliab.*, 25(1), 71–75 (1985).

[4] Goel, L.R., P. Srivastava and R. Gupta, A two-unit cold standby system with correlated failure and repair. *Int. Jr. Systems Science*, 23(3), 379–391 (1992).

[5] Goel, L.R. and P. Srivastava, A warm standby redundant system with correlated failures and repairs. *Micrelectron. Reliab.*, 32(6), 793–797 (1992).

[6] Goel, L.R., P.K. Tyagi and R. Gupta, Analysis of a standby system with dependent repair time and slow switching device. *Microelectron. Reliab.*, 34(2), 383–386 (1994).

[7] Goel, L.R. and S.Z. Mumtaz, Stochastic analysis of a complex system with an auxiliary unit. *Commun. Statist. Theory Meth.*, 23(10), 3003–3017 (1994).

[8] Goyal, V. and K. Murari, Cost analysis of a two unit standby system with a regular repairman and patience time. *Microelectron. Reliab.*, 24(3), 453–459 (1984).

[9] Gupta, S.M., N.K. Jaiswal and L.R. Goel, Switch failure in a two-unit standby redundant system. *Microelectron. Reliab.*, 23(1), 129–132 (1983).

[10] Gupta, S.M., N.K. Jaiswal and L.R. Goel, Stochastic behaviour of a standby redundant system with three models. *Microelectron. Reliab.*, 23(2), 329–331 (1983).

[11] Gupta, R. and L.R. Goel, Profit analysis of two-unit priority standby system with administrative delay in repair. *Int. Jr. Systems Science*, 20, 1703–1712 (1989).

[12] Gupta, R. and S. Bansal, Profit analysis of a two-unit priority standby system subject to degradation. *Int. Jr. Systems Science*, 22(1), 61–72 (1991).

[13] Gupta, R. and A. Chaudhary, A two-unit priority standby system subject to random shocks and Releigh failure time distributions. *Microelection. Reliab.*, 32(2) 1713–1723 (1992).

[14] Gupta, R, A. Chaudhary and R. Goel, Profit analysis of a two-unit priority standby system subject to degradation and random shocks. *Microelection. Reliab.*, 33(6), 1073–1079 (1993).

[15] Gupta, R. and A. Chaudhary, Stochastic analysis of a priority unit standby system with repair machine failure. *Int. Jr. System Science*, 26, 2435–2440 (1995).

[16] Gupta, R, S.Z. Mumtaz and R. Goel, A two dissimilar unit multi-component system with correlated failure and repairs. *Microelectron. Reliab.* 37(5), 845–849 (1997).

[17] Jaiswal, N.K. and J.V. Krishna, Analysis of two-dissimilar-unit standby redundant system with administrative delay in repair. *Int. Jr. Systems Science*, 11 (4), 495–511 (1980).

[18] Kapur, P.K. and K.R. Kapoor, Interval reliability of a two-unit standby redundant system. *Microelectron. Reliab.*, 23(1), 167–168 (1983).

[19] Kumar, A., D. Ray and M. Agarwal, Probabilistic analysis of a two-unit standby redundant system with repair efficiency and imperfect switch over. *Int. Jr. System Science*, 9(7), 731–742 (1978).

[20] Mahmoud, M.I. and M.A.W. Mahmoud, Stochastic behaviour of a two-unit standby redundant system with imperfect switch over and preventive maintenance. *Microelectron. Reliab.*, 23(1), 153–156 (1983).

[21] Mahmoud, M.A.W. and M.A. Esmail, Probabilistic analysis of a two-unit warm standby system subject to hardware and human error failures. *Microelectron. Reliab.*, 36(10), 1565–1568 (1996).

[22] Murari, K. and V. Goyal, Comparison of two-unit cold standby reliability models with three types of repair facilities. *Microelectron. Reliab.*, 24(1), 35–49 (1984).

[23] Naidu, R.S. and M.N. Gopalan, On the stochastic behaviour of a 1-server 2-unit system subject to arbitrary failure, random inspection and two failure modes. *Microelectron. Reliab.*, 24(3), 375–378 (1984).

[24] Nakagawa, T. and S. Osaki, Stochastic behaviour of a two-unit priority standby redundant system with repair. *Microelectron. Reliab.*, 14(2), 309–313 (1975).

[25] Singh, S.K. and B. Srinivasu, Stochastic analysis of a two-unit cold standby system with preparation time for repair *Microelectron. Reliab.*, 27(1), 55–60(1987).

# 8

# Petri Net Modeling and Scheduling of Automated Manufacturing Systems

**MengChu Zhou**
*New Jersey Institute of Technology*

**Huanxin Henry Xiong**
*Unxipros, Inc.*

## 8.1   Introduction

Given a set of production requirements and a physical system configuration, scheduling deals with the allocation of shared resources over time for manufacturing products such that all the production constraints are satisfied, production cost is minimized, and productivity is maximized. Production scheduling problems are known to be very complex and are NP-hard for general cases. Compared with a classical job-shop system, the main characteristics of an automated manufacturing system include multilayer resource-sharing, deadlock, and routing flexibility (13). An automated manufacturing system consists of different kind of resources such as machines, robots, transporters, and buffers. The job processes share all machines and machines share transportation systems, robots, tools, and so on. The complex interaction of the multiple resources and concurrent flow of multiple jobs in an automated manufacturing system can lead to a deadlock situation in which any part flow is inhibited. The occurrence of a deadlock can cripple the entire system. This requires an explicit consideration of deadlock conditions in the scheduling and control methods to prevent or avoid the deadlock states in automated manufacturing. Machine routings specify the machines that are required for each operation of a given job. In an automated manufacturing system, a job may have alternative routings. The routing flexibility results in benefits to the system such as increasing the throughput and handling the machine breakdown situations, while it increases the complexity of scheduling and control of automated manufacturing systems.

Even though scheduling of flow-shops and job-shops has been extensively studied by many researchers (5, 3, 14), most scheduling algorithms ignore the issues such as deadlock, routing flexibility, multiple lot size, and limited buffer size.

Petri net theory has been applied for modeling, analysis, simulation, planning, scheduling, and control of flexible manufacturing systems (FMS) (15, 7, 24, 1, 30, 11–12, 17). A Petri net comprises two types of nodes, namely places and transitions. A place is represented by a circle, and a transition by a bar. A place can be connected to a transition and vice versa by arcs. In order to study dynamic behavior of the modeled system, each place contains a nonnegative integer number of tokens. At any given time instance, the distribution of tokens on places, called Petri net marking, defines the current state of the modeled system. A significant advantage of Petri net-based methods is its representation capability. Petri nets can explicitly and concisely model concurrent and asynchronous activities, multilayer resource sharing, routing flexibility, limited buffers, and precedence constraints in manufacturing systems. The changes of markings in the net describe the dynamic behaviors of the system. Petri nets provide an explicit way for considering deadlock situations in automated manufacturing such that a deadlock-free scheduling and control system can be designed.

Shih and Sekiguchi (19) presented a timed Petri net and beam search method to schedule an FMS. Beam search is an artificial intelligence technique for efficient searching in decision trees. When a transition in a timed Petri net is enabled, if any of its input places is a conflicted input place, the scheduling system calls for a beam search routine. The beam search routine then constructs partial schedules within the beam depth. Based on the evaluation function, the quality of each partial schedule is evaluated and the best is returned. The cycle is repeated until a complete schedule is obtained. This method based on partial schedules does not guarantee global optimization.

Shen et al. (18) presented a Petri net-based branch and bound method for scheduling the activities of a robot manipulator. To cope with the complexity of the problem, they truncate the original Petri net into a number of smaller sized subnets. Once the Petri net is truncated, the analysis is conducted on each subnet individually. However, due to the existence of the dependency among the subnets, the combination of local optimal schedules does not necessarily yield a global optimal or even near-optimal schedule for the original system. Zhou, Chiu, and Xiong (28) also employed a Petri net-based branch and bound method to schedule flexible manufacturing systems. In their method, instead of randomly selecting one decision candidate from candidate sets (enabled transition sets in Petri net-based models), they select the one based on heuristic dispatching rules such as SPT. The generated schedule is transformed into a marked graph for cycle time analysis.

Lee and DiCesare (11) presented a scheduling method using Petri nets and heuristic search. Once the Petri net model of the system is constructed, the scheduling algorithm expands the reachability graph from the initial marking until the generated portion of the reachability graph touches the final marking. Theoretically, an optimal schedule can be obtained by generating the reachability graph and finding the optimal path from the initial marking to the final one. But the entire reachability graph may be too large to generate even for a simple Petri net due to exponential growth of the number of states. Thanks to the proposed heuristic functions, only a portion of the reachability graph is generated. Three kinds of heuristic functions are presented. The first one favors markings that are deeper in the reachability graph. The second one favors a marking that has an operation ending soon. The last one is a combination of the first and the second ones. These three heuristic functions do not guarantee the admissible condition (16), thus the proposed heuristic search algorithm does not guarantee to terminate with an optimal solution. No deadlock issues are discussed in their demonstrated examples because they always put an intermediate place that serves the role of a buffer with unlimited capacity between two operations.

Hatono et al. (6) employed the stochastic Petri nets to describe the uncertain events of stochastic behaviors in FMS, such as failure of machine tools, repair time, and processing time. They develop a rule base to resolve conflicts among the enabled transitions. The proposed method cannot handle the routing flexibility and deadlock situation.

This chapter presents a Petri net-based method for deadlock-free scheduling of automated manufacturing systems. Two hybrid heuristic search strategies that combine the heuristic best-first (BF) strategy with the controlled backtracking (BT) strategy based on the execution of the Petri nets are presented. The searching scheme is controllable, i.e., if one can afford the computation complexities required by a pure best-first strategy, the pure best-first search can be used to locate an optimal schedule. Otherwise, the hybrid BF-BT or BT-BF combination can be implemented. The deadlock states are explicitly defined in the Petri net framework. Hence, no more equations are required to describe deadlock avoidance constraints to derive deadlock-free schedules (Ramaswamy and Joshi 1996 [20]). The issues such as deadlock, routing flexibility, multiple lot size, and limited buffer size are explored.

Section 8.2 contains the discussion of the fundamentals of Petri nets. The conventional methods for Petri net modeling of manufacturing systems are given and illustrated through an example. Section 8.3 presents two Petri net-based hybrid heuristic search strategies. The hybrid heuristic search strategies combine the heuristic best-first strategy with the controlled backtracking strategy based on the execution of the Petri nets. Section 8.4 shows the evaluations of two different search strategies through scheduling a semiconductor test facility with multiple lot sizes for each job. Section 8.5 presents an example for deadlock-free scheduling of a manufacturing system with routing flexibility. Deadlock arises from explicit recognition of buffer space resources. Finally, Section 8.6 draws some conclusions.

## 8.2 Modeling Manufacturing Systems Using Petri Nets

Petri nets were named after Carl A. Petri who created in 1962 a netlike mathematical tool for the study of communication with automata. Further development made Petri nets a promising graphical and mathematical modeling tool applicable to many systems that are characterized as concurrent, asynchronous, distributed, parallel, nondeterministic, and/or stochastic. Petri nets have been used extensively to model and analyze manufacturing systems. Recent overviews of applications of Petri nets in manufacturing areas can be found in Zurawski and Zhou (31) and David and Alla (4).

A Petri net is defined as a bipartite directed graph containing places, transitions, and directed arcs connecting places to transitions and transitions to places. Pictorially, places are depicted by circles, and transitions as bars or boxes. A place is an input (output) place of a transition if there exists a directed arc connecting this place (transition) to the transition (place). A place can contain tokens pictured by black dots. It may hold either none or a positive number of tokens. At any given time instance, the distribution of tokens on places, called Petri net marking, defines the current state of the modeled system. Thus a marked Petri net can be used to study dynamic behavior of the modeled discrete event system.

Formally, a Petri net is defined as $Z = (P, T, I, O, m_0)$ (15, 29), where

- $P = \{p_1, p_2, \ldots, p_n\}$, $n > 0$ is a finite set of places;
- $T = \{t_1, t_2, \ldots, t_s\}$, $s > 0$ with $P \cup T \neq \varnothing$ and $P \cap T = \varnothing$ is a finite set of transitions;
- $I: P \times T \to \{0, 1\}$ is an input function or direct arcs from $P$ to $T$;
- $O: P \times T \to \{0, 1\}$ is an output function or direct arcs from $T$ to $P$;
- $m: P \to \{0, 1, 2, \ldots\}$ is a $|P|$ dimensional vector with $m(p)$ being the token count of place $p$. $m_0$ denotes an initial marking.

In order to simulate the dynamic behavior of a system, a state or marking in a Petri net is changed according to the following transition (firing) rules:

(1) A transition $t$ is enabled if $m(p) \geq I(p, t)$ for any $p \in P$.
(2) An enabled transition $t$ can fire at marking $m'$, and its firing yields a new marking, $m(p) = m'(p) + O(p, t) - I(p, t)$, for arbitrary $p$ from $P$.

The marking $m$ is said to be reachable from $m'$. Given $Z$ and its initial marking $m_0$, the reachability set is the set of all marking reachable from $m_0$ through various sequences of transition firings and is denoted by $R(Z, m_0)$. For a marking $m \in R(Z, m_0)$, if no transition is enabled in $m$, $m$ is called a deadlock marking, and the corresponding system is in a deadlock state.

A Petri net as a graphical and mathematical tool possesses a number of properties. Some of the important properties are as follows.

A Petri net $Z$ is said to be $K$-bounded or simply bounded if the number of tokens in each place does not exceed a finite number $K$ for any marking reachable from $m_0$. $Z$ is said to be safe if it is 1-bounded. For a bounded Petri net, there are a limited number of markings reachable from the initial marking $m_0$ through firing various sequences of transitions.

$Z$ is said to be live if, no matter what marking has been reached from $m_0$, it is possible to ultimately fire any transition of the net by progressing through some further firing sequence. This means that a live Petri net guarantees deadlock-free operation, no matter what firing sequence is chosen.

$Z$ is said to be reversible if, for each marking $m$ in $\in R(Z, m_0)$, $m_0$ is reachable from $m$. Therefore, in a reversible net one can always get back to the initial marking.

The boundedness, liveness, and reversibility of Petri nets have their significance in manufacturing systems. Boundedness or safeness implies the absence of capacity overflows. Liveness implies the absence of deadlocks. This property guarantees that a system can successfully produce without being deadlocked. Reversibility implies cyclic behavior of a system and repetitive production in flexible manufacturing. It means that the system can be initialized from any reachable state.

In this chapter, a place represents a resource status or an operation. A transition represents either start or completion of an event or operation process, and the stop transition for one activity will be the same as the start transition for the next activity according to Zhou and DiCesare (29) and Lee and DiCesare (11). Token(s) in a resource place indicates that the resource is available and no token indicates that it is not available. A token in an operation place represents that the operation is being executed and no token shows none being performed. A certain time may elapse between the start and the end of an operation. This is represented by associating a time delay with the corresponding operation place.

In this chapter, a bottom-up method (15, 9, 8) is used to synthesize the system for scheduling. First, the system is partitioned into subsystems according to the job types, then submodels are constructed for each subsystem, and a complete net model for the entire process is obtained by merging Petri nets of the sub-systems through the places representing the shared resources.

For each subsystem (job type), a Petri net is constructed following the general methodology (29):

(1) Identify the operations and resources (machines/buffers) required;
(2) Order operations by the precedence relations if they exist;
(3) For each operation in order, create and label a place to represent its status, add a transition (start activity) with an output arc(s) to the places, add a transition (stop activity) with an input arc(s) from the places;
(4) For each kind of resources (machines/buffers), create and label a place. If an operation place is a starting activity to require the resource(s), add input arc(s) from that resource place to the starting transition of that operation. If an operation is the ending one to use the resources, add output arc(s) from the ending transition to the resource place(s);
(5) Specify the initial marking, and associate time dealys with operation places.

Consider an example of Petri net modeling for scheduling. A manufacturing system has two machine $M_1$ and $M_2$ and one robot $R$. There are two jobs $J_1$ and $J_2$. Each job has two processes. Table 8.1 shows the job requirements.

The first operation of Job 1 can be carried out at Machine 1 with the help of the robot and needs 4 unit time. The second operation of Job 1 can be carried out at Machine 2 with the help of the robot and

**TABLE 8.1**  Job Requirements for Example 1

| Operations/Jobs | $J_1$ | $J_2$ |
|:---:|:---:|:---:|
| 1 | $(M_1R, 4)$ | $(M_1, 1)$ |
| 2 | $(M_2R, 1)$ | $(M_2, 4)$ |

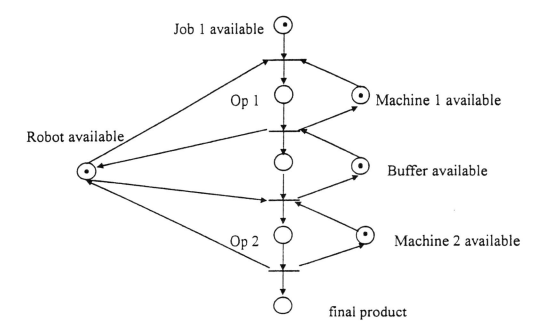

Job 1 available

Op 1

Machine 1 available

Robot available

Buffer available

Op 2

Machine 2 available

final product

**FIGURE 8.1** The Petri net model for subsystem of Job 1.

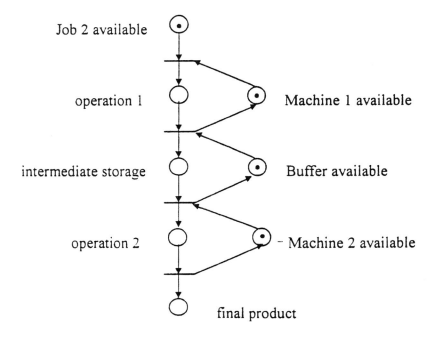

Job 2 available

operation 1

Machine 1 available

intermediate storage

Buffer available

operation 2

Machine 2 available

final product

**FIGURE 8.2** The Petri net model for subsystem of Job 2.

needs 1 unit time. The first operation of Job 2 can be carried out at Machine 1 and needs 1 unit time. The second operation of Job 2 can be carried out at Machine 2 and needs 4 unit time. Job 2 needs the robot for holding. The size of the intermediate buffer for each job is 1. Figure 8.1 shows the Petri net model of subsystem Job 1 and Figure 8.2 shows the Petri net model of subsystem Job 2. The Petri net

**TABLE 8.2**    Interpretation of Places and Transitions in Figure 8.3

| Places | Transitions |
|---|---|
| $p_1$: Job 1 available | $t_1$: Operation 1 of Job 1 starts |
| $p_2$: Job 2 available | $t_2$: Operation 1 of Job 2 starts |
| $p_3$: Operation 1 of Job 1 | $t_3$: Operation 1 of Job 1 finishes |
| $p_4$: Operation 1 of Job 2 | $t_4$: Operation 1 of Job 2 finishes |
| $p_5$: Job 1 ready for the second operation | $t_5$: Operation 2 of Job 1 starts |
| $p_6$: Job 2 ready for the second operation | $t_6$: Operation 2 of Job 2 starts |
| $p_7$: Operation 2 of Job 1 | $t_7$: Operation 2 of Job 1 finishes |
| $p_8$: Operation 2 of Job 2 | $t_8$: Operation 2 of Job 2 finishes |
| $p_9$: Final product of Job 1 | |
| $p_{10}$: Final product of Job 2 | |
| $p_{11}$: Buffer of Job 1 available | |
| $p_{12}$: Buffer of Job 2 available | |
| $p_{13}$: Machine 1 available | |
| $p_{14}$: Machine 2 available | |
| $p_{15}$: Robot available | |

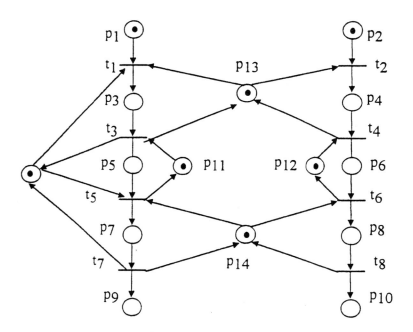

**FIGURE 8.3**    The whole Petri net model.

model for the whole system is obtained by merging the places representing Machine 1 and Machine 2 in two submodels and shown in Figure 8.3. The interpretation of places and transitions is shown in Table 8.2.

The evolution of the system can be completely tracked by the reachability graph of a Petri net. Figure 8.4 shows a partial portion of the reachability graph for the Petri net model shown in Figure 8.3. In the reachability graph, each transition firing sequence of $t_1 t_3 t_2 t_5 t_4 t_7 t_6 t_8$ and $t_2 t_4 t_1 t_6 t_3 t_8 t_5 t_7$ gives a path from the initial marking to the final one. But they generate different performance of schedules. As shown in Figure 8.5, the schedules based on $t_1 t_3 t_2 t_5 t_4 t_7 t_6 t_8$ and $t_2 t_4 t_1 t_6 t_3 t_8 t_5 t_7$ lead to makespan of 9 and 6, respectively. The notation $O_{i,j,k}$ in Figure 8.5 represents the $j$-th operation of the $i$-th job being performed at the $k$-th machine. Furthermore, if the lot size of Job 1 is 2, i.e., there are two tokens in place $p_1$ in the initial state, the transition firing sequence $t_2 t_4 t_1 t_6 t_3 t_8 t_1 t_8$ leads the system from initial state (2 1 0 0 0 0 0 0 0 0 1 1 1 1 1) into a deadlock

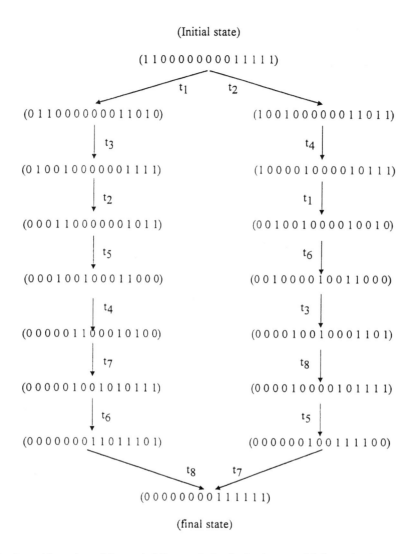

(Initial state)

(1 1 0 0 0 0 0 0 0 0 1 1 1 1 1)

$t_1$          $t_2$

(0 1 1 0 0 0 0 0 0 0 1 1 0 1 0)          (1 0 0 1 0 0 0 0 0 0 1 1 0 1 1)

$t_3$          $t_4$

(0 1 0 0 1 0 0 0 0 0 0 1 1 1 1)          (1 0 0 0 0 1 0 0 0 0 1 0 1 1 1)

$t_2$          $t_1$

(0 0 0 1 1 0 0 0 0 0 0 1 0 1 1)          (0 0 1 0 0 1 0 0 0 0 1 0 0 1 0)

$t_5$          $t_6$

(0 0 0 1 0 0 1 0 0 0 1 1 0 0 0)          (0 0 1 0 0 0 0 1 0 0 1 1 0 0 0)

$t_4$          $t_3$

(0 0 0 0 0 1 1 0 0 0 1 0 1 0 0)          (0 0 0 0 1 0 0 1 0 0 0 1 1 0 1)

$t_7$          $t_8$

(0 0 0 0 0 1 0 0 1 0 1 0 1 1 1)          (0 0 0 0 1 0 0 0 0 1 0 1 1 1 1)

$t_6$          $t_5$

(0 0 0 0 0 0 0 1 1 0 1 1 1 0 1)          (0 0 0 0 0 0 1 0 0 1 1 1 1 0 0)

$t_8$          $t_7$

(0 0 0 0 0 0 0 0 1 1 1 1 1 1)

(final state)

**FIGURE 8.4**   A partial portion of the reachability graph for the Petri net model shown in Figure 8.3.

state (0 0 1 0 1 0 0 0 0 1 0 1 0 1 0), in which any further part flow is inhibited. Figure 8.6 shows the evolution of the system states for the transition firing sequence $t_2 t_4 t_1 t_6 t_3 t_1 t_8$ which leads the system into a deadlock state.

## 8.3   Scheduling Algorithms Based on Petri Nets and Hybrid Heuristic Search

### Best First Search and Backtracking Search

An optimal event sequence is sought in a timed Petri net framework to achieve minimum or near minimum makespan. In the Petri net model of a system, firing of an enabled transition changes the token distribution (marking). A sequence of firings results in a sequence of markings, and all possible behaviors of the system can be completely tracked by the reachability graph of a net. The search space for the

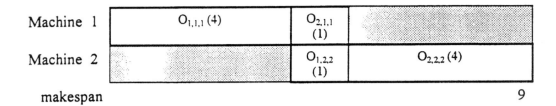

makespan                                                                        9

(a) transition firing sequence $t_1 t_3 t_2 t_5 t_4 t_7 t_6 t_8$

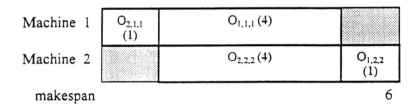

makespan                                                                        6

(b) transition firing sequence $t_2 t_4 t_1 t_6 t_3 t_8 t_5 t_7$

**FIGURE 8.5**   Schedules represented by two different transition firing sequences.

optimal event sequence is the reachability graph of the net, and the problem is to find a firing sequence of the transitions in the Petri net model from the initial marking to the final one. A heuristic search algorithm is developed by combining the Petri net execution and a bestfirst graph search algorithm $A^*$ (16). The most important aspect of the algorithm is the elimination from further consideration of some subsets of markings which may exist in the entire reachability graph. Thus the amount of computation and the memory requirements are reduced.

**Algorithm 1 (Best-First):**

1. Put the start node (initial marking) $m_0$ on OPEN.
2. If OPEN is empty, exit with failure.
3. Remove from OPEN and place on CLOSED a marking $m$ whose $f$ is the minimum.
4. If marking $m$ is a goal node (final marking), exit successfully with the solution obtained by tracing back the pointers from marking $m$ to marking $m_0$.
5. Otherwise find the enabled transitions of marking $m$, generate the successor markings for each enabled transition, and attach to them pointers back to $m$.
6. For every successor marking $m'$ of marking $m$:
   (a) Calculate $f(m')$.
   (b) If $m'$ is on neither OPEN nor CLOSED, add it to OPEN. Assign the newly computed $f(m')$ to marking $m'$.
   (c) If $m'$ already resides on OPEN or CLOSED, compare the newly computed $f(m')$ with the value previously assigned to $m'$. If the old value is lower, discard the newly generated marking. If the new value is lower, substitute it for the old and direct its pointer along the current path. If the matching marking $m'$ resides on CLOSED, move it back to OPEN.
7. Go to step 2.

The function $f(m)$ in Algorithm 1 is the sum of two terms $g(m)$ and $h(m)$. $f(m)$ is an estimate cost (makespan) from the initial marking to the final one along an optimal path which goes through the

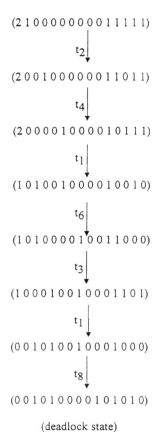

(2 1 0 0 0 0 0 0 0 0 1 1 1 1 1)

$t_2$

(2 0 0 1 0 0 0 0 0 0 1 1 0 1 1)

$t_4$

(2 0 0 0 0 1 0 0 0 0 1 0 1 1 1)

$t_1$

(1 0 1 0 0 1 0 0 0 0 1 0 0 1 0)

$t_6$

(1 0 1 0 0 0 0 1 0 0 1 1 0 0 0)

$t_3$

(1 0 0 0 1 0 0 1 0 0 0 1 1 0 1)

$t_1$

(0 0 1 0 1 0 0 1 0 0 0 1 0 0 0)

$t_8$

**FIGURE 8.6**   The evolution of the system states for the
transition firing sequence $t_2 t_4 t_1 t_6 t_3 t_1 t_8$ which leads the
system into a deadlock state.

(0 0 1 0 1 0 0 0 0 1 0 1 0 1 0)

(deadlock state)

marking $m$. The first term, $g(m)$, is the cost of a firing sequence from the initial marking to the current
one. The second term, $h(m)$ is an estimate cost of a firing sequence from current marking $m$ to the final
marking, called *heuristic function*. The following *heuristic function* is used:

$$h(m) = \max_i \{\xi_i(m), \quad i = 1, 2, \dots, N\}$$

where $\xi_i(m)$ is the sum of operation times of those remaining operations for all jobs which are planned
to be processed on the $i$th machine when the current system state is represented by marking $m$. $N$ is the
total number of machines. The purpose of a heuristic function is to guide the search process in the most
profitable direction by suggesting which transition to fire first.

For the above heuristic function, $h(m)$ is a lower bound to all complete solutions descending from
the current marking, i.e.,

$$h(m) \leq h^*(m), \; \forall m$$

where $h^*(m)$ is the optimal cost of paths going from the current marking $m$ to the final one. Hence,
the $h(m)$ is admissible, which guarantees for an optimal solution (Pearl 1984).

At each step of the best-first search process, we select the most promising of the markings we have
generated so far. This is done by applying an appropriate heuristic function to each of them. We then
expand the chosen marking by firing all enabled transitions under this marking. If one of successor

markings is a final marking, we can quit. If not, all those new markings are added to the set of markings generated so far. Again the most promising marking is selected and the process continues.

Once the Petri net model of the system is constructed, given initial and final markings, an optimal schedule can be obtained using the above algorithm. But for a sizable multiple lot size scheduling problem, it is very difficult or impossible to find the optimal solution in a reasonable amount of time and memory space. To reduce the computation complexity, various combinations of best-first search and backtracking (BT) can be implemented, which cut down the computation complexity at the expense of narrowing the evaluation scope (16). This chapter presents two search strategies by combining the heuristic best-first strategy with the controlled backtracking strategy based on the execution of the Petri nets. The backtracking method applies the last-in-first-out policy to node generation instead of node expansion. When a marking is first selected for exploration, only one of its enabled transitions is chosen to fire, and thus only one of its successor markings is generated. This newly generated marking is again submitted for exploration. When the generated marking meets some stopping criterion, the search process back-tracks to the closest unexpanded marking which still has unfired enabled transitions.

### Algorithm 2 (Backtracking):

1. Put the start node (initial marking) $m_0$ on OPEN.
2. If OPEN is empty, exit with failure.
3. Examine the topmost marking from OPEN and call it $m$.
4. If the depth of $m$ is equal to the depth-bound or if all enabled transitions under marking $m$ have already been selected to fire, remove $m$ from OPEN and go to step 2; otherwise continue.
5. Generate a new marking $m'$ by firing an enabled transition not previously fired under marking $m$. Put $m'$ on top of OPEN and provide a pointer back to $m$.
6. Mark $m$ to indicate that the above transition has been slected to fire.
7. If marking $m'$ is a goal node (final marking), exit successfully with the solution obtained by tracing back the pointers from marking $m'$ to marking $m_0$.
8. If $m'$ is a deadlock marking, remove it from OPEN.
9. Go to step 2.

The best-first search strategy examines, before each decision, the entire set of available alternative markings, those newly generated as well as all those suspended in the past. The backtracking search strategy is committed to maintaining in storage only a single path containing the set of alternative markings leading to the current marking. It proceeds forward heedlessly to find a feasible schedule without considering the optimality. Since only the markings on the current firing sequence are stored, it requires less memory.

## Hybrid Heuristic Search Algorithms

The need to combine BF and BT strategies is a result of computational considerations. For multiple lot size scheduling problems, if we cannot afford the memory space and computation time required by a pure BF strategy, we can employ a BF-BT combination that cuts down the storage requirement and computation time at the expense of narrowing the evaluation scope.

In the following Algorithm 3 (26), the heuristic best-first search strategy is applied at the top of reachability graph of the timed Petri net model and a backtracking search strategy at the bottom. We begin with BF search until a depth-bound $dep_0$ is reached. Then BT search is employed using the best present marking as a starting node. If it fails to find a solution, we return to get the second best marking on OPEN as a new root for a BT search, and so on.

### Algorithm 3 (Hybrid BF-BT):

1. Put the start node (initial marking) $m_0$ on OPEN.
2. If OPEN is empty, exit with failure.
3. Remove from OPEN and place on CLOSED a marking $m$ whose $f$ is the minimum.

4. If marking $m$ is a goal node (final marking), exit successfully with the solution obtained by tracing back the pointers from marking $m$ to marking $m_0$.
5. If the depth of marking $m$ is greater than the depth-bound $dep_0$, go to Step 9; otherwise continue.
6. Find the enabled transitions of the marking $m$, generate the successor markings for each enabled transition, and attach to them pointers back to $m$.
7. For every successor marking $m'$ of marking $m$:
   (a) Calculate $f(m')$.
   (b) If $m'$ is on neither OPEN nor CLOSED, add it to OPEN. Assign the newly computed $f(m')$ to marking $m'$.
   (c) If $m'$ already resides on OPEN or CLOSED, compare the newly computed $f(m')$ with the value previously assigned to $m'$. If the old value is lower, discard the newly generated marking. If the new value is lower, substitute it for the old and direct its pointer along the current path. If the matching marking $m'$ resides on CLOSED, move it back to OPEN.
8. Go to Step 2.
9. Take marking $m$ as a root node for BT search, put it on OPEN0.
10. If OPEN0 is empty, go to Step 2.
11. Examine the topmost marking from OPEN0 and call it $m'$.
12. If all enabled transitions under marking $m'$ have been selected to fire, remove it from OPEN0 and go to Step 10.
13. Generate a successor marking $m''$ for one enabled transition not previously firing, calculate $g(m'')$, put $m''$ on top of OPEN0 and provide a pointer back to $m'$.
14. If marking $m''$ is a goal node (final marking), exit successfully with the solution obtained by tracing back the pointers from marking $m''$ to the initial marking $m_0$.
15. If $m''$ is a deadlock marking, remove it from OPEN0.
16. Go to Step 10.

An opposite approach is starting a backtracking search on the top of the reachability graph followed by heuristic best-first ending. This strategy is implemented in Algorithm 4. We begin BT until a depth-bound $dep_0$ is reached. Then we employ the heuristic BF search from the current marking until it returns the final marking. If the BF search fails to find a solution, we return to backtracking and again use BF upon reaching the depth-bound $dep_0$.

## Algorithm 4 (Hybrid BT-BF)

1. Put the start node (initial marking) $m_0$ on OPEN0.
2. If OPEN0 is empty, exit with failure.
3. Examine the topmost marking from OPEN0 and call it $m$.
4. If all enabled transitions under marking $m$ have already been selected to fire, remove $m$ from OPEN0 and go to Step 2; otherwise continue.
5. If the depth of marking $m$ is greater than the depth-bound $dep_0$, go to Step 10; otherwise continue.
6. Generate a new marking $m'$ by firing an enabled transition not previously fired under marking $m$. Put $m'$ on top of OPEN0 and provide a pointer back to $m$.
7. Mark $m$ to indicate that the above transition has been fired.
8. If $m'$ is a deadlock marking, remove it from OPEN0.
9. Go to step 2.
10. Take marking $m$ from BT search as the start node $m_0$ and put it on OPEN.
11. If OPEN is empty, go back to Step 2 and return to backtracking search.
12. Remove from OPEN and place on CLOSED a marking $m$ whose $f$ is the minimum.
13. If marking $m$ is a goal node (final marking), exit successfully with the solution obtained by tracing back the pointers from marking $m$ to marking $m_0$. Otherwise continue.
14. Find the enabled transitions of marking $m$, generate the successor markings for each enabled transition, and attach to them pointers back to $m$.

**TABLE 8.3**   The Number of Each Type of Facility

| Facility | Type | Quantity |
|----------|------|----------|
| Tester | T1, T2, T3 | 1 |
| Hander | H1 | 1 |
| Hander | H2 | 2 |
| Hardware | Ha1 | 2 |
| Hardware | Ha2 | 1 |

**TABLE 8.4**   Workcenters for Final Test

| Workcenter | Resource Combination |
|------------|----------------------|
| M1 | T1 + H1 + Ha1 |
| M2 | T2 + H2 + Ha2 |
| M3 | T3 + H2 + Ha1 |

**TABLE 8.5**   Job Requirements

| Operations/Jobs | $J_1$ | $J_2$ | $J_3$ | $J_4$ |
|-----------------|-------|-------|-------|-------|
| 1 | $(M_1, 2)$ | $(M_3, 4)$ | $(M_1, 3)$ | $(M_2, 3)$ |
| 2 | $(M_2, 3)$ | $(M_1, 2)$ | $(M_3, 5)$ | $(M_3, 4)$ |
| 3 | $(M_3, 4)$ | $(M_2, 2)$ | $(M_2, 3)$ | $(M_1, 3)$ |

Figure 8.7 shows the Petri net model for this system. The three test stages of the job $i$ are modeled by $p_{i1}, t_{i1}, \ldots, t_{i6}$, and $p_{i7}$, for $i = 1, 2, 3$, and 4, respectively. The places and transitions are interpreted in the same figure. The modeling is briefed as follows. First, model a Petri net model for each job based on their sequence and the use of machines. Then merge these models to obtain a complete Petri net model for the whole system through shared resource places that model the availability of machines. The lot size is represented by the number of tokens in the places $p_{11}$– $p_{41}$. For example in Figure 8.7, the lot sizes for job $J1$–4 are 2, 1, 2, and 3, respectively.

## Scheduling Results

For the above example, we make a comparison between Algorithms 3 and 4. We set the different depth bound to see the relations between the optimality and computation complexity.

The three sets of lot size (5, 5, 2, 2), (8, 8, 4, 4), and (10, 10, 6, 6) are tested. We employ both Algorithms 3 and 4. The scheduling results of makespan, number of generated markings, and computation time are shown in Tables 8.7, 8.8 and 8.9 for the lot size (5, 5, 2, 2), (8, 8, 4, 4), and (10, 10, 6, 6), respectively. The optimal makespans for different cases obtained from pure BF search are also shown in these tables.

Both Algorithms 3 (BF-BT) and 4 (BT-BF) cut down the computation complexity by narrowing the evaluation scope at the expense of losing the optimality. The relations of computation complexity (number of generated markings and computation time) reduced versus optimality lost are shown in Figures 8.8, 8.9, and 8.10 for three different sets of lot size (5, 5, 2, 2), (8, 8, 4, 4), and (10, 10, 6, 6), respectively. In these figures, the percentage of optimality lost, which is the comparison of the makespan, is defined as

$$Percentage\ of\ Optimality\ Lost = \frac{Makespan\ (Hybrid) - Makespan\ (BF)}{Makespan\ (BF)} \cdot 100\%$$

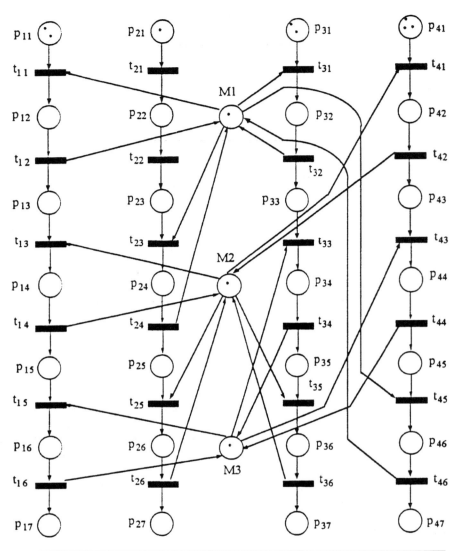

| $p_{i1}$ | Job i available | $t_{i1}$ | Start Job i's first stage test |
|---|---|---|---|
| $p_{i2}$ | Job i's first stage test | $t_{i2}$ | End Job i's first stage test |
| $p_{i3}$ | Waiting for Job i's second stage test | $t_{i3}$ | Start Job i's second stage test |
| $p_{i4}$ | Job i's second stage test | $t_{i4}$ | End Job i's second stage test |
| $p_{i5}$ | Waiting for Job i's third stage test | $t_{i5}$ | Start Job i's third stage test |
| $p_{i6}$ | Job i's third stage test | $t_{i6}$ | End Job i's third stage test |
| $p_{i7}$ | Job i's test complete | $i = 1, 2, 3,$ and $4$ | |
| $M_j$ | Machine j is available for $j = 1, 2,$ and $3$ | | |

**FIGURE 8.7**   A Petri net model of a test facility.

**TABLE 8.7**    Scheduling Results of the Example for Lot Size (5, 5, 2, 2)

| Depth for BF Search | Makespan | | Number of Markings | | CPU Time (sec) (Sun SPARC 20) | | Optimal Makespan Pure BF |
|---|---|---|---|---|---|---|---|
| | BF-BT | BT-BF | BF-BT | BT-BF | BF-BT | BT-BF | |
| 20 | 94 | 88 | 571 | 248 | 0.65 | 0.38 | 58 |
| 40 | 85 | 80 | 1607 | 484 | 4 | 0.8 | 58 |
| 50 | 79 | 70 | 2132 | 1247 | 6 | 3.6 | 58 |
| 60 | 74 | 64 | 2775 | 1520 | 8 | 6.5 | 58 |
| 80 | 64 | 62 | 3308 | 1687 | 11 | 7 | 58 |

**TABLE 8.8**    Scheduling Results of Example 6 for Lot Size (8, 8, 4, 4)

| Depth for BF Search | Makespan | | Number of Markings | | CPU Time (sec) (Sun SPARC 20) | | Optimal Makespan Pure BF |
|---|---|---|---|---|---|---|---|
| | BF-BT | BT-BF | BF-BT | BT-BF | BF-BT | BT-BF | |
| 40 | 168 | 163 | 3888 | 585 | 24 | 1.4 | 100 |
| 60 | 154 | 140 | 5234 | 1590 | 38 | 7 | 100 |
| 80 | 140 | 121 | 7699 | 2873 | 49 | 18 | 100 |
| 100 | 127 | 112 | 8819 | 4545 | 90 | 36 | 100 |
| 120 | 108 | 104 | 9233 | 8045 | 104 | 76 | 100 |

**TABLE 8.9**    Scheduling Results of the Example for Lot Size (10, 10, 6, 6)

| Depth for BF search | Makespan | | Number of Markings | | CPU Time (sec) (Sun SPARC 20) | | Optimal Makespan Pure BF |
|---|---|---|---|---|---|---|---|
| | BF-BT | BT-BF | BF-BT | BT-BF | BF-BT | BT-BF | |
| 80 | 206 | 209 | 6,281 | 1,254 | 64 | 5 | 134 |
| 100 | 198 | 181 | 12,341 | 2,315 | 240 | 16 | 134 |
| 120 | 180 | 162 | 16,602 | 8,495 | 480 | 139 | 134 |
| 140 | 169 | 150 | 20,155 | 11,368 | 540 | 390 | 134 |
| 160 | 153 | 148 | 21,797 | 18,875 | 660 | 560 | 134 |

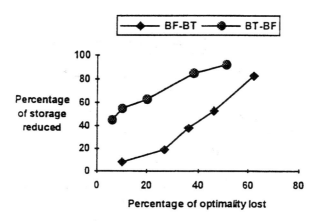

**FIGURE 8.8(a)**    Percentage of storage reduced versus percentage of optimality lost for lot size (5, 5, 2, 2).

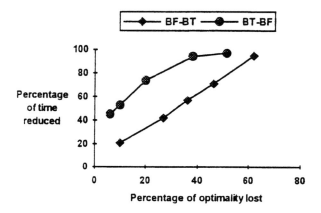

**FIGURE 8.8(b)**    Percentage of computation time reduced versus percentage of optimality lost for lot size (5, 5, 2, 2).

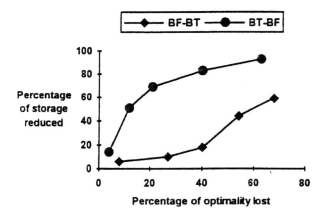

**FIGURE 8.9(a)**    Percentage of storage reduced versus percentage of optimality lost for lot size (8, 8, 4, 4).

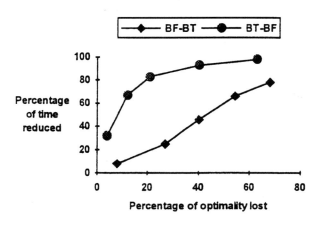

**FIGURE 8.9(b)**    Percentage of computation time reduced versus percentage of optimality lost for lot size (8, 8, 4, 4).

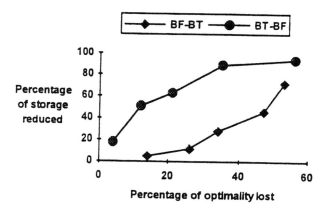

FIGURE 8.10(a)   Percentage of storage reduced versus percentage of optimality lost for lot size (10, 10, 6, 6).

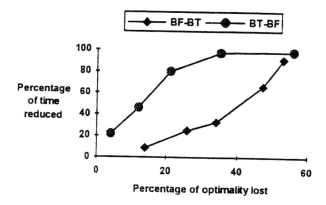

FIGURE 8.10(b)   Percentage of computation time reduced versus percentage of optimality lost for lot size (10, 10, 6, 6).

where hybrid is either BF-BT or BT-BF. The percentage of computation complexity reduced, which is the comparison of the storage (number of generated markings) and computation time, is defined as

$$Percentage\ of\ Storage\ Reduced = \frac{Storage\ (BF) - Storage\ (Hybrid)}{Storage\ (BF)} \cdot 100\%$$

and

$$Percentage\ of\ Time\ Reduced = \frac{CPU\ (BF) - CPU\ (Hybrid)}{CPU\ (BF)} \cdot 100\%$$

From the testing results the following conclusions are drawn. The hybrid heuristic search which employs the heuristic best-first search at the bottom of the Petri net reachability graph (Algorithm 4) performs much better than the one which employs the heuristic best-first search at the top of the Petri net reachability graph (Algorithm 3). This is due to two reasons. One is that the performance of heuristic best-first search is at its best when its guiding heuristic is more informed, and this usually happens at the bottom of the search graph (16). Thus BT-BF search greatly reduces the computation complexity comparing with BF-BT search which employs the heuristic best-first search at the top of the search graph.

Another reason is that there are fewer firing transitions for the markings at the bottom of Petri net reachability graph than those at the top. This is because at the late stages of a scheduling task, the reduced number of remaining operations reduces the number of choices. Hence, the number of alternatives considered in each decision for BT-BF search is less than the one for BF-BT search. However, the important decisions with respect to the quality of a schedule may happen at the early stages of the scheduling activity, this increases the likelihood of missing the critical candidates for BT-BF search which employs backtracking search instead of best-first search at the early stage.

## 8.5  Scheduling Manufacturing Systems with Routing Flexibility

The order in which a job visits different machines is predetermined in the classical job-shop scheduling problem. Routing flexibility is a new feature of flexible manufacturing. In an automated manufacturing system, each operation of a job may be performed by any one of several machines. We consider an automated manufacturing system with three multipurpose machines $M_1$, $M_2$, and $M_3$. There are four jobs, $J_1$, $J_2$, $J_3$, and $J_4$. The first three jobs have three processes each and the last one, $J_4$, has only two processes. Table 8.10 shows the job requirements. The operation times are shown in Table 8.11, where $O_{i,j,k}$ represents the *j*th operation of the *i*th job performed by the *k*th machine.

We note that a job can be carried out more than one routing in Table 8.10. For instance, the first process of job $J_1$ can be performed at either $M_1$ or $M_2$, and the second at either $M_2$ or $M_3$. The Petri net model for $J_1$ is shown in Figure 8.11. The complete Petri net model for the system can be obtained by merging the places representing the shared machines in the Petri net model of each job type.

The hybrid heuristic search algorithm is used to solve the above problem considering different lot sizes. In each case, the depth bound is set to the half of the depth of a reachability graph. The latter is computed by multiplying the number of transitions in the Petri net model and the lot size. The hybrid search is compared with the standard depth-first search (16), and heuristic dispatching rules.

We employ the following benchmark dispatching rules:

1.  A heuristic that chooses the fastest machine which can perform an operation if more than one machine exits, and then the shortest processing time (SPT) rule is used to sequence the operations among the parts waiting in the input buffer of a machine.
2.  A heuristic that chooses a machine whose input buffer currently has the shortest queue, and then SPT rule is used to sequence the operations among the parts waiting in the input buffer of a machine.

**TABLE 8.10**  Job Requirements for the Example with Routing Flexibility

| Operations/Jobs | $J_1$ | $J_2$ | $J_3$ | $J_4$ |
|---|---|---|---|---|
| 1 | $M_1/M_2$ | $M_2$ | $M_1/M_2$ | $M_1$ |
| 2 | $M_2/M_3$ | $M_1/M_3$ | $M_3$ | $M_2/M_3$ |
| 3 | $M_1$ | $M_3$ | $M_1/M_2/M_3$ | N/A |

**TABLE 8.11**  Operation Times

| Operation | Time | Operation | Time | Operation | Time | Operation | Time |
|---|---|---|---|---|---|---|---|
| $O_{1,1,1}$ | 10 | $O_{2,1,2}$ | 5 | $O_{3,1,1}$ | 4 | $O_{4,1,1}$ | 11 |
| $O_{1,1,2}$ | 12 | $O_{2,2,1}$ | 9 | $O_{3,1,2}$ | 8 | $O_{4,2,2}$ | 9 |
| $O_{1,2,2}$ | 7 | $O_{2,2,3}$ | 13 | $O_{3,2,3}$ | 6 | $O_{4,2,3}$ | 9 |
| $O_{1,2,3}$ | 10 | $O_{2,3,3}$ | 8 | $O_{3,3,1}$ | 6 | | |
| $O_{1,3,1}$ | 5 | | | $O_{3,3,2}$ | 2 | | |
| | | | | $O_{3,3,3}$ | 7 | | |

**TABLE 8.12**   The Scheduling Results Using Different Methods

| Lot Size | Makespan | | | |
|---|---|---|---|---|
| | Depth-First | Dispatching (i) | Dispatching (ii) | Hybrid (BT-BF) |
| (1,1,1,1) | 49 | 46 | 37 | 34 |
| (5,5,5,5) | 313 | 204 | 161 | 152 |
| (10,10,10,10) | 713 | 399 | 311 | 296 |
| (20,20,20,20) | 1513 | 789 | 613 | 597 |
| (30,30,30,30) | 2313 | 1179 | 921 | 874 |
| (40,40,40,40) | 3113 | 1569 | 1223 | 1172 |
| (50,50,50,50) | 3913 | 1959 | 1525 | 1468 |

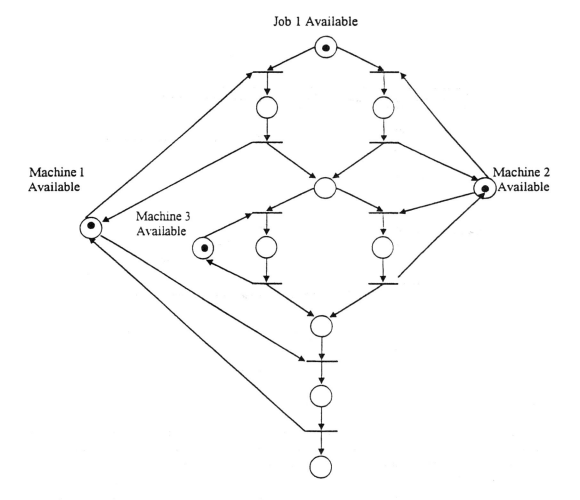

**FIGURE 8.11**   The Petri net model for subsystem Job 1.

Several different lot sizes are tested using the hybrid heuristic BT-BF, depth-first search and dispatching rules a and b. The results of the comparison are given in Table 8.12. In all the cases tested, the presented hybrid method generates schedules with the shortest makespan. The depth-first search generates the worst results since the depth as a heuristic function is not admissible and it is not a good function for this case. The heuristic dispatching methods perform worse than the hybrid search, but better than the depth-first search. This is because the depth-first search explores a path using the totally uninformed knowledge. The dispatching rules

**TABLE 8.13**    The Scheduling Results for Finite Buffer Capacity

| | Makespan | | |
| --- | --- | --- | --- |
| Buffer Size | Dispatching a | Dispatching b | Hybrid (BT-BF) |
| 2 | deadlock | deadlock | 885 |
| 4 | deadlock | deadlock | 883 |
| 6 | deadlock | deadlock | 880 |
| 8 | deadlock | 1032 | 875 |
| 10 | 1189 | 1014 | 874 |
| 12 | 1179 | 981 | 874 |
| 14 | 1179 | 975 | 874 |
| 16 | 1179 | 963 | 874 |
| 18 | 1179 | 942 | 874 |
| 20 | 1179 | 930 | 874 |
| 22 | 1179 | 921 | 874 |
| 24 | 1179 | 921 | 874 |

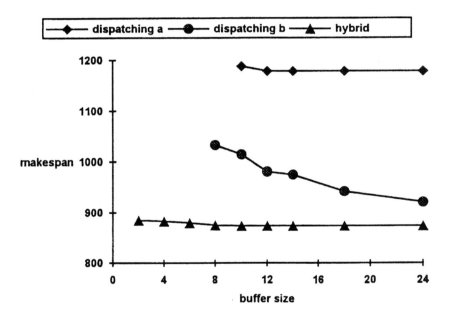

**FIGURE 8.12**    The comparison of makespan for the varying buffer capacity in the lot size case (30,30,30,30).

seek the solutions using the local heuristics, while the hybrid method using the global information by ordering the decision candidates based on the performance indices. The heuristic dispatching rule that chooses a machine whose input buffer currently has the shortest queue performs better than the one that chooses the fastest machine. This is expected because the heuristic that chooses a machine whose input buffer currently has the shortest queue is a dynamic rule, while the heuristic that chooses the fastest machine is a static one. Dynamic rules change priority indices with time and queue characteristics, whereas static ones keep priority indices constant as jobs travel through the plant.

For the computation results shown in Table 8.12, it is supposed that the buffer size for each machine is unlimited. Deadlock is completely avoided because sufficient in-process storage is provided. However, it causes excessive work in-process and leads to an inefficient manufacturing system. The presented hybrid method always generates a deadlock free schedule if it exists because of the explicit representation of deadlock states in the Petri net framework and backtracking capability in the search procedure. However, deadlock could happen when the dispatching rules are employed. It is because the commonly used dispatching rules are "single pass" rules, namely, that once a decision is made by applying a rule, it will not reconsider the alternative courses of action.

For the above example, let us take the lot size case (30,30,30,30) and consider finite buffer size for each machine. Varying the buffer size *N*, scheduling results are obtained by applying the hybrid method and two dispatching methods, as shown in Table 8.13 and Figure 8.12. It is clear that 1) until buffer size reaches 10(8), dispatching rule a(b) generates no dead-lock free schedule; 2) Hybrid algorithm generates deadlock-free schedule regardless of buffer size and constantly performs better than the dispatching rules; 3) After the buffer size reaches 12, 22, and 10, dispatching rules a, b, and hybrid algorithm achieve their best schedules of makespan 1179, 921, and 874, respectively. Thus the hybrid algorithm represents 34.9% and 5.38% better than rules *a* and *b* while it needs 20% and 120% less in buffer capacity.

# 8.6  Conclusions

This chapter investigates different hybrid heuristic search strategies for scheduling of automated manufacturing systems in a Petri net framework. Petri nets are a graphical and mathematical modeling tool applicable to many systems. Petri nets can explicitly and concisely model the concurrent and asynchronous activities, multilayer resource sharing, routing flexibility, limited buffers, and precedence constraints in automated manufacturing systems. Using the concept of markings, the evolution of the system can be completely tracked by the reachability graph of the net. Associating time delays with places or transitions in a Petri net allows it to describe a system whose functioning is time dependent. The problem of deadlock has been ignored by most researchers in scheduling and control when they employ such methods as mathematical programming, heuristics dispatching and knowledge-based, and control theoretic methods. Petri nets can provide an explicit and convenient way for considering deadlock situations in automated manufacturing systems such that a deadlock-free scheduling and control system can be designed.

A Petri net-based method is presented for deadlock-free scheduling of automated manufacturing systems in this chapter. Two hybrid heuristic search strategies that combine the heuristic best-first (BF) strategy with the controlled backtracking (BT) strategy based on the execution of the Petri nets are discussed. The searching scheme is controllable, i.e., if one can afford the computation complexities required by a pure best-first strategy, the pure best-first search can be used to locate an optimal schedule. Otherwise, the hybrid BF-BT or BT-BF combination can be implemented. The deadlock states are explicitly defined in the Petri net framework. No equations are needed to describe deadlock avoidance constraints to derive deadlock-free schedules. The issues such as deadlock, routing flexibility, multiple lot size, and limited buffer size are explored.

Even though the employed heuristic function is admissible, a more effective admissible heuristic function is desired to reduce the search effort. For hybrid search schemes, instead of employing BT on the top and BF on the bottom or vice versa, a more effective way should be employing BT and BF interchangeably based on the current state. This requires a comprehensive analysis of proposed schemes.

Many industrial problems are sizable. For example, a semiconductor manufacturing facility may contain over a hundred machines of different or same kinds. These large-size industrial systems may have to accommodate different priority jobs and face rescheduling when machines break down or are repaired. Both modeling and scheduling solutions are challenging. More research issues have to be addressed to resolve such sizable industrial problems using Petri nets (32).

## References

1. Z. Banaszak and B. Krogh, "Deadlock avoidance in flexible manufacturing systems with concurrently competing process flows," *IEEE Trans. on Robotics and Automation*, vol. 6, no. 6, pp. 724–734, 1990.

2.  T. R. Chen, *Scheduling for IC Sort and Test Facilities via Lagrangian Relaxation*, Ph.D. dissertation, University of California at Davis,1994.
3.  J. Carlier and E. Pinson, "An algorithm for solving the job-shop problem," *Management Science,* vol. 35, no. 2, pp. 164–176, 1989.
4.  R. David and H. Alla, "Petri nets for modeling of dynamic systems—a survey," *Automatica,* vol. 30, no. 2, pp. 175–202, 1994.
5.  S. France, *Sequencing and Scheduling: An Introduction to the Mathematics of the Job-Shop.* New York: Wiley, NY, 1982.
6.  I. Hatono, K. Yamagata, and H. Tamura, "Modeling and on-line scheduling of flexible manufacturing systems using stochastic Petri nets" *IEEE Trans. on Software Engineering,* vol. 17, no. 2, pp. 126–132, 1991.
7.  H. P. Hillion and J. M. Proth, "Performance evaluation of job–shop systems using timed event-graphs," *IEEE Trans. on Automatic Control,* vol. 34, no. 1, pp. 3–9, 1989.
8.  M. D. Jeng and F. DiCesare, "A review of synthesis techniques for Petri nets with applications to automated manufacturing systems," *IEEE Trans. on Systems, Man, and Cybernetics,* vol. 23, no. 1, pp. 301–312, 1993.
9.  B. H. Krogh and C. L. Beck, "Synthesis of place/transition nets for simulation and control of manufacturing systems," in *Proceedings of IFIP Symposium on Large Systems,* Zurich, pp. 1–6, Aug. 1986.
10. C. Y. Lee, R. Uzsoy, and L. A. Martin-Vega, "Efficient algorithms for scheduling semiconductor burn-in operations," *Operations Research,* vol. 40, no. 4, pp. 764–795, 1992.
11. D. Y. Lee and F. DiCesare, "Scheduling FMS using Petri nets and heuristic search," *IEEE Trans. on Robotics and Automation,* vol. 10, no. 2, pp. 123–132, 1994.
12. D. Y. Lee and F. DiCesare, "Petri net-based heuristic scheduling for flexible manufacturing systems," in *Petri Nets in Flexible & Agile Automation,* M. C. Zhou (Ed.), pp. 149–187, Boston, MA: Kluwer Academic, 1995.
13. V. J. Leon and S. D. Wu, "Characteristics of computerized scheduling and control of manufacturing systems," *Computer Control of Flexible Manufacturing Systems,* S. Joshi and G. Smith (Ed.), Chapman & Hall, London: pp. 63–73, 1994.
14. P. B. Luh and D. J. Hoitomt, "Scheduling of manufacturing systems using the Lagrangian relaxation technique," *IEEE Trans. on Automatic Control,* vol. 38, no. 7, 1993, pp. 1066–1079, 1993.
15. Y. Narahari and N. Viswanadham, "A Petri net approach to the modeling and analysis of flexible manufacturing systems," *Ann. Operations Research,* vol. 3, pp. 449–472, 1985.
16. J. Pearl, *Heuristics: Intelligent Search Strategies for Computer Problem Solving.* Reading, MA: Addison-Wesley, 1984.
17. J.-M. Proth and I. Minis, "Planning and scheduling based on Petri nets," in *Petri Nets in Flexible and Agile Automation,* M. C. Zhou (Ed.), pp. 109–148, Boston, MA: Kluwer Academic, 1995.
18. L. Shen, Q. Chen, and J. Luh, "Truncation of Petri net models for simplifying computation of optimum scheduling problems," *Computers in Industry,* 20, pp. 25–43, 1992.
19. H. Shih and T. Sekiguchi, "A timed Petri net and beam search based on-line FMS scheduling system with routing flexibility," in *Proceedings of the 1991 IEEE Int. Conf. on Robotics and Automation,* Sacramento, CA, pp. 2548–2553, April 1991.
20. S. E. Ramaswamy and S. B. Joshi, "Deadlock-free schedules for automated manufacturing workstations," *IEEE Trans. on Robotics and Automation,* vol. 12, no. 3, pp. 391–400, 1996.
21. R. Uzsoy, C. Y. Lee, and L. A. Martin-Vega, "A review of production planning and scheduling models in the semiconductor industry part I: system characteristics, performance evaluation and production planning," *IIE Trans. on Scheduling and Logistics,* vol. 24, no. 4, pp. 47–60, 1992.
22. R. Uzsoy, C. Y. Lee, and L. A. Martin-Vega, "A review of production planning and scheduling models in the semiconductor industry part II: shop-floor control," *IIE Trans. on Scheduling and Logistics,* vol. 26, no. 5, pp. 44–55, 1994.

23. R. Uzsoy, L. A. Martin-Vega, C. Y. Lee, and P. A. Leonard, "Production scheduling algorithms for a semiconductor test facility," *IEEE Trans. on Semiconductor Manufacturing,* vol. 4, pp. 271–280, 1991.

24. N. Viswanadham, Y. Narahari, and T. Johnson, "Deadlock prevention and deadlock avoidance in flexible manufacturing systems using Petri net models," *IEEE Trans. on Robotics and Automation,* vol. 6, no. 6, pp. 713–723, 1990.

25. R. A. Wysk, N. Yang, and S. Joshi, "A detection of deadlocks in flexible manufacturing cells," *IEEE Trans. on Robotics and Automation,* vol. 7, no. 6, pp. 853–859, 1991.

26. H. H. Xiong, M. C. Zhou, and R. J. Caudill, "A hybrid heuristic search algorithm for scheduling flexible manufacturing systems," in *Proceedings of 1996 IEEE Int. Conf. on Robotics and Automation,* Minneapolis, MN, pp. 2793–2797, Apr. 1996.

27. H. H. Xiong, *Scheduling and Discrete Event Control of Flexible Manufacturing Systems Based on Petri Nets.* Ph.D. dissertation, New Jersey Institute of Technology, 1996.

28. M. C. Zhou, H. Chiu, and H. H. Xiong, "Petri net scheduling of FMS using branch and bound method," *Proc. of 1995 IEEE Int. Conf. on Industrial Electronics, Control, and Instrumentation,* Orlando, FL, pp. 211–216, Nov. 1995.

29. M. C. Zhou and F. DiCesare, *Petri Net Synthesis for Discrete Event Control of Manufacturing Systems.* Boston, MA: Kluwer Academic, 1993.

30. M. C. Zhou, F. DiCesare, and A. Desrochers, "A hybrid methodology for synthesis of Petri nets for manufacturing systems," *IEEE Trans. on Robotics and Automation,* vol. 8, no. 3, pp. 350–361, 1992.

31. R. Zurawski and M. C. Zhou, "Petri nets and industrial applications: a tutorial," *IEEE Trans. on Industrial Electronics,* vol. 41, no. 6, pp. 567–583, 1994.

32. Zhou, M. C. and K. Venkatesh, *Modeling, Simulation and Control of Flexible Manufacturing Systems: A Petri Net Approach.* Singapore: World Scientific, 1998.

# 9

# Improvement of Damping Characteristics in Machine Tools in Manufacturing Systems

Satoshi Ema
*Gifu University*

E. Marui
*Gifu University*

In this study, we introduce an impact damper developed to improve the damping capability of long, thin cutting tools, such as drills or boring tools. The impact damper has the following features:

(1) small in size and simple in construction;
(2) easy to mount on the main vibratory systems; and
(3) no need to adjust parameters of an impact damper to the vibratory characteristics of the main vibratory systems.

Furthermore, a construction, a mechanism, and characteristics of the impact damper are described and their application to cutting tools is presented.

## Nomenclature

$C_L$      clearance (range of free mass motion)
$f$        frequency of free damped vibration of leaf spring system with free mass
$f_n$      natural frequency of leaf spring system without free mass
$f_{n2}$     natural frequency of drill in drill point restricted mode
$g$        acceleration of gravity
$K$      equivalent spring constant of leaf spring

$M$        main mass (equivalent mass of leaf spring)

$m$        free mass (equivalent mass of impact damper)

$N$        workpiece rotation per second

$T$        thickness of leaf spring

$Y$        displacement amplitude of leaf spring

$Y_D$      critical amplitude where impact damper no longer functions

$Y_0$      initial displacement of leaf spring

$\Delta Y$       amplitude decrement between neighboring cycles, $Y_i - Y_{i+1}$

$\Delta Y_{av}$  average amplitude decrement

$\gamma$        corresponding logarithmic decrement of leaf spring with free mass

$\gamma_2$       corresponding logarithmic decrement of drill point restricted mode

$\gamma_0$       logarithmic decrement of leaf spring without free mass

$\mu$        mass ratio, $m/M$

# 9.1   Introduction

In order to suppress chatter vibration and to improve cutting stability, many types of dampers have been used, and viscous materials and frictional force in contacting parts have been utilized. Studies on a Lanchester damper and a viscoelastic damper for boring bars or turning tools have been carried out for many years with beneficial results [1–7]. However, in a Lanchester damper consisting of a mass, a damper and a spring, their parameters must be adjusted to main vibratory systems. On the other hand, boring tools or drills with great ratios of overhang lengths to diameters often lead to chatter vibration due to their extremely low damping capabilities [8]. However, it is next to impossible from a construction standpoint to equip an absorber in a long, thin cutting tool such as boring tools or drills. In order to improve the damping capability of these cutting tools, we developed an impact damper with the following features [9–10]:

(1) small in size and simple in construction;

(2) easy to mount on the main vibratory systems; and

(3) no need to adjust parameters of an impact damper to the vibratory characteristics of the main vibratory systems.

Figure 9.1 shows a cutting tool (a drill) with an impact damper. An impact damper (free mass) is equipped on the middle of the drill overhang. The free mass is a ring shape and has a hole in the middle slightly larger than the diameter of the drill. Therefore, the free mass can move freely within this clearance. Thus, the impact damper has a very simple construction consisting of a free mass and a clearance. As chatter vibration occurs in a drilling and the drill deflects in the direction perpendicular to its axis, the free mass collides with the drill (main mass). Vibratory energy is consumed by this collision, and as a result the chatter vibration can be prevented. This is damping mechanism of the impact damper.

We have developed two types of impact dampers in this study. One is an impact damper which functions effectively only when cutting tools vibrate in the direction of gravity. The other is an impact damper which functions effectively not only when cutting tools vibrate in the direction of gravity but also when they vibrate in the direction perpendicular thereto. Hereafter, the former is called the impact damper A and the latter is called the impact damper B. Vibratory systems with these impact dampers can be represented as equivalent vibratory systems illustrated in Figure 9.2. The left in the figure shows a vibratory system with the impact damper A and the right shows a vibratory system with the impact damper B. In addition, the impact dampers allow a free mass to be equipped on the main vibratory system, but in the model vibratory system presented in Figure 9.2, the free mass exists inside the main mass.

The experiment was carried out in the use of these impact dampers, in which the main vibratory system was vibrated in the direction of gravity and the direction perpendicular to it. In order to investigate the effects of the free mass $m$ and the clearance $C_L$ on the damping capability of the impact dampers, $m$ and $C_L$ were varied widely. Furthermore, in order to investigate the damping effects of the impact dampers

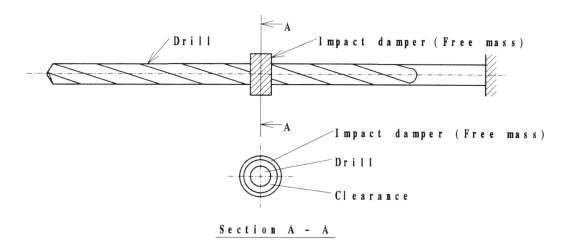

**FIGURE 9.1**   Cutting tool with impact damper.

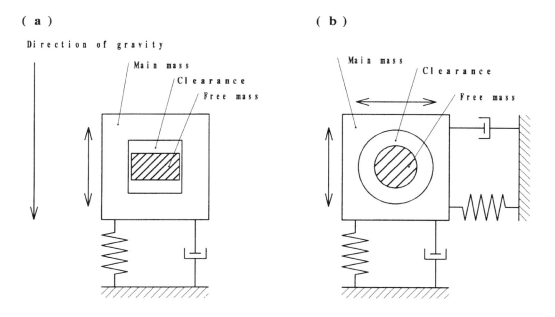

**FIGURE 9.2**   Impact dampers; (a) Impact damper A and (b) Impact damper B.

when vibratory characteristics of the main vibratory system are different (mainly the main mass and the spring constant), the thickness of leaf springs was also varied widely. Based on the results obtained from these fundamental experiments, we present the following damping characteristics of the impact dampers:

(1) range where the impact dampers function effectively (critical amplitude $Y_D$);
(2) frequency of the vibratory systems with the impact dampers; and
(3) degree of the damping capability of the impact dampers (average amplitude decrement $\Delta Y_{av}$ and corresponding logarithmic decrement $\gamma$).

Furthermore, the impact dampers were applied to drills and boring tools so that the damping capability of these tools can be improved considerably and chatter vibrations occurring in these tools can be prevented or suppressed.

## 9.2   Experimental Apparatus and Method

Figures 9.3 and 9.4 show outlines of the experimental apparatuses used to investigate the damping characteristics of impact dampers. Figure 9.3 is the experimental apparatus for the impact damper A and Figure 9.4 is for the impact damper B. In the impact damper A, the leaf spring was vibrated in the direction of gravity. In the impact damper B, the leaf spring was vibrated in the direction of gravity and the direction perpendicular thereto. The case in which the leaf spring vibrates in the direction of gravity is called vertical vibration, while the case in which the leaf spring vibrates in the direction perpendicular to gravity is called horizontal vibration. In this experiment, the leaf springs with a width of 30 mm and an overhang of 165 mm as shown in Figures 9.3 and 9.4 were used as the main vibratory system. In vertical vibration, the leaf spring was fixed on the table by two bolts. For horizontal vibration, the leaf spring shown in Figure 9.4 was fixed on the table by means of an auxiliary block in a state in which the leaf spring was rotated 90 degrees around the center axis. To diminish the influence of the fixing condition of the leaf spring clamping part on the damping characteristics of the system, the thickness of the clamping part was kept at 20 mm, which was thicker than the thickness of the overhang, and the clamping torque was kept at a constant value of 29.4 N·m.

In the impact damper A shown in Figure 9.3, a ring-shaped free mass indicated by the hatching was attached to the end of the leaf spring by the bolt. After the free mass and an auxiliary sleeve were installed between the leaf spring and the bolt, the bolt and the auxiliary sleeve were tightly clamped by two nuts. As the auxiliary sleeve was slightly taller than the free mass (20 mm in height), the free mass can move freely within a certain range. Therefore, this range of free mass motion is a clearance $C_L$ indicated in the figure. A radial clearance between the free mass and the auxiliary sleeve was maintained at 0.1 mm. In the impact damper B shown in Figure 9.4, a ring-shaped free mass indicated by the hatching was attached to the end of the leaf spring by the bolt. The diameter ($d$) of the bolt beneath the head is slightly smaller than the hole diameter of the free mass (12 mm), so that the free mass can move within a certain range. Therefore, this range of free mass motion, that is, the difference in diameter ($12 - d$) is a clearance $C_L$.

After an initial upward deflection (displacement) $Y_0$ or an initial horizontal deflection $Y_0$ was given to the leaf spring by the adjusting screw and the rod indicated in Figures 9.3 and 9.4, the rod was removed instantaneously by a hammer. The vibratory displacement of the leaf spring at this time was detected using an eddy current-type gap detector and was stored in a microcomputer through an A/D converter. In this manner, the free damped vibration of the leaf spring with an impact damper could be observed.

In this experiment, in order to investigate the effects of the free mass $m$ and the clearance $C_L$ on the damping capability of the impact damper, $m$ and $C_L$ were varied widely. The amount of the free mass $m$ was controlled by varying the diameter of the free mass $D$ in both impact dampers A and B. The size of the clearance $C_L$ was controlled by varying the length of the auxiliary sleeve in the impact damper A and by varying the diameter of the bolt $d$ beneath the bolt head in the damper B. Furthermore, the thickness of the leaf spring $T$ was varied widely in order to investigate the damping effects of the impact damper when vibratory characteristics of the main vibratory system were different (mainly the main mass and the spring constant). In Table 9.1, the thickness of the leaf spring $T$ and the initial displacement of the leaf spring $Y_0$ used in the experiment are listed. In Table 9.2, the free mass $m$ and the clearance $C_L$ are listed. The leaf springs were made of carbon steel for machine structural use (0.45% C) and the free mass was made of brass (70% Cu and 30% Zn).

## 9.3   Experimental Results

### Method for Evaluating Damping Capability

Figures 9.5 and 9.6 show the result of the vertical vibration, and Figures 9.7 and 9.8 show the result of the horizontal vibration. The other experimental conditions are appended in the captions: impact damper B, thickness of the leaf spring $T = 8$ mm, free mass $m = 0.1$ kg, clearance $C_L = 0.1$ mm, and initial displacement $Y_0 = 0.5$ mm. Figures 9.5 and 9.7 show the relationship between the vibration cycle $i$ and

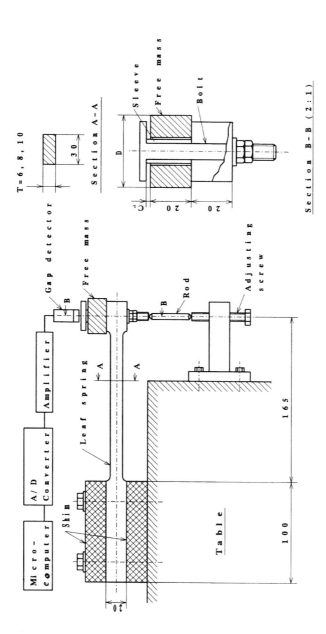

**FIGURE 9.3**  Experimental apparatus for impact damper A.

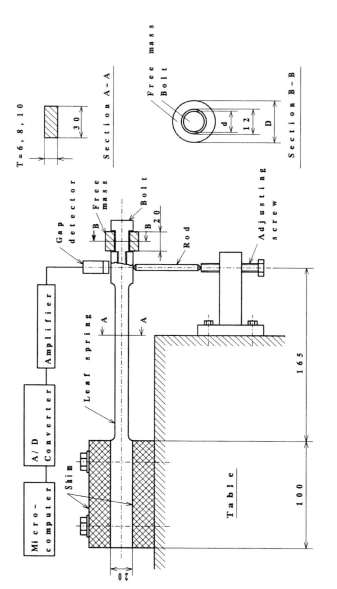

**FIGURE 9.4** Experimental apparatus for impact damper B (setting for vertical vibration).

**TABLE 9.1**  Thickness and Initial Displacement of Leaf Springs Used

| Damper | Thickness $T$ (mm) | | | Initial Displacement $Y_0$ (mm) | | | | Vibratory Direction | |
|---|---|---|---|---|---|---|---|---|---|
| A | 6 | 8 | 10 | 0.2 | 0.3 | 0.4 | 0.5 | Vertical | |
| B | 6 | 8 | 10 | – | 0.3 | 0.4 | 0.5 | Vertical | Horizontal |

**TABLE 9.2**  Mass and Clearance of Impact Dampers Used

| Damper | Mass $m$ (kg) | | | | Clearance $C_L$ (mm) | | | | | | | | | |
|---|---|---|---|---|---|---|---|---|---|---|---|---|---|---|
| A | 0.064 | – | 0.143 | 0.250 | 0.1 | 0.2 | 0.3 | 0.4 | – | 0.6 | – | 0.8 | – | 1.0 |
| B | 0.064 | 0.100 | 0.143 | – | 0.1 | 0.2 | 0.3 | 0.4 | 0.5 | 0.6 | 0.7 | 0.8 | 0.9 | 1.0 |

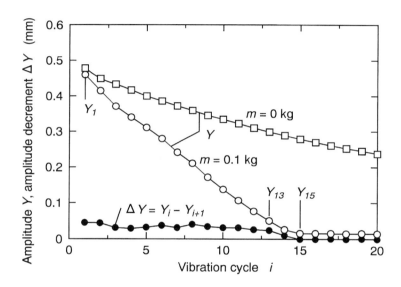

**FIGURE 9.5**  Amplitude reduction of leaf spring (impact damper B, vertical vibration, $T = 8$ mm, $m = 0.1$ kg, $C_L = 0.1$ mm, $Y_0 = 0.5$ mm).

the amplitude $Y$ of a leaf spring obtained from free damped vibrations, and Figures 9.6 and 9.8 show the state of amplitude reduction which can be visually presented by plotting alternately the amplitudes of continuous vibration cycles as coordinates on the ordinate and the abscissa. In these figures, the amplitude $Y$ (marked by circles) when a free mass exists and the amplitude $Y$ (marked by squares) when a free mass does not exist, are presented together.

When there is no free mass, as the amplitude of the leaf spring decreases exponentially as shown in Figures 9.5 and 9.7, and as inclinations of lines combining the origin with each point plotted in Figures 9.6 and 9.8 are constant, it is obvious that the amplitude reduction of the leaf spring is caused by viscous damping of the system. From such free damped vibrations without a free mass, the natural frequency $f_n$ and the logarithmic decrement $\gamma_0$ of the leaf spring were obtained. An equivalent spring constant of the leaf spring $K$ was obtained by applying a static bending load at the position (which coincides with the overhang of 165 mm) where an initial displacement was given to make the leaf spring a free damped vibration. Furthermore, the equivalent mass of the leaf spring $M$ was calculated from the natural frequency $f_n$ and the equivalent spring constant $K$. The vibrational characteristics of the main vibratory system, that is, the equivalent mass $M$, the equivalent spring constant $K$, the natural frequency $f_n$ and the logarithmic decrement $\gamma_0$ of leaf springs are summarized in Table 9.3. In addition, the logarithmic decrement $\gamma_0$ of the

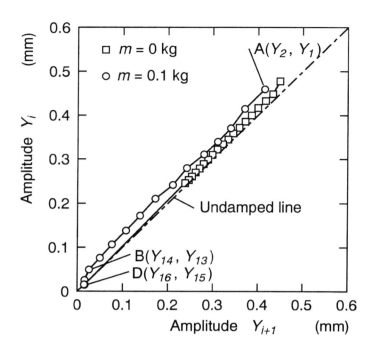

**FIGURE 9.6**   Damping curve of leaf spring (impact damper B, vertical vibration, $T = 8$ mm, $m = 0.1$ kg, $C_L = 0.1$ mm, $Y_0 = 0.5$ mm).

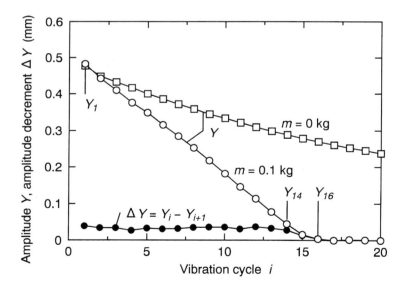

**FIGURE 9.7**   Amplitude reduction of leaf spring (impact damper B, horizontal vibration, $T = 8$ mm, $m = 0.1$ kg, $C_L = 0.1$ mm, $Y_0 = 0.5$ mm).

same leaf spring is affected by the initial displacement, but a mean value of logarithmic decrements is presented in the table.

On the other hand, the amplitude decrements of the leaf spring with a free mass differ considerably from those of the leaf spring without a free mass. In the case of the vertical vibration shown in Figure 9.5, the amplitude linearly decreases from the first cycle (amplitude $Y_1$) to the 13th cycle (amplitude $Y_{13}$). However,

**TABLE 9.3** Vibration Characteristics of Leaf Springs Used

| Damper | $T$ (mm) | $M$ (kg) | $K$ (kN/m) | $f_n$ (Hz) | $\gamma_0$ |
|---|---|---|---|---|---|
| | 6 | 0.234 | 75 | 90 | 0.006 |
| A | 8 | 0.248 | 175 | 135 | 0.016 |
| | 10 | 0.258 | 326 | 179 | 0.023 |
| | 6 | 0.263 | 75 | 85 | 0.013 |
| B | 8 | 0.284 | 178 | 126 | 0.015 |
| | 10 | 0.289 | 326 | 169 | 0.020 |

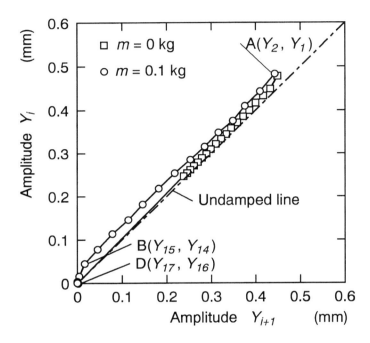

**FIGURE 9.8** Damping curve of leaf spring (impact damper B, horizontal vibration, $T$ = 8 mm, $m$ = 0.1 kg, $C_L$ = 0.1 mm, $Y_0$ = 0.5 mm).

only a very small amplitude is maintained after the 15th cycle. In Figure 9.6, every point indicated by point A (determined by the amplitudes at the first and second cycles) as far as point B (determined by the amplitudes at the 13th and 14th cycles), is separated from a dotted line showing that the system has no damping and lies on a line with an inclination larger than one. However, elongation of this line never reaches the origin. Since every point after point D (determined by the amplitudes at the 15th and 16th cycles) lies on a line passing through the origin when there is no free mass, the damping capability of the leaf spring in this amplitude range may depend on viscous damping.

With the horizontal vibration shown in Figure 9.7, the amplitude linearly decreases from the first cycle (amplitude $Y_1$) to the 14th cycle (amplitude $Y_{14}$), but the amplitude reaches almost zero after the 16th cycle. The amplitude reduction in the horizontal vibration shown in Figure 9.8 is similar to that in the vertical vibration shown in Figure 9.6. Accordingly, the impact damper functions effectively in the amplitude range corresponding to point A till point B in the vertical and horizontal vibrations, but does not function in the amplitude range after point D. The amplitude range between point B and point D seems to be in a transitional stage. As described above, since the amplitude reduction state of the leaf spring with a free mass considerably differs from that of the leaf spring without a free mass, the damping capability of the impact damper essentially differs from that of the main vibratory system caused by viscous damping. Therefore, the damping characteristics of the impact damper was assessed by the following manner.

From point A and point B shown in Figures 9.6 and 9.8, the amplitude reductions do not depend on the vibration direction of the leaf spring, but have the same tendency. Then, an amplitude decrement between neighboring cycles ($\Delta Y = Y_i - Y_{i+1}$) was obtained. Solid circles in Figures 9.5 and 9.7 present the amplitude decrement $\Delta Y$. As this amplitude decrement influences the magnitude of amplitudes of the leaf spring an average value $\Delta Y_{av}$ of the amplitude decrements $\Delta Y$ was obtained. This $\Delta Y_{av}$ is an amplitude decrement in a cycle from point A to point B and called an average amplitude decrement. Furthermore, the amplitude $Y_D$ at point $D$ where the impact damper no longer functions, was obtained. This $Y_D$ is called a critical amplitude. Impact dampers with a large average amplitude decrement $\Delta Y_{av}$ and a small critical amplitude $Y_D$ are desirable.

## Critical Amplitude

The effects of the clearance $C_L$ and the initial displacement $Y_0$ on the critical amplitude $Y_D$ are shown in Figure 9.9. The experimental conditions are appended in the chapter: impact damper B, thickness of the leaf spring $T = 8$ mm and free mass $m = 0.1$ kg. It is clear from the figure that the critical amplitude where the impact damper ceases to function is hardly affected by the clearance and the initial displacement. Furthermore, the effects of the clearance and the initial displacement on the critical amplitude were investigated for all combinations of the impact damper, the vibration direction, the leaf spring and the free masses. As a result, it was confirmed that the critical amplitude was hardly affected by the clearance and the initial displacement. Then, an average value of the critical amplitudes obtained when the clearance and the initial displacement were varied widely, was considered to be the critical amplitude for the impact damper, the leaf spring, and the free mass used.

Figures 9.10 and 9.11 show the average critical amplitude plotted to the natural frequency of each leaf spring $f_n$. Figure 9.10 shows the result of the vertical vibration when the impact damper A was used, and Figure 9.11 the result of the vertical and horizontal vibrations when the impact damper B was used. With respect to the vertical vibration, both the impact dampers A and B, the critical amplitude decreases as the natural frequency of the system is increased. Furthermore, the critical amplitude has the same tendency as the curve calculated by an equation in the figures. This equation was introduced from the following assumption.

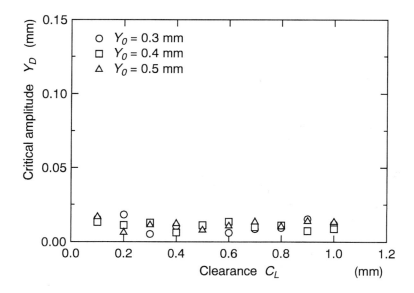

**FIGURE 9.9**  Effect of clearance and initial displacement on critical amplitude (impact damper B, vertical vibration, $T = 8$ mm, m = 0.1 kg).

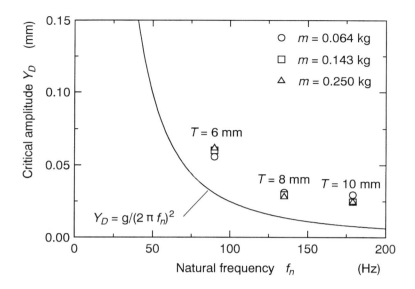

**FIGURE 9.10**   Relationship between critical amplitude and natural frequency (impact damper A).

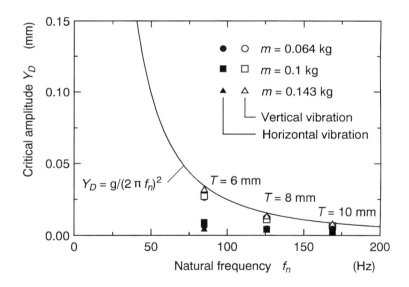

**FIGURE 9.11**   Relationship between critical amplitude and natural frequency (impact damper B).

When the leaf spring starts a free damped vibration from the upward displacement as shown in Figure 9.3, the leaf spring and the free mass begin to vibrate toward the acting direction of the acceleration of gravity. In this case, if the downward acceleration of the leaf spring is faster than the acceleration of gravity $g$, the free mass is separated from the upper surface of the leaf spring and starts a free-falling motion. Now, when the leaf spring maintains a steady vibration with a natural frequency $f_n$ and an amplitude $Y_s$, a vibratory acceleration of the leaf spring can be presented as $Y_s(2\pi f_n)^2$. If the vibratory acceleration of the leaf spring is faster than the acceleration of gravity $g$, the free mass is separated from the leaf spring and starts a free-falling motion. That is to say, if the vibratory acceleration of the leaf spring is faster than the acceleration of gravity $g$, the free mass is separated from the main mass and starts a collision. Accordingly, the critical amplitude $Y_D$ where the impact damper does not function can be presented

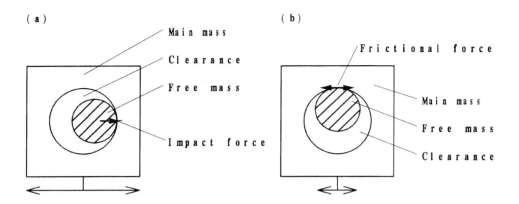

**FIGURE 9.12** Damping mechanism of impact damper varying with amplitude in horizontal vibration; (a) Large amplitude and (b) Small amplitude.

as the next equation.

$$Y_D = \frac{g}{(2\pi f_n)^2} \qquad (9.1)$$

It is clear from equation (1) that the critical amplitude $Y_D$ when the leaf spring vibrates in the direction of gravity is not affected by the free mass $m$, the clearance $C_L$ and the initial displacement $Y_0$ and is determined by the natural frequency $f_n$ and the acceleration of gravity $g$. It is also clear that the greater the natural frequency of the leaf spring is, the more effectively the impact damper functions.

On the other hand, with respect to the horizontal vibration in the use of the impact damper B, the critical amplitude has no relation to the natural frequency and remains low for the following reasons. As shown in Figure 9.12(a), the impact damper functions effectively when the amplitude of the leaf spring is large, because the vibratory energy is absorbed by the collision between the free mass and the main mass. As the amplitude of the leaf spring decreases (Figure 9.12(b)), the free mass begins to contact the main mass in a certain period of the vibration. At this time, as the vibratory direction of the horizontal vibration deviates 90 degrees from the direction of gravity, Coulomb's friction occurs in the contacting area of the free mass and the main mass due to the relative motion between them. This frictional force always acts as a damping force on the system while the main mass vibrates. Therefore, the critical amplitude when the main mass vibrates in the horizontal direction has no relation to the natural frequency as shown in Figure 9.11 and is smaller and the critical amplitude of the vertical vibration.

## Frequency

The frequency of the leaf spring with a free mass was obtained between point A and point B where the impact damper functions effectively shown in Figures 9.6 and 9.8. When the leaf spring with the impact damper A vibrates in the vertical direction, the effects of the clearance $C_L$ and the initial displacement $Y_0$ on the frequency $f$ are shown in Figure 9.13. The other experimental conditions are appended in the caption. The frequency $f$ is hardly affected by the clearance and the initial displacement used in the experiment; however, it is lower than the natural frequency of this leaf spring ($f_n = 135$ Hz) indicated by a heavy line. Accordingly, from these experimental results when the clearance and the initial displacement were varied widely, an average value of the frequencies was obtained for the impact damper, the vibratory direction, the leaf spring and the free mass used. Furthermore, the average value $f$ of the frequencies was non-dimensioned by the natural frequency $f_n$ of each leaf spring and was plotted to the mass ratio $\mu = m/M$.

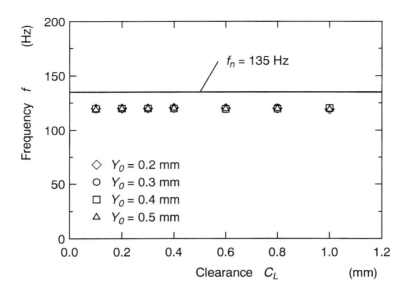

**FIGURE 9.13** Frequency of leaf spring with impact damper (impact damper A, $T = 8$ mm, $m = 0.064$ kg).

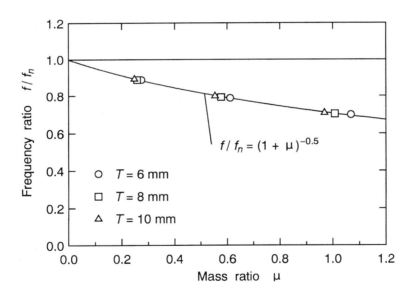

**FIGURE 9.14** Relationship between frequency ratio and mass ratio (impact damper A).

Figure 9.14 shows the frequency ratio $f/f_n$ when the leaf spring with the impact damper A vibrates in the vertical direction. The frequency ratio $f/f_n$ decreases with the increase of the mass ratio $\mu$. Furthermore, the frequency ratio coincides well with the curve calculated by the following equation.

$$\frac{f}{f_n} = \frac{1}{(1 + \mu)^{0.5}} \tag{9.2}$$

This equation is introduced on the assumption that the free mass is fixed (appended) to a main mass.

On the other hand, when the impact damper B was used, especially when the leaf spring vibrated in the horizontal direction, the frequency dispersed. Accordingly, not only an average value of the frequencies but

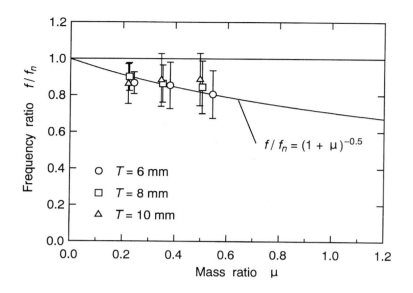

**FIGURE 9.15**   Relationship between frequency ratio and mass ratio (impact damper B, vertical vibration).

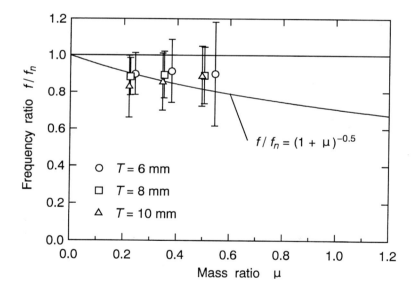

**FIGURE 9.16**   Relationship between frequency ratio and mass ratio (impact damper B, horizontal vibration).

also a standard deviation of them were obtained when the impact damper B was used. Figures 9.15 and 9.16 show the frequency ratio $f/f_n$ of the leaf spring with the impact damper B and present the results of the vertical vibration and the horizontal vibration, respectively. The symbol and longitudinal bar appended to it in the figures present the average value and the standard deviation of the frequency ratio. The frequency ratio $f/f_n$ of the vertical vibration shown in Figure 9.15 disperses, but decreases with the increase of the mass ratio $\mu$. Furthermore, the frequency ratio almost coincides with the curve calculated from equation (2).

When the leaf spring vibrates in the horizontal vibration shown in Figure 9.16, as the frequency ratio $f/f_n$ is smaller than unity, the frequency $f$ is lower than the natural frequency $f_n$. However, there is no obvious reduction tendency of the frequency ratio to the mass ratio as shown in the vertical vibration. The scatter of the frequency ratio (standard deviation) increases with the mass ratio. These results may be attributed to the fact that the frequency could not be obtained precisely, because the leaf spring reached a critical amplitude by a few vibration cycles as the free mass or the clearance increased. Accordingly, even though the leaf spring vibrates in any direction, the frequency of the system with a free mass can be estimated by the natural frequency and the mass ratio, but the estimation accuracy for the horizontal direction is low.

## Average Amplitude Decrement

Figure 9.17 is a sample showing he relationship between the average amplitude decrement $\Delta Y_{av}$ and the clearance $C_L$ in the experimental conditions presented in the caption. There is a clearance where the average amplitude decrement becomes maximum for each initial displacement (indicated by solid symbols). Thus, when the appropriate clearance is chosen with regard to the initial displacement, the damping effect of the impact damper becomes greatest. The optimum clearance $C_{Lopt}$ obtained in this manner for all combinations of the leaf spring and the free mass is summarized in Tables 9.4 and 9.5.

In the impact damper A shown in Table 9.4, the optimum clearance $C_{Lopt}$ to obtain the greatest damping effect disperses in the range from 0.2 to 1.0 mm. In the impact damper B shown in Table 9.5, the optimum

**TABLE 9.4** Optimum Clearance of Impact Damper A

| $T$ (mm) | $m$ (kg) | $\mu = m/M$ | Optimum Clearance $C_{Lopt}$ (mm) | | | |
|---|---|---|---|---|---|---|
| | | | $Y_0 = 0.2$ | 0.3 | 0.4 | 0.5 mm |
| 6 | 0.064 | 0.274 | 0.2 | 0.5 | – | – |
| | 0.143 | 0.611 | 0.2 | 0.4 | – | – |
| | 0.250 | 1.068 | 0.2 | 0.4 | – | – |
| 8 | 0.064 | 0.258 | 0.4 | 0.8 | 1.0 | 1.0 |
| | 0.143 | 0.577 | 0.2 | 0.3 | 0.6 | 0.8 |
| | 0.250 | 1.008 | 0.2 | 0.4 | 0.4 | 0.8 |
| 10 | 0.064 | 0.248 | 0.4 | 0.4 | 0.8 | 0.8 |
| | 0.143 | 0.554 | 0.2 | 0.3 | 0.6 | 0.8 |
| | 0.250 | 0.969 | 0.2 | 0.3 | 0.6 | 0.6 |

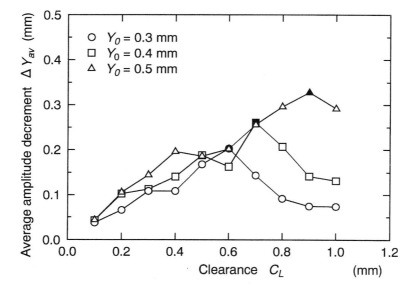

**FIGURE 9.17** Average amplitude decrement (impact damper B, vertical vibration, $T = 8$ mm, $m = 0.1$ kg).

**TABLE 9.5**   Optimum Clearance of Impact Damper B

| T (mm) | m (kg) | μ = m/M | Optimum Clearance $C_{Lopt}$ (mm) | | | | | |
|---|---|---|---|---|---|---|---|---|
| | | | Vertical Vibration | | | Horizontal Vibration | | |
| | | | $Y_0 = 0.3$ | 0.4 | 0.5 | 0.3 | 0.4 | 0.5 mm |
| | 0.064 | 0.243 | 0.8 | 0.9 | 1.0 | 0.8 | 0.9 | 1.0 |
| 6 | 0.100 | 0.380 | 0.7 | 0.9 | 1.0 | 0.8 | 0.9 | 1.0 |
| | 0.143 | 0.544 | 0.6 | 0.7 | 0.8 | 0.7 | 0.8 | 0.9 |
| | 0.064 | 0.225 | 0.7 | 0.9 | 1.0 | 0.8 | 0.9 | 1.0 |
| 8 | 0.100 | 0.352 | 0.6 | 0.7 | 0.9 | 0.7 | 0.8 | 1.0 |
| | 0.143 | 0.504 | 0.6 | 0.8 | 0.9 | 0.6 | 0.8 | 0.9 |
| | 0.064 | 0.221 | 0.6 | 0.9 | 1.0 | 0.6 | 0.8 | 1.0 |
| 10 | 0.100 | 0.346 | 0.6 | 0.7 | 0.9 | 0.5 | 0.8 | 1.0 |
| | 0.143 | 0.495 | 0.6 | 0.7 | 0.9 | 0.6 | 0.8 | 0.9 |

clearance $C_{Lopt}$ disperses in the range from 0.5 to 1.0 mm. However, for any impact damper, any leaf spring and any vibration direction, as the initial displacement increases, the optimum clearance increases when the free mass is constant. As the free mass decreases, the optimum clearance tends to increase when the initial displacement is constant; but its effects are not so remarkable. In addition, the optimum clearance is about two times greater than the initial displacement $Y_0$.

## Corresponding Logarithmic Decrement

From the experimental results described above, the damping characteristics of the impact dampers used in this study were almost clarified. However, as the assessment standards of the damping characteristics differed depending on whether the impact damper exists or not, the damping capabilities of both cases could not be compared. Accordingly, an amount corresponding to the logarithmic decrement $\gamma$ (called corresponding logarithmic decrement) was calculated using the initial displacement $Y_0$, the critical amplitude $Y_D$ and the number of vibration cycles until the leaf spring reaches the critical amplitude $Y_D$. Further, the corresponding logarithmic decrement $\gamma$ obtained in this manner was divided by the logarithmic decrement $\gamma_0$ obtained for each initial displacement and each leaf spring without the impact damper.

Figure 9.18 is a sample showing the corresponding logarithmic decrement ratio $\gamma/\gamma_0$ obtained when the impact damper B, the vertical vibration, the thickness $T = 8$ mm and the free mass $m = 0.1$ kg were used. The corresponding logarithmic decrement ratio $\gamma/\gamma_0$ disperses with the variations in the initial displacement and the clearance. This can be attributed to the fact that the slight difference (error) in estimating the critical amplitude $Y_D$ remarkably affects the corresponding logarithmic decrement $\gamma$. Then, in order to evaluate how the impact damper improves the damping capability of the main vibratory system, the maximum and minimum values of the corresponding logarithmic decrement ratio (marked by solid symbols) were obtained for all combinations for all combinations of the impact damper, the vibratory direction, the leaf spring and the free mass used in this experiment. These results are listed in Tables 9.6 and 9.7. It is clear from tables that using the impact damper, the damping capability of the main vibratory systems can be improved considerably. Namely, the damping capability of the main vibratory systems can be improved at least fivefold when the impact damper A was used and at least eightfold for both vertical and horizontal vibrations when the impact damper B was used.

## 9.4   Application of Impact Damper

When the impact damper was applied to cutting tools, in order to investigate how degree the damping capability of cutting tools can be improved, how degree the chatter vibration can be prevented and which considerations should be taken in the actual setting, the following experiments for drills and boring tools were carried out.

**TABLE 9.6**   Minimum and Maximum Values of Corresponding Logarithmic Decrement Ratio of Impact Damper A

| $T$ (mm) | Vibratory Direction | Corresponding Logarithmic Decrement Ratio $\gamma/\gamma_0$ | | | | | |
|---|---|---|---|---|---|---|---|
| | | $m = 0.064$ kg | | $m = 0.143$ kg | | $m = 0.250$ kg | |
| | | Min. | Max. | Min. | Max. | Min. | Max. |
| 6 | Vertical | 13.4 | 89.0 | 22.0 | 166.2 | 24.4 | 242.5 |
| 8 | Vertical | 6.7 | 58.4 | 10.5 | 75.7 | 12.6 | 86.6 |
| 10 | Vertical | 5.0 | 30.5 | 10.4 | 65.4 | 10.8 | 65.2 |

**TABLE 9.7**   Minimum and Maximum Values of Corresponding Logarithmic Decrement Ratio of Impact Damper B

| $T$ (mm) | Vibratory Direction | Corresponding Logarithmic Decrement Ratio $\gamma/\gamma_0$ | | | | | |
|---|---|---|---|---|---|---|---|
| | | $m = 0.064$ kg | | $m = 0.1$ kg | | $m = 0.143$ kg | |
| | | Min. | Max. | Min. | Max. | Min. | Max. |
| 6 | Vertical | 12.0 | 51.0 | 13.6 | 62.5 | 14.2 | 50.8 |
| | Horizontal | 12.1 | 52.8 | 21.6 | 67.1 | 23.8 | 65.4 |
| 8 | Vertical | 13.7 | 37.3 | 16.3 | 53.4 | 18.6 | 62.1 |
| | Horizontal | 8.2 | 31.7 | 12.5 | 41.8 | 17.5 | 43.7 |
| 10 | Vertical | 12.0 | 34.1 | 14.6 | 33.8 | 14.0 | 50.4 |
| | Horizontal | 8.8 | 21.9 | 9.9 | 29.6 | 10.4 | 33.9 |

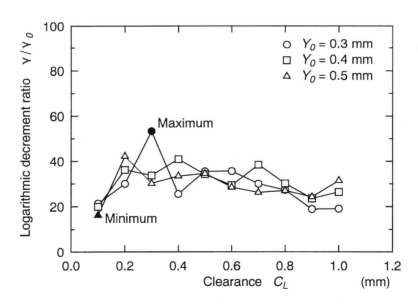

**FIGURE 9.18**   Corresponding logarithmic decrement ratio (impact damper B, vertical vibration, $T = 8$ mm, $m = 0.1$ kg.)

## Drills

An application of the impact damper to drills is described below. The drill used in the experiment was an one on the market with a diameter of 10 mm and an overhang of 260 mm. It has been clarified by the authors [8] that when chatter vibration occurs in a drill, its vibration mode becomes not the drill point free mode shown in Figure 9.19(b) but the drill point restricted mode shown in Figure 9.19(c),

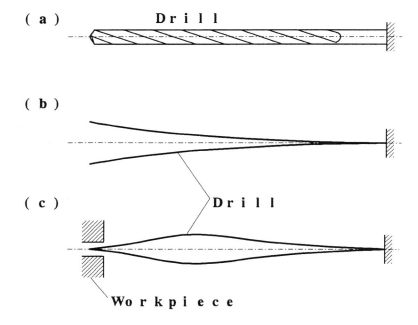

**FIGURE 9.19**  Vibration mode of drill; (a) Drill whose end is fixed on a chuck, (b) Drill point free mode, and (c) Drill point restricted mode.

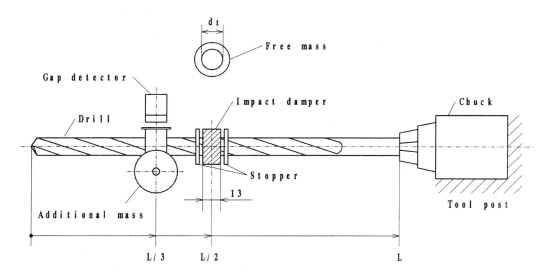

**FIGURE 9.20**  Configuration of impact damper and method for equipping it on drill.

in which the drill point movement is restricted in a machined hole. Then, as illustrated in Figure 9.20, the impact damper (free mass) was equipped at a position of about a half of the overhang away from the end of drill where the drill deflection becomes greatest during chatter vibration. The free mass is a ring with a width of 13 mm and has a hole in the middle which diameter $(d_1)$ is slightly larger than the drill diameter $D$. Therefore, the free mass can move freely within this clearance $(C_L = d_1 - D)$. A workpiece was mounted on the chuck on the lathe main spindle, and a drill was fixed to the tool post. The drill displacement in drilling was measured using an eddy current-type gap detector indicated in

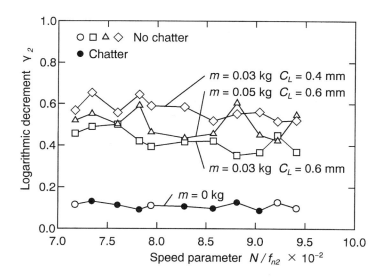

**FIGURE 9.21** Suppression effect of chatter vibration in drill by impact damper.

the figure. Units of additional mass up to ten which weighs 3.8 g per unit were attached to the drill. In this manner, the state of chatter vibration occurrence was investigated when a speed parameter $N/f_{n2}$ consisting the rotation of the workpiece per second $N$ and the natural frequency of the drill $f_{n2}$ was varied slightly.

Three kinds of impact dampers were used in the experiment. Their free mass and clearance are as follows: $m = 0.03$ kg and $C_L = 0.4$ mm, $m = 0.03$ kg and $C_L = 0.6$ mm, and $m = 0.05$ kg and $C_L = 0.6$ mm. The drill used was made of high speed tool steel, the free mass was of brass (70% Cu and 30% Zn), and the workpiece was of rolled steel for general structure ($\sigma_B = 402$ Mpa). The cutting conditions were a feed rate 0.029 mm/rev. and a workpiece rotation 1200 rpm.

Figure 9.21 shows the suppression effect of chatter vibration in the use of the impact dampers. The ordinate in the figure indicates the corresponding logarithmic decrement of the drill $\gamma_2$, while the abscissa indicates the speed parameter $N/f_{n2}$. The open symbols in the figure indicate that no chatter vibration occurred while the solid symbols indicate its occurrence. The chatter vibration occurs without an impact damper ($m = 0$ kg), since the damping of the system is extremely low. However, when the three kinds of impact dampers were used, no chatter vibration whatsoever occurred in any speed parameter. This is due to the fact that the corresponding logarithmic decrement of the drill $\gamma_2$ was improved considerably by using the impact damper.

As shown in the foregoing, it is clarified that the chatter vibration can be suppressed completely, if the damping characteristic of a drill is improved by using an impact damper. However, since a drill is a tool for boring holes in materials, it is impossible to attach an impact damper to its end. When an impact damper is positioned at the shank end (base) of a drill, its damping effect cannot be expected. Accordingly, for practical use, it is advisable for the impact damper to be positioned one-third or half the overhang from the end of the drill.

## Boring Tools

An application of the impact damper to boring tools is described below. The boring tool used in the experiment was an one on the market with a shank diameter of 22 mm and an overall length of 180 mm illustrated in Figure 9.22. The impact damper was attached at the position of 40 mm away from the end of boring tool. The free mass is a ring with a width of 12 mm and has a hole ($d_1$) in the middle slightly larger than the diameter of an auxiliary sleeve $D$. Therefore, the free mass can move freely within this

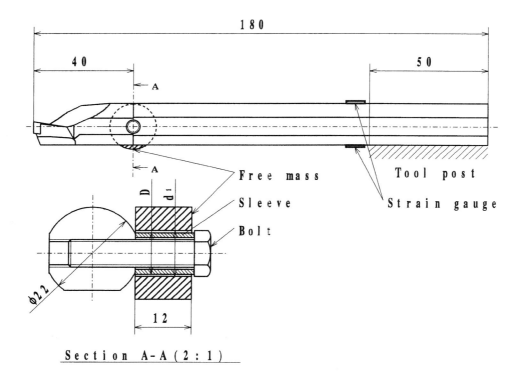

**FIGURE 9.22**   Configuration of impact damper and method for equipping it on boring tools.

clearance ($C_L = d_1 - D$). A workpiece was mounted on the chuck on the lathe main spindle, and the boring tool was fixed to the tool post. The workpiece was a ring with a diameter of 140 mm and a thickness of 6 mm.

Four kinds of impact dampers consisting of the free mass $m = 0.008$ and 0.024 kg and the clearance $C_L = 0.4$ and 0.8 mm were used in the experiment. The vibratory displacement of the end of the boring tool in direction of the main cutting force was measured by strain gauges indicated in Figure 9.22. The free mass used was made of brass (70% Cu and 30% Zn), the cutting tools (throw-away tip) was of cemented carbide alloy (grade K10), and the workpiece was of rolled steel for general structure ($\sigma_B = 402$ Mpa). The cutting conditions were a cutting depth 0.2 mm and a feed rate 0.029 mm/rev. and workpiece rotations 80, 130, and 200 rpm.

Figure 9.23 shows the suppression effect of chatter vibration in the use of the impact dampers. The ordinate in the figure indicates the amplitude of the end of the boring tool, while the abscissa indicates the workpiece rotation. The chatter vibration occurs without an impact damper ($m = 0$ kg), the amplitude increases with the increase of the workpiece rotation. However, when the impact damper of a free mass of $m = 0.008$ kg was equipped on the boring tool, the amplitude reduced two-third smaller than the amplitude obtained when $m = 0$ kg. Furthermore, when the impact damper of a free mass of $m = 0.024$ kg was equipped on the boring tool, the amplitude reduced one-third smaller than the amplitude obtained when $m = 0$ kg. This is due to the fact that the logarithmic decrement of the boring tool was improved considerably by using the impact damper.

As showing in the foregoing, it is clarified that the chatter vibration can be suppressed considerably, if the damping characteristic of a boring tool is improved by using an impact damper. To utilize effectively the damping capability of the impact damper, the impact damper must be equipped on the end of a boring tool which amplitude is large. However, since a boring tool is a tool for magnifying and finishing holes in materials, it is impossible to attach an impact damper larger than the diameter of the hole.

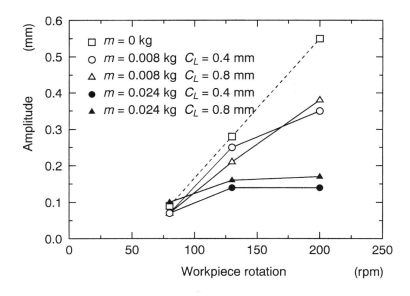

**FIGURE 9.23**  Suppression effect of chatter vibration in boring tool by impact damper.

Accordingly, for practical use, it is required that a free mass and a clearance are chosen adequately and an impact damper does not interfere with a hole machined in materials.

## 9.5  Conclusions

Damping characteristics of an impact damper consisting of a ring-shaped free mass and a clearance and its application to cutting tools were investigated, and the following was clarified.

(1) Frequency of a vibratory system with an impact damper is approximately decided by the natural frequency of the system and the mass ratio.

(2) Where the system vibrates in the direction of gravity, the critical amplitude at which the impact damper no longer functions is decided by the natural frequency and the acceleration of gravity. On the other hand, where the system vibrates in the direction perpendicular to gravity, the impact damper functions effectively until the amplitude of the system reaches almost zero, because the frictional force occurring between the main mass and the free mass acts as a damping force.

(3) Use of the impact damper can improve considerably the damping capability of the main vibratory systems. Namely, the damping capability can be improved at least five-fold when the impact damper A was used and at least eight-fold for both vertical and horizontal vibrations when the impact damper B was used.

(4) It is possible to apply the impact damper to actual cutting tools. However, when the impact damper is put into practice, such considerations as features of cutting tools are not spoiled are required.

### Acknowledgment

The authors would like to express their appreciation to the kind permission from Elsevier Science Ltd., The Boulevard, Langford Lane, Kidlington OX5 1GB, UK. Some parts of this article are translated from the article which appeared in Ema, S., and Marui, E., 1994, "A Fundamental Study on Impact Dampers," *International Journal of Machine Tools and Manufacture*, Vol. 34, No. 3, pp. 407–421, and Ema, S., and Marui, E., 1996, "Damping Characteristics on an Impact Damper and Its Application," *International Journal of Machine Tools and Manufacture*, Vol. 36, No. 3, pp. 293–306.

# References

[1]  Tobias, S. A., 1965, *Machine-Tool Vibration,* Blackie, pp. 333–338.

[2]  Hahn, R. S., 1951, "Design of Lanchester Damper for Elimination of Metal-Cutting Chatter," *Trans. of ASME,* Vol. 73, April, pp. 331–335.

[3]  Kato, S., E. Marui, and H. Kurita, 1969, "Some Considerations on Prevention of Chatter Vibration in Boring Operations," *Trans. of ASME Ser. B,* Vol. 91, Aug., pp. 717–729.

[4]  Kim, K. J. and J. Y. Ha, 1987, "Suppression of Machine Tool Chatter Using a Viscoelastic Dynamic Damper," *ASME Journal of Engineering for Industry,* Vol. 109, Feb., pp. 58–65.

[5]  Masri, S. F. and T. K. Caughey, 1966, "On the Stability of the Impact Damper," *ASME Journal of Applied Mechanics,* Vol. 33, Sept., pp. 586–592.

[6]  Sadek, M. M. and B. Mills, 1970, "Effect of Gravity on the performance of an Impact Damper: Part 1. Steady-State Motion," *Journal Mechanical Engineering Science,* Vol. 12, No. 4, pp. 268–277.

[7]  Thomas, M. D., W. A. Knight, and M. M. Sadek, 1975, "The Impact Damper as a Method of Improving Cantilever Boring Bars," *ASME Journal of Engineering for Industry,* Vol. 97, August, pp. 859–866.

[8]  Ema, S., H. Fujii, and E. Marui, 1988, "Chatter Vibration in Drilling," *ASME Journal of Engineering for Industry,* Vol. 110, Nov., pp. 309–314.

[9]  Ema, S. and E. Marui, 1994, "A Fundamental Study on Impact Dampers," *International Journal of Machine Tools and Manufacture,* Vol. 34, No. 3, pp. 407–421.

[10] Ema, S. and E. Marui, 1996, "Damping Characteristics on an Impact Damper and Its Application," *International Journal of Machine Tools and Manufacture,* Vol. 36, No. 3, pp. 293–306.

# Index